T0206415

Introduction to Cosmic Inflation and Dark Energy

Series in Astronomy and Astrophysics

The *Series in Astronomy and Astrophysics* includes books on all aspects of theoretical and experimental astronomy and astrophysics. Books in the series range in level from textbooks and handbooks to more advanced expositions of current research.

Series Editors:
M Birkinshaw, University of Bristol, UK
J Silk, University of Oxford, UK
G Fuller, University of Manchester, UK

Recent books in the series

Dark Sky, Dark Matter
J M Overduin and P S Wesson

Dust in the Galactic Environment, 2nd Edition
D C B Whittet

The Physics of Interstellar Dust
E Krügel

Very High Energy Gamma-Ray Astronomy
T C Weekes

Numerical Methods in Astrophysics: An Introduction
P Bodenheimer, G P Laughlin, M Rózyczka, H W Yorke

An Introduction to the Physics of Interstellar Dust
Endrik Krugel

Astrobiology: An Introduction
Alan Longstaff

Fundamentals of Radio Astronomy: Observational Methods
Jonathan M Marr, Ronald L Snell, and Stanley E Kurtz

Stellar Explosions: Hydrodynamics and Nucleosynthesis
Jordi José

Cosmology for Physicists
David Lyth

Cosmology
Nicola Vittorio

Cosmology and the Early Universe
Pasquale Di Bari

Fundamentals of Radio Astronomy: Astrophysics
Ronald L. Snell, Stanley E. Kurtz, and Jonathan M. Marr

Introduction to Cosmic Inflation and Dark Energy
Konstantinos Dimopoulos

Introduction to Cosmic Inflation and Dark Energy

Konstantinos Dimopoulos

CRC Press
Taylor & Francis Group
Boca Raton London New York

CRC Press is an imprint of the
Taylor & Francis Group, an **informa** business

First edition published 2021
by CRC Press
6000 Broken Sound Parkway NW, Suite 300, Boca Raton, FL 33487-2742

and by CRC Press
2 Park Square, Milton Park, Abingdon, Oxon, OX14 4RN

Library of Congress Cataloging-in-Publication Data
Names: Dimopoulos, K. (Konstantinos) author.
Title: Introduction to cosmic inflation and dark energy / Konstantinos
    Dimopoulos.
Description: First edition. | Boca Raton : CRC Press, 2020. | Series:
    Series in astronomy and astrophysics | Includes bibliographical
    references and index.
Identifiers: LCCN 2020024113 | ISBN 9780815386759 (hardback) | ISBN
    9781351174862 (ebook)
Subjects: LCSH: Cosmology--History. | Dark energy (Astronomy) | Big bang
    theory.
Classification: LCC QB981 .D56 2020 | DDC 523.1/8--dc23
LC record available at https://lccn.loc.gov/2020024113

ISBN: 978-0-8153-8675-9 (hbk)
ISBN: 978-1-351-17486-2 (ebk)

Typeset in LMRoman
by Nova Techset Private Limited, Bengaluru & Chennai, India

Visit the Taylor & Francis Web site at
http://www.taylorandfrancis.com

and the CRC Press Web site at
http://www.crcpress.com

Visit the eResources: www.routledge.com/9780815386759

Contents

Preface vii
1 Introduction 1
2 Dynamics and Content of the Universe 4
3 History of the Universe 46
4 Inflation Basics 67
5 CMB Primordial Anisotropy and Structure in the Universe 91
6 The Inflationary Paradigm 109
7 Models of Inflation 137
8 Beyond Slow-Roll Inflation 171
9 Dynamic Dark Energy 204
10 Epilogue 253
A A Taste of General Relativity 257
B Correlators of the Curvature Perturbation 260
C Light Scalar Field Superhorizon Spectrum 262
D Field Equation and Energy-Momentum of a Free Scalar Field 265
References 267
Index 271

Preface

Sometime in 2017, I was asked to review an introductory book on general relativity and cosmology. The book was very well written, especially the part on general relativity. However, to my amazement, hardly anything was mentioned on the topics of cosmic inflation and dark energy. I sent my report to the publishing company emphasising this point. They asked me if I would be interested in writing a book. Did I have a proposal for one? At this point, I realised that most of the existing introductory books on cosmology have very little material on cosmic inflation and dark energy, even though a sizeable fraction of cosmological research activity is concentrated in these fields. Sure, there are a number of excellent textbooks for specialists, but not much for senior undergraduate or early graduate students. And there is so much that one can say on the subject at this level. I felt that this situation must be remedied. This is how I first came up with the idea of this book.

I have been teaching cosmology to undergraduate and graduate students for more than 15 years now. I am grateful to my undergraduate cosmology students, whose feedback (either from end-of-year questionnaires or because of many insightful questions during the lectures) throughout the years enabled me to improve and refine my lecture notes, upon which a large part of this book is based. I should also include here my master's and doctorate students. Of particular help was my former master's student, Evelina Petkova, who read through the first draft of the background cosmology chapters; my current doctorate student, Samuel Sánchez López, who read through the entire draft; and my former doctorate student, Charlotte Owen, who also read through the whole draft and additionally provided me with a couple of graphs (Figures 6.2 and 7.6) well after her graduation and despite working full-time. Their detailed comments and suggestions were invaluable. I should also thank my former doctorate student, Leonora Donaldson-Wood, who created the plots in Figures 9.2 and 9.8 as part of two of our published papers. Special thanks go to my friend and collaborator, Carsten van de Bruck, who kindly offered to read through the manuscript and came back to me with pointers and suggestions.

My knowledge and understanding of cosmology and in particular of cosmic inflation and dark energy is shaped by more than two decades of active research, interactions with numerous collaborators, and participation and presentations to a large number of conferences (many of the graphs I have designed in the book have their origins in old presentation slides). My understanding was particularly boosted and shaped through interaction and collaboration with my friend and mentor, David H. Lyth, while I should also thank my former supervisor, Anne-Christine Davis, and my prominent collaborator, George Lazarides. The Physics Department of Lancaster University should also be thanked for providing me with the opportunity to write this book, in the form of a sabbatical leave during which the bulk of the book was written. I would also like to thank CRC Press for their continuous support.

Finally, none of this would have been possible without the loving support of my wife, Marilita Papastathi. Apart from reading through most of the manuscript (with her unique point of view), she has been constantly encouraging me and prompting me to move forward, allowing me space to work, and putting up with my preoccupation with this demanding project. I think that her biggest effect was believing in me so, in turn, I was persuaded to believe in myself. I thank her from the bottom of my heart.

1

Introduction

Cosmology can be defined as the study of the Universe as a system, much like astrophysics can be defined as the study of stars as systems (a star is called "astro" ($\check{\alpha}\sigma\tau\rho o$) in Greek). The term cosmology derives from the Greek word Cosmos ($K\acute{o}\sigma\mu o\varsigma$), which literally means *beauty*. In fact, the terms cosmology and cosmetics have the same root! The above definition of cosmology, however, begs the question: What is the Universe? *The Oxford English Dictionary* defines the Universe as "The whole of existing things" (including space and time, I should add) or "all things (including the Earth, the heavens, and all the phenomena of space) considered as constituting a systematic whole."

Why do people study cosmology? Well, one answer could be that cosmology leads to an understanding of how Nature works on the largest possible scales. This is a focal point for human curiosity, especially for these humans who call themselves physicists. Now, for physicists, there are also practical advantages in studying cosmology. Indeed, cosmology is the meeting point of different areas of physics such as astrophysics, general relativity (the study of gravity) and particle physics. Cosmology facilitates communication of knowledge between these areas. For astrophysics in particular, cosmology sets the stage for astrophysical processes: for example, star and galaxy formation, abundance of elements in the Universe, age of stellar systems, and so on. With respect to particle physics, cosmology assists in the study of fundamental theory. This is because in the early Universe extreme conditions were manifested, at energies so high that are impossible to create in the lab. Phenomena occurring in such conditions may leave traces behind, which, when observed, can offer insights on particle and high-energy physics, for example, on the unification of fundamental forces. In a sense, cosmology uses the entire Universe as a giant laboratory aiming to reveal fundamental physics on energies well beyond the reach of Earth-based experiments such as colliders. Yakov B. Zel'dovich, the prominent soviet cosmologist, famously said "the very early Universe is an accelerator for poor people." The observational side of cosmology is clearly astronomical, using telescopes, balloons, and satellites, while the theoretical part is mainly particle theory and employs what is known or conjectured about fundamental physics. Moving past physicists, there are profound implications of cosmology of philosophical and even existential nature. For example, it is clear that understanding our place in the Cosmos is a step towards understanding ourselves.

Throughout the history of humankind virtually every civilisation had a cosmogony myth, that is a theory for the origin of the Universe. Gradually, with the advance of science, our understanding of the Cosmos was stripped from folklore, religion, and "common sense" prejudices. In the last hundred years or so, a unified picture emerged to the point that physicists, and especially cosmologists, feel that, in broad terms, almost the entire history of the Universe is now known, with most unresolved questions concentrated in a fraction of the first second of its existence. This is a tremendous achievement, of course, but we need to be reminded that cosmologists are only a few thousand people, embedded in the billions of humanity, who may still be largely ignorant of this state of affairs. Note also that, what we cosmologists think we "know" might be subject to change, even radical change, in the future. Still, we have to remember that the objective of science is not really the discovery of "the truth" but rather the formulation of a model that describes the world successfully and can be put to use in order to predict, predetermine, and exploit the world. Since there is no way to prove that what we have discovered is "the truth" (we could only disprove it), the latter is, in a sense, irrelevant to science. This means that the hubris of claiming the understanding of the history of the Universe is swept away by realising that we have merely formulated a model that seems to satisfy and explain most of the observations. This model in now called the standard model of cosmology.

The standard model of cosmology is comprised of two pieces: the Hot Big Bang and cosmic inflation. Now, I should say here that the most undisputed part of the standard model of cosmology is the Hot Big Bang, which describes the bulk of the Universe history. Cosmic inflation amounts to a period of superluminal expansion of space preceding the Hot Big Bang and is the part of the standard model of cosmology, which is used to determine the initial conditions of the Hot Big Bang, which are rather fine-tuned. Observations overwhelmingly support the basic picture of cosmic inflation, and this is why the vast majority of cosmologists now accept it (some reluctantly) as part of the standard model of cosmology. Efforts for alternative scenarios to cosmic inflation are ongoing but this is science and such activity is always pursued in all areas of science. Modified gravity theories, for example, all attempt to go beyond Einstein's general relativity even though the latter is one of the most complete and successful theories. Cosmic inflation, of course, has not been around for as long as general relativity. Neither is it a theory of the scope and completeness of general relativity. Nor has it been tested on the same level as general relativity. Still, it is constantly being tested by cosmological observations at ever higher precision and so far remains successful.

But cosmic inflation is not only successful; it is also compelling. For, apart from setting the initial conditions of the Hot Big Bang, it also provides an elegant explanation of the origin of structures in the Universe, such as galaxies and galactic clusters. It is, therefore, relevant not only for very early times but also for the later stages of the Universe's history, when galaxies formed. Finally, there is another piece of the puzzle, which has to do with the latest part of the Universe history (i.e., today) that is surprisingly connected to cosmic inflation. This is the recent discovery of the mysterious dark energy, which seems to comprise nearly 70% of the content of the Universe at present. The characteristics of this dark energy are not too different from the characteristics of the substance responsible for cosmic inflation, even though the latter occurred a tiny fraction of a second after the beginning of the Universe. This primordial dark energy gave way to the usual radiation and matter content of the Hot Big Bang. But recently, dark energy seems to have reappeared (or it could be another dark energy substance), and it is taking over the evolution of the Universe again. Conversely, it can be said that the Universe is currently engaging in a bout of late-time cosmic inflation, similar to the primordial cosmic inflation, which set up the initial conditions of the Hot Big Bang.

I feel that this connection between cosmic inflation and dark energy has not been highlighted enough. In this book, I provide an overview of cosmic inflation and of dark energy and I also discuss how these could be directly connected, beyond just being two similar but unconnected periods of the Universe's history. But first, I present a brief account of the Hot Big Bang itself, which is the backbone of the standard model of cosmology, an understanding of which is essential before discussing cosmic inflation and dark energy.

Before we begin though, I need to explain a little bit the rationale behind this book. This is not a general introduction to cosmology, although Chapters 2 and 3 summarise the basics, because this is imperative in order to set the stage and motivate cosmic inflation and dark energy. However, there are many other high-quality introductory books to cosmology, so I deliberately chose not to attempt another go on the subject, assured that if readers so wish they can find more information on basic cosmology elsewhere. Thus, worthy subjects such as, e.g., leptogenesis or reionisation are absent from my book. I put a lot of emphasis on concepts, especially regarding the notion of the Universe expansion in Chapter 2 and how cosmic inflation solves the main fine-tunings of Hot Big Bang cosmology in Chapter 4. However, I steered away from going into much detail regarding the observations. Observations are really important in cosmology in general and in particular in research on cosmic inflation and dark energy. However, I am not an observationalist, and so had I attempted to delve into the observations I would need to rely almost exclusively on other better sources and I would not do them justice. Having said that, as with all of science, cosmic inflation and dark energy aim to explain observations and are tested by them, sometimes rather stingily. So, I analysed and summarised what I think are the most important observational data and how they come about in constraining cosmic inflation and dark energy. For example, Chapter 5 is entirely devoted to the structure of CMB primordial anisotropy and large scale structure formation, both of which are seeded by the primordial density perturbations generated by cosmic inflation. In Chapter 6, the book slightly changes gear. First of all, natural units are introduced. The inflationary paradigm is presented, which uses a scalar field to drive the Universe expansion. Even though we mostly talk about cosmic inflation, much of the findings in Chapter 6 are also applied in Chapter 9, where we discuss late-time dark energy. The level rises a little when we move to Chapters 7, 8, and 9, which are the "meat" of the book. Chapters 7 and 8 are devoted to cosmic inflation. They discuss some of the main models but also some additional modes of inflation eventually connecting with the beginning of time and of the existence of a classical Universe. Chapter 9 is larger than the rest (only Chapter 2 is comparable) and deals with current dark energy, ending with a projection to the future; a discussion of the ultimate fate of the Universe.

The book is geared at the level of senior undergraduate and early graduate students in cosmology, but it can also be useful for specialists in neighbouring areas, for example fundamental theorists or astrophysicists. I refrain from highly technical proofs and where it is unavoidable (for example, when discussing the particle production process in Chapter 4) I offer a heuristic picture which hopefully captures the essence of the argument. Consequently, prior knowledge of general relativity or Quantum Field Theory is not needed. However, as a hint towards deeper understanding, I have included appendices A, B, C, and D, where the level is even higher. The appendices address a number of selected topics such as the origin of the Friedmann equations and the cosmological constant problem in general relativity, elements on correlators of the curvature perturbation, particle production of a light scalar field in de Sitter space(time) and a formal derivation of the field equation and the energy-momentum tensor of a scalar field in curved spacetime. The appendices provide a glimpse of a more thorough grounding upon which all the discussions in the book are based, suggesting that the heuristic arguments used occasionally are not the end of the line. Still, the level is not quite the same as that of a specialist even in the appendices. For example, I refrain from using the Einstein summation convention or conformal time. In places I offer a few philosophical thoughts, which is hard to resist given the subject matter, but I think I managed to keep this at a minimum.

I believe this is, so far, a unique book, in that in focuses on cosmic inflation and dark energy, which are subjects dealt in detail only in textbooks for specialists, while they are typically mentioned in passing in most introductory books in cosmology. The subject of this book touches the frontier of research, and as such, the discussion is more fluid than, say, a textbook in electromagnetism. The observational data is also subject to continuous change and refinement, which means that there is danger of elements of this book becoming obsolete in the near future. Having said that, I am certain that the idea of cosmic inflation is not going to go away. Similarly, the observations of the current accelerated expansion, which are behind the idea of dark energy, are undeniable. I feel it is time these issues are presented in more detail to the younger generation of physicists. Also, it is an opportunity to explain to specialists in related fields what the fuss is all about regarding cosmic inflation and dark energy.

2

Dynamics and Content of the Universe

2.1 The Universe observed 4
    Cosmography • Standard candles
2.2 The Universe expansion 8
    Hubble-Lemaitre law • Observable Universe • Scale
    factor and Hubble parameter • Cosmological redshift
2.3 Cosmic dynamics 14
    Cosmological principle • Friedmann equation •
    Continuity equation • Equation of state • Acceleration
    equation
2.4 Cosmic geometry 18
    The curvature parameter • Critical density
2.5 Dynamics and curvature 20
    The Friedmann universes • Closed, matter-dominated
    Universe
2.6 The cosmological constant 22
    Dynamics with Λ • The problem with Λ
2.7 A first brush with dark energy 27
2.8 Dynamics with spatial flatness 29
    Basic equations • Matter-radiation equality • Phantom
    dark energy
2.9 The metric 32
2.10 The cosmological horizon revisited 33
2.11 The age of the Universe 36
2.12 Dark matter 38
    Observational evidence for dark matter • Properties of
    non-baryonic dark matter • Candidates for dark matter
    • Problems of CDM • Direct observational evidence for
    dark matter

2.1 The Universe observed

In this chapter, we begin our brief crash course in the Hot Big Bang, which is the backbone of the standard model of cosmology. We briefly review the state of affairs of the Universe, its present morphology and content and its dynamics.

2.1.1 Cosmography

In charting our Universe, we can outline the observed structures as follows:

- **Solar System**
  The radius of the Earth is about 6400 km. The average distance of the Earth from the Sun is an astronomical unit defined as:

  $$1\,\text{AU} = 1.5 \times 10^8\,\text{km} \simeq 8.3\,\text{lmin}$$

  where a light-minute is defined as $1\,\text{lmin} \equiv c \times (1\,\text{min})$, with $c = 3.0 \times 10^5\,\text{km/sec}$ being light-speed. The radius of the (photosphere of the) Sun is about $7.0 \times 10^5\,\text{km} \simeq$ 109 Earths. The radius of our Solar System is about $30\,\text{AU} = 4.2\,\text{lhr}$, where a light-hour is defined as $1\,\text{lhr} \equiv c \times (1\,\text{hr})$.

- **Stars**
  Our Sun is a typical star although a star can be a thousand times larger or as small as a few tens of km in diameter (compact stars). The nearest star is Proxima Centauri (part of a triple system) at a distance of $4.2\,\text{ly} = 1.3\,\text{pc}$, where a parsec is defined as

  $$1\,\text{pc} \equiv 3.26\,\text{ly},$$

  where a light-year is defined as $1\,\text{ly} \equiv c \times (1\,\text{y})$. A parsec is the distance from Earth where a source (e.g., a star) would have parallax equal to one second, with parallax being half of the apparent angular displacement of the source on the celestial sphere per semester. Also, a parsec is the typical interstellar distance.

- **Open Star Clusters**
  These are of irregular shape and are comprised of a few tens to several thousands of stars. Typically, their radii are a few tens of light-years.

- **Globular Star Clusters**
  These are spherical conglomerations of about 10^5 stars. They are gravitationally bound systems with typical radii of a few tens of parsecs. Globular clusters are found in the halos of galaxies. Spiral galaxies like the Milky Way may have hundreds of such clusters (at least 160 for the Milky Way, while the Andromeda galaxy has more than 500), but elliptical galaxies may have thousands. They orbit their host galaxy at typical distances of the order of a few tens of kpc (kilo-parsec), where $1\,\text{kpc} \equiv 1000$ pc.

- **Galaxies**
  Galaxies are gravitationally bound systems of stars, stellar remnants, interstellar gas and dust (and dark matter). They are comprised typically of 10^{11} (a hundred billion) stars but can be as small as 10^8 stars (dwarf galaxies) or as large as 10^{14} stars (giant galaxies). They are broadly divided morphologically into spiral, elliptical, and irregular galaxies, the Milky Way being a spiral galaxy. Their size ranges as 1–$100\,\text{kpc}$. The typical intergalactic distance is about $1\,\text{Mpc}$ (megaparsec), where $1\,\text{Mpc} \equiv 1000\,\text{kpc}$.

- **Galactic Clusters and Superclusters**
  Galaxies form groups, which in turn form clusters of about hundreds to thousands of galaxies. They are the largest gravitationally bound systems observed. The typical dimensions of galactic clusters are of the order of 10 Mpc.
  Galactic superclusters comprise of a few dozens of galactic clusters and have typical dimensions of several tens of megaparsecs. They are not gravitationally bound.

- **Large scale structure**
  The distribution of galactic clusters and superclusters is not uniform, but it forms giant sheets (or walls) and filaments, which delineate a bubbly structure, called the large scale structure of the Universe (also called the cosmic web). Inside the bubbles, the density of galaxies is relatively smaller, so space is called a void. Thus, bubbles enclose roughly spherical voids with typical dimensions of order 100 Mpc. A glimpse of large scale structure as revealed by galaxy redshift observations is shown in Fig. 2.1.

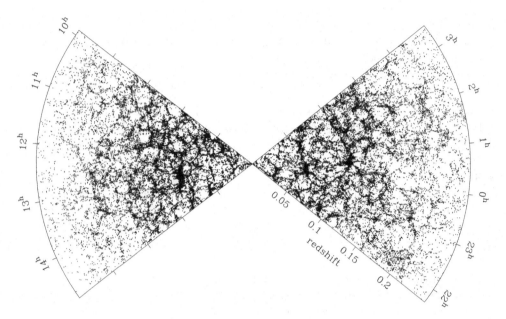

FIGURE 2.1   Observations from the 2dF Galaxy Redshift Survey reveal the distribution of the large scale structure. The survey is towards the poles of the Milky Way, away from the galactic plane. Each dot is a galaxy, and the survey reaches out to distances of about 1 Gpc (for a relation between distance and redshift, see Sec. 2.2.4). The galaxy distribution features bubbles and voids but becomes smooth on very large scales as larger structures spanning the entire field of vision are not seen. © : "the 2dF Galaxy Redshift Survey team" (www.2dfgrs.net). Reproduced with permission from Ref. [C+01].

- **Last Scattering Surface**
  Structure in the Universe is not of fractal form, i.e., one does not find even larger structures when considering larger scales. It seems that at scales larger than the bubbles and voids of the large scale structure the matter distribution becomes uniform. However, there is another signal we observe coming from distances much larger than the sizes of bubbles and voids. This is the light of the cosmic microwave background (CMB) radiation, which was emitted soon after the beginning of the Universe. Now, the further we observe the earlier the time when the signals observed were emitted because they need time to reach us. Thus, when we observe the CMB we look at a time close to the beginning of time, i.e., close to the largest possible distance we can observe. This corresponds to the dimensions of the *observable Universe* (see Sec. 2.2.2), which has a radius of about 14 Gpc (giga-parsec), where 1 Gpc $\equiv$ 1000 Mpc. The observable Universe contains roughly 10^{11} (a hundred billion) to 10^{12} (a trillion) galaxies. The surface where the CMB is emitted is engulfing the field of vision and is called the last scattering surface, for reasons which will become apparent when we discuss the CMB.

2.1.2   Standard candles

In order to chart the Universe, we need to know how big and how far away the structures we observe are. Because there is no single method that allows this for the entire range of scales, we have to rely on a multitude of overlapping methods, which calibrate one another and together constitute the *cosmic distance ladder* (see Fig. 2.2). Therefore, the cosmic distance ladder is a sequence of techniques for the estimation of distances of celestial bodies used to chart the

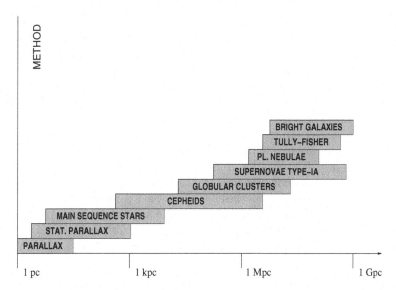

FIGURE 2.2   Schematic representation of the cosmic distance ladder. Some of the most prominent techniques are depicted, as well as their range of applicability. Overlapping ranges allow the calibration of the cosmic distance ladder.

observable Universe. Each technique in the sequence corresponds to a range of distances, greater than but overlapping with the previous technique in the sequence.

One way to estimate distances is by considering the luminosity distance r. The total power L (sometimes called absolute luminosity) radiated by some source is related to the apparent luminosity ℓ of the source as $L = \ell \times (4\pi r^2)$, where ℓ is defined as the power per unit area normal to the line of sight. If L is known and ℓ measured, one may calculate the luminosity distance as $r = \sqrt{L/4\pi\ell}$.

Objects of known absolute luminosity L are known as *standard candles*. They are used to calibrate the cosmic distance ladder. In order for a given type of object to be useful as a standard candle, it must be identifiable over a significant range of distances and have little variance (meaning it remains the same) over cosmological timescales. Prominent examples of standard candles are the following:

- **Cepheids**
  These are variable stars with a known relation between their brightness variation period and their average absolute luminosity. The prototype of these stars is δ-Cephei. Thus, by observing the period of their variation one can obtain L. Then, observing ℓ one can estimate the luminosity distance. Using Cepheids, distances can be probed within 10 Mpc or so.

- **Supernovae type Ia**
  These are white dwarfs in close binary systems that flare up when they collapse after accreting enough mass from their companion star to surpass the Chandrasekhar limit $M_C = 1.44\, M_\odot$, where $M_\odot$ is the solar mass. This means that their absolute luminosity is almost universal with a small distortion due to metallicity and other factors, which can be calibrated based on their fading times (Phillips relation). These supernovae are as bright as the host galaxy, which means they can be seen from huge distances up to 500 Mpc or so, thus reaching the nearest galactic supercluster (Virgo).

- **Galaxies**

  Galaxies themselves can be used as standard candles in various ways. One example considers spiral galaxies for which there is a specific relation between absolute luminosity and galactic rotational velocity v_{rot}. This is the *Tully-Fisher* relation: $L \propto v_{\text{rot}}^4$. Thus, if L of a given spiral galaxy is known (through some other method) then it can be used to estimate L for another spiral galaxy, provided the ratio of the rotational velocities is estimated. Note, however, that at high distances, the method becomes inapplicable because of errors in the measurement of the corresponding parameters and also because the variance in the galactic properties becomes significant over large time-scales. Distance determination using this method can reach 150 Mpc.

  Another example is considering the brightest galaxy in a galactic cluster and assuming that its absolute luminosity is universal. This method is motivated by the observed fact that the apparent luminosity of galaxies within a cluster has a sharp upper limit. Using this method, we can reach distances of about 8 Gpc, but because galaxies evolve in sizeable time intervals, the method is unreliable for $r > 1\,\text{Gpc}$.

2.2 The Universe expansion

2.2.1 Hubble-Lemaitre law

Arguably, the birth of modern cosmology occurred in 1929 with the observation by astronomer Edwin Hubble that distant galaxies are receding from us with speed proportional to their distance. The constant of proportionality H_0 is called the Hubble constant:

$$v = H_0 r \,. \tag{2.1}$$

The above relation is called the *Hubble-Lemaître law*. It was derived earlier by Friedmann and individually by Lemaître, but it was Hubble's observations that "breathed fire" to the equation.

Supernovae type Ia observations may provide an estimate of H_0, which means that, if the recession speed is estimated then the distance r can be obtained. Indeed, using the Hubble-Lemaître law, we can estimate distances up to several hundreds of megaparsecs. Traditionally, the Hubble constant is written in strange units (because they are convenient to astronomers) as

$$H_0 = 100\,h\,\frac{\text{km}}{\text{sec}\,\text{Mpc}} \,, \tag{2.2}$$

with "little h" determined by observations. Latest estimates suggest $h = 0.674 \pm 0.005$. The above choice of units makes distance estimation straightforward. For example, a galaxy which is receding with speed $v = 10^4\,\text{km/s}$ is at a distance of $100/h$ Mpc.

However, the Hubble-Lemaître law had a much deeper impact than being just another method of estimating cosmological distances. Indeed, it gave rise to the notion, the observational fact, that our Universe is expanding. This is a collective behaviour of the Universe at large, corresponding to scales well beyond intergalactic distances. Thus, Hubble's observation revealed that there are global, collective dynamics which supersede the peculiar motions of individual galaxies. The latter are considered as parts of a coherent whole, whose behaviour sweeps galaxies and galactic clusters into a grander evolution. It was the first evidence of a distinct behaviour of the Universe, which was not simply a sum of its parts. In order to begin understanding this, we first need to ask: Are we at the centre of the Universe? At face value, Hubble-Lemaître law suggests this very thing. After all, distant galaxies are receding from *us*—all of them!

In order to answer this question, we use Hubble-Lemaître law in vector form: $\vec{v} = H_0\,\vec{r}$. Consider two arbitrary galaxies with position vectors $\vec{r}_1$ and $\vec{r}_2$, which are far away enough not to be gravitationally bound either with our own galaxy at the origin or with each other, as depicted in Fig. 2.3. Then, it is straightforward to show

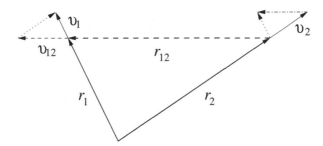

FIGURE 2.3 Hubble-Lematre law applies not only for the observer at the origin but also for the observers with position vectors $\vec{r}_1$ and $\vec{r}_2$.

$$\left.\begin{array}{r} \vec{v}_1 = H_0\,\vec{r}_1 \\ \vec{v}_2 = H_0\,\vec{r}_2 \end{array}\right\} \;\Rightarrow\; \vec{v}_1 - \vec{v}_2 = H_0\,(\vec{r}_1 - \vec{r}_2) \;\Rightarrow\; \vec{v}_{12} = H_0\,\vec{r}_{12}\,. \qquad (2.3)$$

In view also of $-\vec{v}_2 = H_0\,(-\vec{r}_2)$, we see that Hubble-Lemaître law is also true for the observer at galaxy number 2. Observer 2 also finds that distant galaxies are receding from him/her with speed proportional to their distance, with the same proportionality constant H_0. Similarly, because $-\vec{v}_{12} = H_0\,(-\vec{r}_{12})$ and $-\vec{v}_1 = H_0\,(-\vec{r}_1)$ we find that the same is true for the observer at galaxy number 1. Thus, we conclude that the Hubble-Lemaître law is true for all observers in the Universe, wherever they are located. All distant galaxies are receding from each other. We started by suspecting that we live at a special place (the centre of the Universe!) and we ended up reaffirming the Copernican principle, that there is nothing privileged about our location in the Universe, i.e., the Earth is not the centre of the Universe after all.

Therefore, we can talk about a global expansion, which occurs everywhere at the same rate, characterised by H_0. Thus, Universe expansion means that all unbound parts of the Universe move away from each other. Crucially, "move away" does *not* mean travel through space. It is space itself which is expanding between them. To clarify "unbound" here, we should define bound objects. These are objects held together by an attractive force. For example, stars in a globular star cluster or a galaxy are gravitationally bound. Similarly, protons and electrons in atoms are electromagnetically bound. Bound objects like atoms or galaxies do not grow in size because of Universe expansion. This means that the forces of Nature are not affected by Universe expansion.*

A useful metaphor to understand the Universe expansion pictorially is the following. Consider a river-mouth, as depicted in Fig. 2.4. The width of the river increases as it approaches the sea. Suppose we put two pieces of cork floating on the river waters at a given distance apart. As the pieces of cork float towards the sea, the distance between them grows, following the river flow, which opens up. These pieces of cork are unbound and their motion traces the river flow. Now consider connecting the two pieces with a solid rod. Now, the pieces of cork are bound and the distance between them (the length of the rod) stays unchanged as they both float towards the sea. Of course, there is an intermediate situation, when the pieces of cork are linked with a loose spring. If the spring is loose enough, the distance between the rods would grow but not as much as when the two pieces were unbound. Now, the pieces of cork only partially trace the river flow. In a similar way, structures follow the *Hubble flow* of the Universe expansion, with

---

*Of course, this is only a model. We can equally assume that there is no space expansion but that the forces are getting stronger with time so that the size of objects is diminishing. This picture, however, is more difficult to gauge (plus there are some additional assumptions to be made so that this picture works, related for example to the quantum world), so the expansion of space is a better choice.

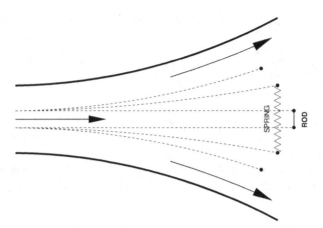

FIGURE 2.4 The black dots correspond to pieces of cork travelling down a river mouth, which can be used as a metaphor for galaxies following the Universe expansion (Hubble flow). The dashed lines depict their trajectories. If the pieces of cork are free, they follow the river flow. If they are attached at the ends of a solid rod, they remain a fixed distance apart (bounded). If they are connected through a loose spring, they partially follow the river flow and they are partially bounded.

galaxies being fully bound and galactic superclusters unbound, while structures in between, such as galactic clusters of different sizes, being semi-bound depending on their size.

2.2.2 Observable Universe

As we have seen, Hubble's observations demonstrated that the Universe is not static but expanding. This means that, if we go backwards in time, then the Universe contracts. In fact, all points in space merge together in finite time. Thus, we conclude that the Universe began with an explosion at all points in space. This proposal sounded preposterous when it first appeared. In fact, astronomer Fred Hoyle in 1949 dismissed the idea in a BBC radio broadcast, calling it one "Big Bang." However, the name stuck, and now the cataclysmic explosion, which marked the beginning of the Universe, is actually called the Big Bang, the onset of time itself. The universality of the Hubble-Lemaître law demonstrates that the rate of expansion is uniform, which means that there are not many "Big Bang" explosions in different locations but that the Big Bang happened once and everywhere.

A finite age for the Universe means that there is a finite distance that signals can travel, because nothing can move faster than light-speed. Using $v \leq c$, Hubble-Lemaître law suggests that the maximum observable distance is

$$D_H \simeq c/H_0 \qquad (2.4)$$

Signals from beyond D_H cannot reach us because the time required is larger than the age of the Universe. D_H is, therefore, the range of causal correlations and is called the *cosmological horizon*. Thus, the cosmological horizon can be defined as the dimensions of a region in the Universe which is causally connected.

The horizon is defined with respect to the observer, which means different observers in the Universe correspond to different locations of a horizon. Also, the horizon grows with time as the range of causal correlations $D_H(t) \simeq ct$. Comparing this expression to the above, we see that $H_0 \sim 1/t_0$, where t_0 is the age of the Universe. It is important to note here that the above expressions are crude order-of-magnitude estimates, which only serve to convey the physical meaning of the horizon. We revisit the horizon later on, in a more rigorous way.

Now, it is evident that there is more of the Universe beyond the horizon, in the same way that there are more lands beyond our horizon on Earth. But any regions of the Universe beyond the horizon are unobservable. Thus, our horizon encompasses our *observable Universe*. We define, therefore, the observable Universe as the region of the Universe spanned by the horizon at present. Equivalently, we can say that the observable Universe is the part of the Universe which is within causal contact with us.

The observable Universe is of finite volume (a sphere of radius D_H). Often, when people discuss the size of the Universe, they mean the observable Universe. The global Universe may well be infinite in size. We really have no information about the actual global Universe apart from the trivial requirement that it should not be wildly different to our observable Universe for modest distances $\sim 10\,D_H$, just beyond the horizon, for otherwise there would be observable effects. For distances much larger than D_H, though, little if anything can be determined. Such regions of the actual global Universe, however, are not (and never have been*) in causal contact with our observable Universe and so their form is actually academic. To simplify things, therefore, we consider that the rest of the global Universe is more of the same, i.e., identical to our observable Universe. Thus, we say that the Universe is a global construction whose properties we conjecture by extrapolating the observations within our observable Universe. We can do this without loss of generality, keeping in the back of our mind, though, that the actual reality can be very different. We revisit this conjecture much later, when we discuss eternal inflation in Sec 8.8.

2.2.3   Scale factor and Hubble parameter

As we have explained, the expansion of the Universe at large scales is uniform because the Hubble-Lemaître law is universal. This means that structures following the Hubble flow grow in size in a self-similar manner, just like features in a magnified photograph (but three-dimensional). To picture this, consider a particular galactic supercluster in the shape of a little bunny. After some time, because of the Universe's expansion, the supercluster becomes larger. It still, however, retains its original bunny shape, because the uniformity of the expansion means that there are no locations or directions that expand faster than others. Were it not so, the shape of the cluster would have been deformed, stretched or bloated. Instead, if the distance from the bunny nose to the tail has doubled, say, other similar features (e.g., the bunny ears) would have doubled in size too. Thus, because the Universe expansion is self-similar, it can be parametrised by a single factor $a(t)$, which is a function of time only. If after some time, features in the large scale structure double in size then then $a(t_f) = 2a(t_i)$, where t_i and t_f are the initial and final times, respectively. We call $a(t)$ the *scale factor* of the Universe. Therefore, the scale factor $a(t)$ is a function of cosmic time, which parametrises the Universe expansion.

We can use $a(t)$ to factor out the Universe expansion by defining a coordinate system that follows the expansion by moving with it. Such coordinates are called *comoving*. Comoving distance x is related to physical distance r as

$$r(t) = a(t)x\,. \qquad (2.5)$$

The expansion does not affect comoving distances, which means x is not varying with the expansion. Our bunny supercluster would remain invariant in the comoving reference frame.

To better visualise the Universe expansion it is helpful to think of it like this. Imagine a scaffolding, which extends to infinity in all three dimensions as depicted in Fig. 2.5. Then, consider that the characteristic lengthscale of the scaffolding (the bay) is growing in time, so the entire lattice is enlarged. This depicts the grid of the comoving coordinates and the growth

---

*Unless extensions beyond the Hot Big Bang are considered, like cosmic inflation.

of the bays is the Universe expansion. As you can see, there is no centre to the expansion. Any figure that we can draw by connecting neighbouring joins together retains its shape because the growth of the scaffolding is self-similar. If we go backwards in time, the size of the bays contracts. When the size of the bays diminishes to zero, all distances connecting any arbitrary joins in the scaffolding become zero. This is the Big Bang, and it happens simultaneously at all points in space.

The vector form of Eq. (2.5) is $\vec{r}(t) = a(t)\vec{x}$, where $\vec{x}$ and $\vec{r}$ are the comoving and the physical position vectors, respectively. Taking the time derivative we obtain

$$\vec{v} = \frac{d\vec{r}}{dt} = \frac{da}{dt}\vec{x} + a\frac{d\vec{x}}{dt} = \left(\frac{\dot{a}}{a}\right)\vec{r} + \vec{v}_{\text{pec}} \Rightarrow \vec{v} = \vec{v}_H + \vec{v}_{\text{pec}} , \qquad (2.6)$$

where $\vec{v}_{\text{pec}} \equiv a\,\dot{\vec{x}}$ is the *peculiar velocity* due to proper motion, e.g., of galaxies revolving around each other (nothing to do with the Universe expansion), and $\vec{v}_H \equiv H\,\vec{r}$ is the Hubble flow due to Universe expansion only, with the dot denoting a derivative with respect to the cosmic time. In the above we have defined

$$H(t) \equiv \frac{\dot{a}}{a} , \qquad (2.7)$$

which is called the *Hubble parameter*. The Hubble parameter is, therefore, the fractional growth rate of the scale factor. Since $a(t)$ parametrises the expansion, the Hubble parameter is the rate of the Universe expansion. Now, for negligible peculiar velocity, we obtain $\vec{v} \simeq H\,\vec{r}$. Evaluating this at present, we recover Hubble-Lemaître law by identifying the Hubble constant with the present value of the Hubble parameter $H_0 \equiv H(t_0)$. Because the peculiar velocity of galaxies is of the order $v_{\text{pec}} \sim 100$ km/sec, we see that the Hubble flow dominates peculiar motions for distances larger than a few tens of megaparsecs.

From Eq. (2.5), since physical and comoving distances are positive, we see that $a > 0$, except at the Big Bang itself where physical distances are zero so that $a = 0$. Choosing the dimensions of the comoving coordinates to be dimensions of length, like the physical coordinates, implies that the scale factor is dimensionless.* Note also that the scale factor has only relative values and can be normalised to unity at any convenient time. On many occasions, $a(t)$ is normalised to unity at present, such that $a_0 \equiv a(t_0) = 1$. However, this is not necessarily so, in general. We will keep the normalisation of the scale factor undefined unless explicitly mentioned. This freedom in evaluating the scale factor suggests that it is not a physical quantity and that it can only be related to physical quantities, such as $H(t)$, in ratios, as in Eq. (2.7).

2.2.4   Cosmological redshift

As we have seen, the Hubble expansion of space results in distant galaxies moving away from us. This motion of galaxies with respect to us affects identifiable sets of absorption and emission lines in their light spectra due to the Doppler effect on light waves. In a nutshell, the Doppler effect is the following: If a source moves away from us then light waves spread out thereby lowering the frequency. Consequently, we say that spectral lines get redshifted. Conversely, if a source moves towards us then light waves get crowded together raising the frequency and we say that spectral lines get blueshifted. Because the Hubble-Lemaître law implies that most galaxies are receding from us, we talk about redshift and not blueshift in cosmology. Thus, the cosmological redshift corresponds to the stretching of the wavelengths of light due to the Universe expansion.

---

*It was not always like this. In the past, the comoving coordinates were chosen to be dimensionless, which meant that the scale factor had dimensions of length. It used to be symbolised as $R(t)$.

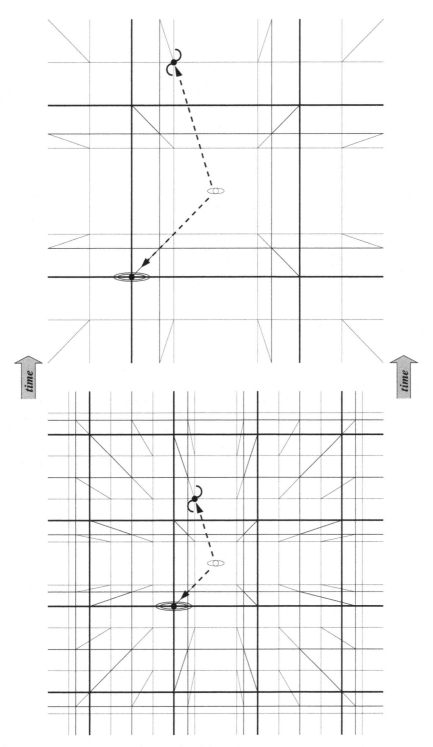

FIGURE 2.5 The grid of comoving coordinates, which extends as a scaffold in all three-dimensional space. The characteristic distance (the bay) of the scaffold is growing with time in all directions, resulting in the growth of physical distances.

The redshift is defined as:

$$z \equiv \frac{\lambda_0 - \lambda}{\lambda}, \tag{2.8}$$

where λ is the emitted wavelength while λ_0 is the observed wavelength. Since the scale factor parameterises the growth of lengthscales in the Universe, we have $\lambda \propto a$. As a result, the above gives

$$\frac{\lambda_0}{\lambda} = \frac{a_0}{a} \Rightarrow \frac{a_0}{a} = z + 1. \tag{2.9}$$

For distances smaller than the horizon, the Doppler shift of light wavelengths gives approximately $z \simeq v/c$, where v is the recession velocity due to the Hubble flow. Then Eq. (2.1) suggests that $D = cz/H_0$, meaning that the redshift is proportional to the distance.[*] However, Hubble-Lemaître law breaks down for $z \gtrsim 1$ due to relativistic corrections (since $v \simeq c$).[†] Also, Hubble-Lemaître law becomes inapplicable anyway because of the evolution of the Hubble parameter $H(t)$; recall that the Hubble constant is $H_0 \equiv H(t_0)$, where t_0 is the present time. Still, we can continue to employ Eq. (2.9), but we have to remember that when $z \gg 1$, redshift is not proportional to distance because the latter is approximately the radius of the horizon at present.

2.3   Cosmic dynamics

2.3.1   Cosmological principle

Hubble-Lemaître law shows that the Universe expansion is uniform, which suggests that the content of the Universe is also uniformly distributed. Indeed, as already discussed, the distribution of galactic clusters and superclusters becomes uniform on scales above 100 Mpc, which is about a few percent of the size of the observable Universe. This is observational support for the *cosmological principle*, which simply states that, on large scales, the Universe is homogeneous and isotropic. Homogeneity means that the Universe looks the same at every point in space and isotropy means that the Universe looks the same towards any spatial direction. Hence, according to the cosmological principle, the Universe is invariant under translations and rotations. Consequently, there are no special points or directions in the Universe, no centre, no edges, and no axis of rotation. Additionally, each region contains enough information to account for a complete global description of the Universe.

Homogeneity and isotropy are distinct properties of the Universe. To understand these terms better, consider a uniform universe permeated by a uniform magnetic field. This Universe would be homogeneous but not isotropic because the magnetic field selects a preferred direction.[‡] Conversely, consider a Universe with a spherically symmetric distribution of matter. If we find ourselves at the centre of the matter distribution, this Universe would look isotropic but would not be homogeneous. There is a mathematical theorem that states that global isotropy enforces homogeneity.

As stated, the cosmological principle applies at large scales, where structure is coarse-grained and becomes smooth. One can view the Universe content as a fluid in thermal equilibrium, which

---

[*]This is really convenient for measuring the distances of galaxies provided H_0 is known.

[†]The exact relativistic expression is:

$$z = \sqrt{\frac{1 + v/c}{1 - v/c}} - 1,$$

which reduces into $z \simeq v/c$ when $v \ll c$. However, when $v \to c$ we have $z \gg 1$.

[‡]If we send an electrically-charged particle, it will either continue to move in a straight line if its velocity is parallel to the magnetic field lines, or it will undergo Larmor circular motion if its velocity is perpendicular to the magnetic field lines, or spiral away, if its velocity is at an intermediate angle with the magnetic field lines. Thus, the motion of this particle is diretion dependent, which means space is not isotropic.

is homogeneous and isotropic at scales much larger than molecular level. In this picture, galaxies are actually test particles, following the flow of the fluid, with their peculiar motions (the fluid equivalent of molecular thermal vibrations) ignored.

The cosmological principle was originally put forward by Sir Isaac Newton himself when discussing universal gravitation. As a conjecture, it was assumed by early works of Friedmann and Lemaître in the beginning of the 20th century in order to simplify the newly introduced Einstein equations and allow the mathematical study of the global Universe dynamics. However, it turned out that we are fortunate enough that the cosmological principle is more than a simplifying assumption. It is supported by the universality of Hubble-Lemaître law and the large scale structure distribution, which seams to be smooth over scales larger than bubbles and voids. It is also strongly supported by the overwhelming isotropy of the observed CMB radiation (discussed later on). Thus, the cosmological principle has been fleshed out into an observational fact.*

2.3.2 Friedmann equation

Alexander Friedmann was the first to study the dynamics of the global Universe in 1922, employing Einstein's theory of general relativity (GR), which appeared a few years earlier. Assuming the cosmological principle to simplify equations, he modelled the content of the Universe as a fluid with density $\rho(t)$, which is a function of time only, like the scale factor $a(t)$. Then, the temporal-temporal component of the Einstein equations becomes

$$H^2 \equiv \left(\frac{\dot{a}}{a}\right)^2 = \frac{8\pi G}{3}\rho - \frac{kc^2}{a^2}, \tag{2.10}$$

where G is Newton's gravitational constant and k is the curvature parameter, which determines the spatial geometry (more on this later). The above equation is named the *Friedmann equation* and is of paramount importance because it demonstrates that, in general, the Universe is not static when $\rho \neq 0$ (for a static Universe $H = 0$ because $a = $ constant). Mirroring GR, the Friedmann equation demonstrates that the dynamics of the Universe are determined by a balance between its geometry and its content, with the rate of expansion H accounting for the geometry in the time dimension. How Eq. (2.10) is obtained in GR is briefly discussed in Appendix A.

2.3.3 Continuity equation

Equation (2.10) is not enough to determine the evolution of the Universe, because it is one equation linking two unknown functions $a(t)$ and $\rho(t)$. In order to obtain another equation involving these functions we reason as follows. Density is mass per spatial volume $\rho = M/V$. The volume, in three-dimensional expanding space, grows as $V \propto a^3$. This is evident if one thinks of the volume of a sphere of radius r: $V = \frac{4\pi}{3}r^3$, where $r \propto a$ as suggested in Eq. (2.5). Similarly, for any three-dimensional volume, e.g., a cube with edge equal to r for which $V = r^3$. Derivating $V \propto a^3$ with respect to time we have $\dot{V}/V = 3\dot{r}/r = 3\dot{a}/a = 3H$, where we used Eqs. (2.5) and (2.7).

---

*For a long time, a misconception, the *cosmic egg fallacy*, plagued the research community. In a nutshell, the proposal was that the Universe originally used to be infinitely compressed in a single point inside an otherwise empty background space: the "cosmic egg" (also called the "primordial atom"). This "egg" exploded with a "Big Bang," and matter has since then been spreading outwards into the empty external space. This demonstrated how "common sense" prejudices led to confusion. One way to debunk the fallacy is by contrasting it with the cosmological principle, because for someone near the edge of the expanding matter distribution, the Universe would look neither homogeneous nor isotropic, as one side would be almost empty of matter. Also, superluminal relative motion (as in cosmic inflation) would be impossible because of relativity.

Thus, for the density we find

$$\frac{\dot{\rho}}{\rho} = \frac{\dot{M}}{M} - \frac{\dot{V}}{V} \quad \Rightarrow \quad \dot{\rho} + 3H\rho = \rho\frac{\dot{M}}{M}. \tag{2.11}$$

The above means that if the mass within a volume which grows with the Universe expansion (called a comoving volume) remains invariant (i.e., $M =$ constant), then $\dot{\rho} + 3H\rho = 0$ and the density is diluted due to the growth of the comoving volume only (this is the case of dust, as explained later). However, in general, we have

$$\dot{\rho} + 3H\left(\rho + \frac{p}{c^2}\right) = 0, \tag{2.12}$$

where we have set $p = -\frac{\rho c^2}{3H}\frac{\dot{M}}{M}$. The above equation is called the *continuity equation* (also called the fluid equation). The quantity $p(t)$ is called isotropic pressure and it expresses work done by the Universe expansion. The continuity equation expresses local energy conservation.

Here we should note that the continuity equation is also valid individually for all independent components ρ_i of the density ρ, where independent components are the ones which do not net decay into one another.

2.3.4 Equation of state

We have obtained a second equation determining the dynamics of the Universe, but we also introduced a third unknown function $p(t)$, besides $a(t)$ and $\rho(t)$. To close the system we consider the Universe content as a collection of barotropic fluids with equations of state of the form $p_i = p_i(\rho_i)$, i.e., each type of density is associated in a unique manner with its pressure. The equations of state of the cosmological perfect fluids are written as

$$p_i = w_i \epsilon_i = w_i \rho_i c^2, \tag{2.13}$$

where w_i is the barotropic parameter and $\epsilon_i \equiv \rho_i c^2$ is the energy density of the i−th fluid. The energy density is defined as energy per volume $\epsilon = E/V$, with energy $E = Mc^2$.

The barotropic parameter of the Universe is defined as

$$w \equiv \frac{\sum_i p_i}{(\sum_i \rho_i)c^2}. \tag{2.14}$$

Practically, the above means that the barotropic parameter of the Universe is the barotropic parameter of the dominant component of the Universe content, as there are only a few occasions when two or more components have comparable contributions to the density (but the present is one such occasion), i.e.,

$$w = p/\epsilon = p/\rho c^2, \tag{2.15}$$

where the energy density of the Universe is $\epsilon \equiv \rho c^2$. For a barotropic fluid with $p \geq 0$, the speed of sound waves c_s is given by

$$c_s^2 = \frac{\partial p}{\partial \rho} = wc^2. \tag{2.16}$$

Demanding that c_s is not superluminal implies that $w \leq 1$. However, negative values of w (and p) are also considered.

The continuity equation for a given independent fluid can be written as $\dot{\rho}_i + 3(1 + w_i)H\rho_i = 0$. Integrating, we find $\rho_i \propto a^{-3(1+w_i)}$, assuming $w_i =$ constant. Considering the dominant component of the Universe content this becomes

$$\rho \propto a^{-3(1+w)} \qquad \text{for } w = \text{constant}. \tag{2.17}$$

Consider a fluid comprised of particles of mass m and energy E_p with $E_p = \sqrt{(mc^2)^2 + (Pc)^2}$, with P being the particle momentum. We can obtain the fluid equation of state if we consider a (comoving) box of volume V filled with this fluid. How much energy is included in the box, and how much pressure is applied to the sides of the box by the motion of the fluid particles? By linking the two we can find the equation of state.

First, consider nonrelativistic particles, i.e., particles that move much slower than lightspeed. These particles hardly move, so the pressure they apply is negligible (they don't hit the box walls). Nonrelativistic matter is simply called *matter* in cosmology. Thus, we find that matter is pressureless and $p_m = 0$. The barotropic parameter of matter is $w_m = p_m/\rho_m c^2 = 0$. This kind of matter is also called dust. Galaxies, whose motion is not relativistic, are considered dust, and if they dominate the Universe, then we have a period of *matter domination* during which the Universe pressure is zero and $w = 0$.

Now consider relativistic particles. These particles travel at lightspeed. One can estimate their pressure as follows: $p = F/A = \Delta P/A\Delta t = \Delta P\, c/A(c\Delta t) = E/V = \epsilon$, where A is the area of the side of the box, Δt is the time it takes for the particles to cross the box (such that $V = A(c\Delta t)$), the force on the box wall is $F = \Delta P/\Delta t$ and we considered that $E = \Delta P c$ since the particles are relativistic so that $E_p \simeq Pc$. Now, only one third of the particles included in the box are impacting on a given area because space is three-dimensional. Relativistic matter is called *radiation* in cosmology. Thus, the equation of state of radiation is $p_r = \frac{1}{3}\epsilon_r = \frac{1}{3}\rho_r c^2$, which means that the barotropic parameter is $w_r = \frac{1}{3}$. When relativistic particles, e.g., photons or neutrinos (or both), dominate the Universe then we have a period of *radiation domination* during which the Universe pressure is $p = \frac{1}{3}\rho c^2$ and $w = \frac{1}{3}$.

Using our finding that $\rho_i \propto a^{-3(1+w_i)}$ for $w_i =$ constant, we obtain $\rho_m \propto a^{-3}$ and $\rho_r \propto a^{-4}$, which means radiation is diluted faster than matter with the expansion of the Universe. Considering the Universe content only as a collection of relativistic and nonrelativistic particles, we see that the contribution of relativistic particles in the density budget of the Universe is increasing as we move to earlier times because $\rho_r/\rho_m \propto 1/a$. Hence, regardless of its current relative abundance, radiation dominates the early Universe.

The different scaling of the densities of matter and radiation with the Universe expansion can also be understood as follows. The energy density is

$$\epsilon_i = \rho_i c^2 = n_i E_p \,, \tag{2.18}$$

where n_i is the number density and E_p is the typical particle energy. The number density is $n_i = N_i/V$, where N_i is the number of particles in a comoving volume V. Now, for an independent fluid (because there is no net decay) $N_i =$ constant, which means $n_i \propto a^{-3}$, since $V \propto a^3$. In the case of matter, $mc^2 \gg Pc$ and $E_p \simeq mc^2 =$ constant. This means that $\rho_m = mn_m \propto a^{-3}$. In the case of radiation, $mc^2 \ll Pc$ and $E_p \simeq Pc = hc/\lambda$, where λ is the de Broglie wavelength and h is the Planck constant. One can understand this by considering photons, whose energy is $E_p = hf$, with $f = c/\lambda$ being the frequency. Because of the Universe expansion, the wavelength grows as all lengthscales as $\lambda \propto a$. Thus, the particle energy is diminished as $E_p \propto 1/a$. Therefore, the density of radiation scales as $\rho_r = n_r E_p/c^2 \propto a^{-3} \times a^{-1} = a^{-4}$. In summary, we have

Matter (nonrelativistic matter) $\quad E_p = mc^2 \Rightarrow \rho_m = mn_m \quad \Rightarrow \rho_m \propto a^{-3}, \quad (2.19)$

Radiation (relativistic matter) $\quad E_p = Pc = hf = hc/\lambda \propto a^{-1} \Rightarrow \rho_r \propto a^{-4}. \quad (2.20)$

Hence, radiation is diluted faster than matter because its wavelength is enlarged (redshifted) by the Universe expansion. Note that radiation was not assumed to be in thermal equilibrium (see next chapter).

2.3.5   Acceleration equation

By combining the Friedmann and the continuity equations it is easy to find the *acceleration equation*

$$\frac{\ddot{a}}{a} = -\frac{4\pi G}{3}\left(\rho + \frac{3p}{c^2}\right). \tag{2.21}$$

The above can be directly obtained by the Einstein equations (see Appendix A). Sometimes, Eqs. (2.10) and (2.21) are both called Friedmann equations. After Friedmann, they were independently obtained by Georges Lemaître in 1927.

In view of the equation of state Eq. (2.15), the acceleration equation is written as $\ddot{a}/a = -\frac{4\pi}{3}(1+3w)G\rho$. Hence, we find that accelerated expansion ($\ddot{a} > 0$) is possible only if the pressure is negative enough: $w < -\frac{1}{3}$, where we considered that $a > 0$.

2.4   Cosmic geometry

2.4.1   The curvature parameter

The curvature parameter in Eq. (2.10) is a dimensionful constant with dimensions of $[\text{length}]^{-2}$. This means that it cannot have a value of ± 1 (despite contrary claims in the literature) unless a unit system is chosen such that length is dimensionless. In general, the value of the curvature parameter is given by the inverse-square of the comoving radius of curvature X_c:

$$|k| = 1/X_c^2. \tag{2.22}$$

This is because a homogeneous and isotropic space can have only three types of global geometry, each with constant comoving radius of curvature (constant because there are no special locations or directions). These three possibilities are:

- **Euclidean (flat) geometry.** This corresponds to $k = 0$ ($X_c = \infty$).
  All the geometrical facts, known from school, are valid. In particular, parallel straight lines remain a fixed distance apart, the angles of a triangle add up to 180°, and the circumference of a circle of radius r is $2\pi r$. The volume is infinite.
- **Spherical (closed) geometry.** This corresponds to $k > 0$ ($k = 1/X_c^2$).
  In this case, parallel straight lines meet at a finite distance, the angles of a triangle add up to more than 180°, and the circumference of a circle of radius r is smaller than $2\pi r$. The volume is finite.
- **Hyperbolic (open) geometry.** This corresponds to $k < 0$ ($k = -1/X_c^2$).
  In this case, parallel straight lines diverge away from each other, the angles of a triangle add up to less than 180°, and the circumference of a circle of radius r is larger than $2\pi r$. The volume is infinite.

In the above, by straight line, we mean the shortest distance between two points (geodesic). Note that, in contrast to its sign, the value of k is not a physical quantity, but it depends on the normalisation of the scale factor. This is true for comoving distances in general.

Because it is difficult to visualise curved three-dimensional (3-D) space, we have to downgrade to two dimensions (2-D), so we consider a 2-D surface. The comoving scaffolding is now an infinite plane with a square grid, whose characteristic distance is growing in time. This is the flat case.

The closed geometry corresponds to the surface of a sphere. There are no edges to the surface but the total area is finite. Similarly, upgrading to 3-D, the total volume of space with closed geometry is finite even though there are no edges. On the surface of a sphere, if one travels in any direction in a straight line then one eventually returns to the point of origin (after completing a great circle on the sphere). Similarly in 3-D closed geometry, only "any direction" now refers to

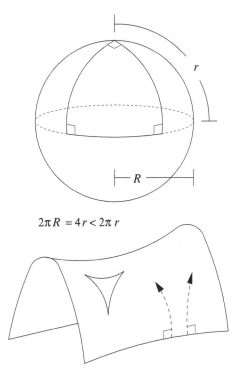

$$2\pi R = 4r < 2\pi r$$

FIGURE 2.6 Depiction of 2-D spherical (above) and hyperbolic (below) geometry. In spherical geometry, a triangle has sum of angles larger than 180° and parallel lines (perpendicular to the equator) meet at a finite distance (r), while the circumference of a circle of radius r is less than $2\pi r$ (for example, the circumference of a great circle is $4r < 2\pi r$). In hyperbolic geometry, a triangle has sum of angles smaller than 180° and parallel lines diverge from each other.

all possible directions in 3-D space. On the 2-D spherical surface, we can introduce a geographical coordinate system. Then it is evident that two neighbouring lines that are orthogonal to the equator meet at the pole. These lines are parallel at the equator and yet they meet at a finite distance, while the triangle they form with the equator features two right angles plus the pitch angle at the pole, so the sum of angles is larger than 180°. The same applies in 3-D.

The expansion, in the case of a 2-D closed geometry, corresponds to the enlargement of the area of an inflating spherical balloon. Note that if we reverse time, the balloon is shrinking until the area and all the distances between points on the surface of the balloon diminish to zero. Here we should point out that a curved geometry is perfectly possible without embedding space in extra dimensions. This means that our 3-D balloon surface is not necessarily expanding in a higher dimensional space.

The flat geometry is simply the closed geometry, with the radius of curvature enlarged to infinity $X_c \to \infty$. Visualising the open geometry is not as straightforward. The hyperbolic 2-D surface resembles a saddle. Indeed, if we draw a circle of radius r on a saddle then its circumference is larger than $2\pi r$. The two non-flat geometries (spherical and hyperbolic) described above are depicted in Fig. 2.6.

The cosmological principle suggests that the Universe is either closed, flat, or open, corresponding to the 3-D geometries we have discussed. This is always true regardless of the Universe content. As we have seen, the volume of the Universe is finite only in the case of spherical geometry (closed Universe). However, we cannot be certain of the actual geometry on scales significantly larger than the horizon, because the actual Universe may not respect in its entirety the cosmological principle. Still, as we have explained, the global Universe considered is merely an extrapolation of the observable Universe, where the cosmological principle holds true. Because the evolution of our observable Universe cannot be affected by scales significantly larger than the horizon, the actual geometry of the Universe is irrelevant, and we can postulate that the cosmological principle is universal without problem.

2.4.2 Critical density

The *critical density* ρ_c is defined as the density that a flat Universe would have, for a given value of the Hubble parameter H. Setting $k = 0$ in Eq. (2.10), we find

$$\rho_c(t) \equiv \frac{3H^2}{8\pi G} \tag{2.23}$$

because $H(t)$ is a function of time. We now define the *density parameter* as the ratio of the actual density of the Universe over the critical density

$$\Omega(t) \equiv \frac{\rho}{\rho_c} \, . \tag{2.24}$$

Particular components have individual density parameters $\Omega_i(t) \equiv \rho_i/\rho_c$. By dividing Eq. (2.10) with H^2 we can rewrite the Friedmann equation as

$$\Omega - 1 = \frac{kc^2}{a^2 H^2} \, . \tag{2.25}$$

In view of the above, it is straightforward to find

$$
\begin{aligned}
\rho > \rho_c &\Rightarrow \quad \Omega > 1 \Leftrightarrow k > 0 \quad : \text{Closed Universe}\,, \\
\rho < \rho_c &\Rightarrow \quad \Omega < 1 \Leftrightarrow k < 0 \quad : \text{Open Universe}\,, \\
\rho = \rho_c &\Rightarrow \quad \Omega = 1 \Leftrightarrow k = 0 \quad : \text{Flat Universe}\,.
\end{aligned}
\tag{2.26}
$$

Thus, the density of the Universe and how it compares with the critical density determines the global geometry. This may by understood by considering that a large density curves space substantially, such that $\rho > \rho_c$ forces space to close onto itself resulting in a closed Universe.

If we consider that the content of the Universe is matter and radiation, because $\rho_m \propto a^{-3}$ and $\rho_r \propto a^{-4}$, the density term diminishes with time faster than the curvature term, $-kc^2/a^2 \propto a^{-2}$, in the right-hand side of the Friedmann equation (2.10). Going backwards in time, this means that the curvature term is negligible compared to the density term in the early Universe. Therefore, the early Universe is effectively flat.

2.5 Dynamics and curvature

2.5.1 The Friedmann universes

We now assume that the content of the Universe is comprised only of matter and radiation. Because ρ_m and ρ_r decrease faster than a^{-2} with the Universe expansion, the curvature term in the Friedmann equation (2.10) is increasingly important and determines the fate of the Universe. Since there are three possibilities for k, we have three possible outcomes, called the Friedmann universes.

- **Closed Universe:** $(k > 0)$
  In this case, the curvature term eventually becomes comparable to the density term in the right-hand side of Eq. (2.10). Momentarily $H = 0$ and the Universe stops expanding (since $\dot{a} = 0$). Indeed, a closed Universe stops expanding in finite time, and afterwards it collapses down to a Big Crunch, which is the opposite to the Big Bang as all

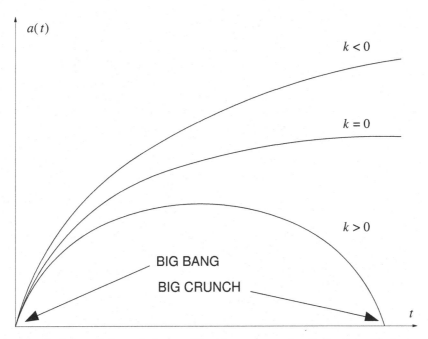

FIGURE 2.7   Evolution of the scale factor $a(t)$ with respect to time in the three Friedmann universes. The beginning of time is taken to be the Big Bang. Very early in the history of the Universe, the evolution is similar and the three curves are near each other. This is because the curvature is negligible and the early Universe is approximately flat. If the Universe is closed ($k > 0$), then it expands to a maximum value, momentarily stops expanding, and then recollapses to a Big Crunch. This case corresponds to a cycloid curve. In contrast, if the Universe is flat ($k = 0$) or open ($k < 0$), it expands forever. The flat case is actually half a cycloid curve stretched to infinity, such that the maximum value is asymptotically approached when $t \to \infty$. In all cases, the curves are concave, meaning they are curved downwards. This is because the expansion is decelerating in all cases, $\ddot{a} < 0$.

distances decrease to zero. Note, however, that there is no time reversal. This is due to the second law of thermodynamics, which demands that the entropy of a closed system (and the Universe is one) cannot decrease.

- **Open Universe:** ($k < 0$)
  In this case, both terms on the right-hand side of Eq. (2.10) are positive, which means that the Universe expands forever because H remains positive. In time, the density term becomes negligible, so that the Friedmann equation reduces to $\dot{a} = \sqrt{-k}c$, suggesting $a \propto t$. Such an empty universe is called the Milne Universe.

- **Flat Universe:** ($k = 0$)
  This is the limiting case between the two cases above. The Universe is asymptotically static because the rate of the expansion H is approaching zero in infinite time. Thus, the Universe expands forever.

From the above, we see that only a flat or open Universe expands forever. A closed Universe exists only for a finite time, so it is finite both in space and time. The three Friedmann universes are shown in Fig. 2.7. Notice that, in all cases, $a(t)$ is curved downwards. This is because the barotropic parameter of the Universe leads to decelerated expansion (both $w_r = \frac{1}{3}$ and $w_m = 0$ are larger than $-\frac{1}{3}$). Note also that near the Big Bang the three curves merge, as the early Universe is approximately flat.

2.5.2   Closed, matter-dominated Universe

It is instructive to look at the closed Universe in a bit more detail. Consider matter domination. As we have mentioned, $k = 1/X_c^2$, where X_c is the constant comoving radius of curvature. The physical radius of the Universe is $r = aX_c = a/\sqrt{k}$. The scale factor grows to a maximum size, corresponding to a maximum radius $r_{\max} = a_{\max}/\sqrt{k}$ and a minimum density $\rho_{\min} = \rho(a_{\max})$. At $r_{\max}$, we have $H = 0$, so the Universe expansion momentarily stops. Then, the Friedmann equation (2.10) suggests $kc^2/a_{\max}^2 = (8\pi G/3)\rho_{\min}$. For matter, we also have $\rho = \rho_{\min}(a_{\max}/a)^3$ because $\rho_m \propto a^{-3}$. Using the above, it is easy to show that the Friedmann equation has the parametric solution

$$
\begin{aligned}
2a &= a_{\max}\,(1 - \cos\tau) \\
2\sqrt{k}\,c\,t &= a_{\max}\,(\tau - \sin\tau)\,,
\end{aligned}
\tag{2.27}
$$

parametrised by $\tau \in (0, 2\pi)$. The above corresponds to a cycloid curve as shown in Fig. 2.7.

Now, the *Schwarzschild radius* is defined as the radius within which a given (nonrotating and uncharged) mass becomes a black hole. The Schwarzschild radius of the Universe is $r_S = 2GM/c^2$, where the total mass is $M \propto r^3 \rho = $ constant (because $r = a/\sqrt{k}$ and $\rho \propto a^{-3}$). Then, it can be shown that the ratio of the Schwarzschild radius to the actual radius of the Universe is $r_S/r \propto 1/(1 - \cos\tau)$. At maximum radius we have $a = a_{\max}$, which means $\tau = \pi$. As time passes $\tau \to 2\pi$, which means that the ratio r_S/r becomes very large, i.e., $r \ll r_S$. Thus, the entire Universe is contained within its Schwarzschild radius. As a result, the closed Universe collapses into a black hole. This eventual collapse of the closed Universe is called the Big Crunch.* The total lifetime of the Universe corresponds to $\tau = 2\pi$ and is given by $t_{\text{tot}}^2 = 3\pi^2/8\pi G\rho_{\min}$, i.e., it is inversely proportional to the density at maximum expansion.

2.6   The cosmological constant

Symmetries in general relativity allow a constant contribution on the right-hand-side of the Friedmann equation (2.10)

$$
H^2 \equiv \left(\frac{\dot{a}}{a}\right)^2 = \frac{8\pi G}{3}\rho - \frac{kc^2}{a^2} + \frac{\Lambda c^2}{3}\,.
\tag{2.28}
$$

The same contribution also appears in the acceleration equation due to energy conservation as can be readily verified using the continuity equation (2.12). The acceleration equation becomes

$$
\frac{\ddot{a}}{a} = -\frac{4\pi G}{3}\left(\rho + \frac{3p}{c^2}\right) + \frac{\Lambda c^2}{3}\,.
\tag{2.29}
$$

Λ is called the *cosmological constant*. In principle, Λ can be anything, positive or negative. It has dimensions of $[\text{length}]^{-2}$ similarly to k, but in contrast to the curvature parameter, it is a physical quantity (it does not depend on the normalisation of $a(t)$).

The cosmological constant can be considered as a constant contribution to the total density of the Universe. Considering that the Universe includes matter and radiation, we can write

$$
\frac{8\pi G}{3}(\rho_m + \rho_r) - \frac{kc^2}{a^2} + \frac{\Lambda c^2}{3} = \frac{8\pi G}{3}(\rho_m + \rho_r + \rho_\Lambda) - \frac{kc^2}{a^2}\,,
\tag{2.30}
$$

---

*According to Douglas Adams, it is also called the Gnab Gib, which is just Big Bang spelled backwards.

where we defined

$$\rho_\Lambda \equiv \frac{\Lambda c^2}{8\pi G}. \tag{2.31}$$

Therefore, $\rho = \rho_m + \rho_r + \rho_\Lambda$. When there is no matter and radiation $\rho_m = \rho_r = 0$, we say we are in the vacuum, when $\rho = \rho_\Lambda$. Thus, we see that the cosmological constant corresponds to constant, nonzero *vacuum density* ρ_Λ.

We can model vacuum density as a barotropic fluid with pressure $p_\Lambda = w_\Lambda \rho_\Lambda c^2$, where w_Λ is the corresponding barotropic parameter. From the continuity equation, we have $\rho_\Lambda \propto a^{-3(1+w_\Lambda)}$. However, $\rho_\Lambda = $ constant, which means that $w_\Lambda = -1$. Putting $p_\Lambda = -\rho_\Lambda c^2$ in Eq. (2.21) and using Eq. (2.31) it is straightforward to recover Eq. (2.29). From now on, we consider Eqs. (2.10) and (2.21) assuming that the Universe content can include a contribution of vacuum density, apart from the usual matter and radiation. As explained below, this has profound consequences to the dynamics of the Universe, if the vacuum density becomes important.[*]

2.6.1   Dynamics with Λ

Vacuum density can be negative if $\Lambda < 0$. If $\Lambda > 0$ then vacuum pressure is negative, since $p_\Lambda = -\rho_\Lambda c^2$. In fact, if positive vacuum density dominates the Universe content, then the acceleration equation (2.21) suggests that $\ddot{a} > 0$ and the Universe expansion is accelerating (since $w_\Lambda < -\frac{1}{3}$, with $\rho_\Lambda > 0$). Here we briefly comment on the effect of Λ on the Universe dynamics.

Considering the weak field limit of GR (linearised gravity), it can be shown that a positive cosmological constant corresponds to a repulsive force which counteracts the attractive gravitational force of regular matter. Exactly the opposite is true for a negative cosmological constant, i.e., a negative cosmological constant corresponds to an attractive force, which reinforces the attractive gravitational force of regular matter. As a result, a closed Universe ($k > 0$) expands forever, when $\Lambda > 0$. Conversely, an open Universe ($k < 0$) recollapses when $\Lambda < 0$. Despite the fact that the geometry of the Universe (spherical, Euclidean, or hyperbolic) is not affected, the fate of the Friedmann universes strongly depends on the value (positive, negative, or zero) of vacuum density.

Cosmologies with $\Lambda < 0$ lead to a Big Crunch if the value of the cosmological constant is sizeable (regardless of k). This would not allow enough time for galaxies to form (and us to exist). Therefore, we concentrate on the case $\Lambda > 0$. If the vacuum density is dominant (even over the curvature) then the Friedmann equation (2.10) reduces to

$$H^2 = \frac{8\pi G}{3}\rho_\Lambda = \frac{\Lambda c^2}{3} = \text{constant}, \tag{2.32}$$

where we also considered Eq. (2.31). The result is quite reasonable, in view of Eq. (2.28), by noting that only the Λ-term survives in the right-hand side. From the definition of H in Eq. (2.7) it is straightforward to obtain

$$H \equiv \dot{a}/a = \text{constant} \Rightarrow a(t) \propto e^{Ht}. \tag{2.33}$$

Thus, space is expanding exponentially. Plotting $a(t)$ in terms of time, we find that the line is curved upwards (see Fig. 2.8), which means that $\ddot{a} > 0$ and expansion is accelerating. This is evident because $w < -\frac{1}{3}$, since $w = w_\Lambda = -1$ when we have vacuum density domination. One way to understand this strange behaviour of space is by considering that, for regular matter, the Universe expansion decreases the energy density. It seems that, in the case of vacuum density, the Universe is expanding explosively, but in vain, in order to lower $\rho_\Lambda = $ constant.

---

[*]We need to be careful not to count the vacuum density twice. If ρ_Λ is included in ρ then we should not use Eq. (2.28), which features Λ explicitly. We should use Eq. (2.10) instead.

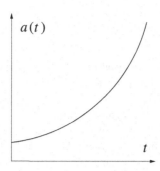

FIGURE 2.8   Evolution of the scale factor $a(t)$ when dominated by a cosmological constant, or equivalently by vacuum density. The curve is convex, meaning curved upwards, because the expansion is accelerating, $\ddot{a} > 0$.

What happens to the Friedmann universes if $\Lambda > 0$? Well, when the vacuum density comes to dominate, all three curves approach the exponential solution in Eq. (2.33), despite the value of k. A closed universe can indeed expand forever.

A comprehensive way to present the various possibilities is shown in Fig. 2.9. Since $\rho_m/\rho_r \propto a$, we ignore radiation and consider the Universe filled with matter and vacuum density only. This corresponds to relatively late times, because as we said, the early Universe is radiation dominated. The total density parameter of the Universe is $\Omega \equiv \rho/\rho_c$, where ρ_c is the critical density. Now, $\rho = \rho_m + \rho_\Lambda$. By dividing with ρ_c, we have $\Omega = \Omega_m + \Omega_\Lambda$, where $\Omega_m \equiv \rho_m/\rho_c$ and $\Omega_\Lambda \equiv \rho_\Lambda/\rho_c$. A closed Universe requires $\Omega > 1$, which implies $\Omega_\Lambda > 1 - \Omega_m$. In the presence of matter and vacuum density only, the acceleration equation (2.21) becomes

$$\frac{\ddot{a}}{a} = -\frac{4\pi G}{3}\left[\rho_m + \rho_\Lambda - \frac{3}{c^2}(p_m + p_\Lambda)\right] = -\frac{4\pi G}{3}(\rho_m - 2\rho_\Lambda), \qquad (2.34)$$

where we considered that $p_m = 0$ and $p_\Lambda = -\rho_\Lambda c^2$. The above suggests that the expansion is accelerating ($\ddot{a} > 0$) if $2\rho_\Lambda > \rho_m$. Dividing with ρ_c we obtain the condition $\Omega_\Lambda > \frac{1}{2}\Omega_m$. The conditions for a closed Universe $\Omega_\Lambda > 1 - \Omega_m$ and an accelerating expansion $\Omega_\Lambda > \frac{1}{2}\Omega_m$ explain the diagram in Fig. 2.9. Note also that, if $\Lambda \gg 0$ ($\Omega_\Lambda > 1$) there is not even a Big Bang (the Universe bounces at the beginning of the expansion).

Recent observations confirm that the Universe is close to spatially flat: $\Omega_0 = 1.00 \pm 0.01$. Hence, $\Omega_\Lambda \simeq 1 - \Omega_m$, and we are on the line denoting a flat Universe in Fig. 2.9. Additionally, data from high-redshift supernovae type-Ia suggest that the Universe is accelerating at present: $\Omega_\Lambda > \frac{1}{2}\Omega_m$. The two datasets correspond to regions in the $\Omega_m - \Omega_\Lambda$ plane, which are almost orthogonal, see Fig. 2.10. The best fit point corresponds to $\Omega_m = 0.31 \pm 0.01$, which results in $\Omega_\Lambda = 0.69 \pm 0.02$ today. This estimate has also been recently confirmed by large scale structure observations of the distribution of galactic clusters. Hence, it seems that $\rho_\Lambda \simeq 0.7\rho_0$. The discovery of recent accelerated expansion was a surprise for most cosmologists. In 2011, it led to a Nobel prize for Saul Perlmutter, Brian P. Schmidt, and Adam G. Riess.

2.6.2   The problem with Λ

Einstein's equations in general relativity include a dimensionful constant, Newton's gravitational constant G, which sets the scale for gravity. The natural scale for gravity is the *Planck mass*. The Planck mass M_P is defined as the mass of a particle whose Compton/de Broglie wavelength equals its Schwarzschild radius. Thus,

$$\frac{2\pi\hbar}{M_P c} \sim \frac{2GM_P}{c^2} \quad \Rightarrow \quad M_P \equiv \sqrt{\frac{\hbar c}{G}}, \qquad (2.35)$$

2

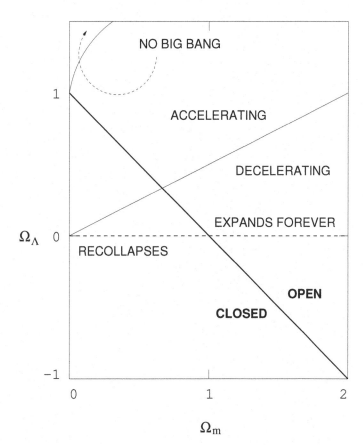

FIGURE 2.9 The $\Omega_m - \Omega_\Lambda$ plane. The horizontal dashed line corresponds to zero cosmological constant. The parameter space above the horizontal dashed line corresponds to a Universe that expands forever regardless of the geometry. The parameter space below the horizontal dashed line corresponds to a Universe that recollapses into a Big Crunch regardless of the geometry. The thick line corresponds to a flat Universe $\Omega_m + \Omega_\Lambda = 1$. The parameter space above the thick line corresponds to an open Universe, since $\Omega_\Lambda > 1 - \Omega_m$. The parameter space below the thick line corresponds to a closed Universe, since $\Omega_\Lambda < 1 - \Omega_m$. The thin straight line corresponds to the condition $\Omega_\Lambda = \frac{1}{2}\Omega_m$, when the expansion is linear with time $a \propto t$ and $\ddot{a} = 0$. The parameter space below the thin straight line corresponds to decelerated expansion since $\Omega_\Lambda < \frac{1}{2}\Omega_m$ implies $\ddot{a} < 0$. The parameter space above the thin straight line corresponds to accelerated expansion since $\Omega_\Lambda > \frac{1}{2}\Omega_m$ implies $\ddot{a} > 0$. Note that if $\Omega_\Lambda > 1$ and Ω_m is not too big, there is not even a Big Bang and the Universe bounces at the beginning of the expansion.

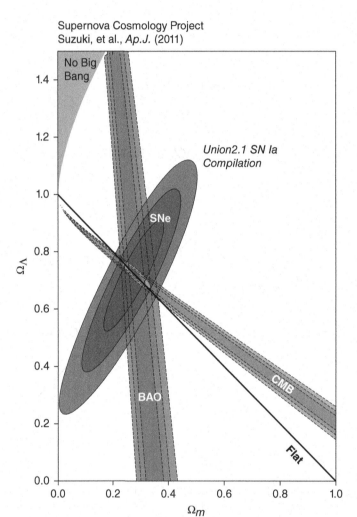

FIGURE 2.10 CMB observations suggest that the Universe is approximately flat, while supernovae (SNe) Ia suggest that it is undergoing accelerated expansion. The two datasets are almost orthogonal and give $\Omega_m \simeq 0.3$ and $\Omega_\Lambda \simeq 0.7$. This is corroborated by large-scale structure observations, which depict a characteristic scale of enhanced density contrast (about 150 Mpc) due to baryon acoustic oscillations (BAO) (see Sec. 5.2). Reproduced with permission from *Astrophysical Journal.* © original source Supernova Cosmology Project Collaboration [S+12].

where $\hbar \equiv h/2\pi$ is the reduced Planck constant and we ignored the factor of π. The associated Planck energy is

$$M_P c^2 = 1.2211 \times 10^{19}\,\text{GeV}\,.$$

In a similar manner, we define the Planck length $\ell_P = \hbar/(M_P c)$ and the Planck density $\rho_P \equiv M_P/\ell_P^3 = c^5/\hbar G^2$.

Assuming that gravity is characterised by a single scale only, we would expect that the vacuum density is similar to the Planck density $|\rho_\Lambda| \simeq \rho_P$, which suggests $|\Lambda| \simeq c^3/\hbar G$. However, this value is huge and would not allow the Universe we observe to form. If $\Lambda < 0$, this value would result in a Big Crunch almost immediately after the Big Bang, while if $\Lambda > 0$, the Universe would engage in eternal exponential expansion again right after the Big Bang, not allowing structures such as galaxies to ever form. Indeed, the formation of structure needs $|\rho_\Lambda| \lesssim \rho_0 \sim 10^{-120}\rho_P$, where ρ_0 is the average density of the Universe at present. Note, however, that in theoretical physics if an expected contribution is observed not to be there, its absence needs explaining.[*]

In addition to the above, there is another reason to expect a huge value of the cosmological constant. In quantum field theory, all particles and fields are producing a nonzero contribution to the constant vacuum density, whose magnitude is determined by the cutoff of the theory and sometimes is referred to as *zero-point energy*. This has no effect on QFT, so particle physicists happily ignore it. However, it can have a huge gravitational effect, as we have seen. Now, QFT is an effective theory, and this is why it features a cutoff, which is again set by the Planck density. Thus, we find again $\rho_\Lambda \simeq \rho_P$, which as we said is unacceptable. This is the infamous *cosmological constant problem*. If, in addition, supersymmetry is assumed, then there is a massive improvement to the situation, because unbroken supersymmetry sets the vacuum density at zero. However, supersymmetry is always broken by the presence of nonzero energy density, which still introduces a vacuum density roughly 10^{60} times larger than ρ_0, which is still entirely unacceptable.

Until recently, the "solution" to the cosmological constant problem was simply setting $\Lambda = 0$ due to an *unknown* symmetry, postulating, for example, that there is as much positive vacuum density as there is negative vacuum density. However, as mentioned already, recent observations suggest that the expansion of the Universe is starting to accelerate at present. This led many cosmologists to consider that $\Lambda \neq 0$ after all and that $\rho_\Lambda \sim \rho_0$. Because $\rho_\Lambda \sim 10^{-120}\rho_P$ seems highly unlikely though, this has been called (by Laurence Krauss) "the worst fine-tuning in Physics" (see also Appendix A). For those who do not accept Λ as the explanation behind the recent accelerated expansion, a different substance needs to be introduced to account for the observations. This substance is called *dark energy*.

2.7 A first brush with dark energy

The term "dark energy" was coined by Michel Turner and his then-student Dragan Huterer in 1998 [HT99] and is meant to signify its elusive nature and distingiush it from dark matter (see later).[†] Dark energy can be defined as follows:

> **Dark energy:** a substance with a barotropic parameter negative enough so that it causes accelerated expansion if dominant.

In view of the acceleration equation (2.21), the barotropic parameter of dark energy is $w_{\text{DE}} < -\frac{1}{3}$. Thus, since $w_\Lambda = -1$, vacuum density can be thought of as a particular type of dark energy.

---

[*]We discuss this issue also in Secs. 8.8 and 9.1.1.

[†]Attracting research funding from the US Department of Energy is purely coincidental.

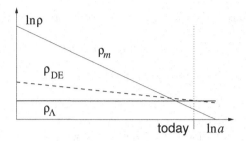

FIGURE 2.11   Log-log plot of the density of matter ρ_m (thin line) and the vacuum density ρ_Λ (thick line) or the density of some other dark energy substance ρ_{DE}, with constant barotropic parameter $-1 < w < -\frac{1}{3}$ (dashed line). Today, dark energy dominates, but going into the past we see that matter density takes over and the contribution of dark energy in the density budget of the Universe is exponentially suppressed (it is a log-log diagram). This is because the density of matter decreases much faster than that of dark energy.

If not vacuum density, dark energy can explain the observed recent accelerated expansion with $\Lambda = 0$. However, whenever the Universe undergoes accelerated expansion, the content of the Universe needs to be dominated by some form of dark energy. Cosmic inflation is also a period of accelerated expansion, albeit in the very early Universe. Thus, some form of dark energy is required to drive inflation too.

If constant, the barotropic parameter of the late-time dark energy has been constrained by observations as [A+18a]

$$w_{\text{DE}} = -1.028 \pm 0.031 \,, \tag{2.36}$$

at 2-σ. Also, if $w_{\text{DE}} > -1$, then observations suggest $w_{\text{DE}} < -0.95$ [A+18a]. For a constant barotropic parameter, we know that $\rho_{\text{DE}} \propto a^{-3(1+w_{\text{DE}})}$. Because $w_{\text{DE}} \approx -1$, this implies that the density of dark energy varies little with time. In contrast, $\rho_m \propto a^{-3}$. This means that, towards the past, the density of matter grows much faster than ρ_{DE} (see Fig. 2.11). As dark energy becomes dominant at present, we conclude that the importance of dark energy is recent only (remember that radiation is negligible in the late Universe). This is fortunate, because, were dark energy important at earlier times, then it would have inhibited galaxy formation, because it corresponds to a repulsive, antigravity force similar to positive vacuum density (which is merely a particular type of dark energy).

However, it is possible that the barotropic parameter of dark energy is not constant. If dark energy is dynamical, then this introduces a new problem; that of its initial conditions. At the moment, there are multiple efforts to observationally discern dark energy from pure Λ. By Taylor expanding $w_{\text{DE}}(a)$ near the present value of the scale factor a_0, we can consider the following parametrisation used in order to investigate observationally the possible variation of the dark energy barotropic parameter:

$$w_{\text{DE}} = w_{\text{DE}}^0 + w_a \left(1 - \frac{a}{a_0}\right) = w_{\text{DE}}^0 + w_a \frac{z}{1+z} \,, \tag{2.37}$$

where $w_{\text{DE}}^0, w_a = \text{constant}$, with w_{DE}^0 being the value of w_{DE} at present, $w_a = -(dw_{\text{DE}}/da)_0$, i.e., given by the derivative of w_{DE} today, and we used Eq. (2.9). This was introduced by Michel Chevallier and David Polarski in 2001 [CP01] and also by Eric V. Linder in 2003 [Lin03]. It is sometimes called CPL parametrization.

If the dark energy is not the vacuum density then there are several new issues introduced. First, one introduces a new substance with exotic properties, such as a negative enough pressure. This may not sound good, but we know of such substances already. A prominent example is a scalar field. After the observation of the Higgs boson particle at the LHC in 2012 (Peter W. Higgs and Francois Englert were awarded the Nobel prize in 2013), we have evidence of the existence of at least one fundamental scalar field.* As we explain in Chapter 6, under certain circumstances, a scalar field can behave as a barotropic fluid with pressure negative enough to be a dark energy candidate. Such a dark energy can source primordial inflation or even late-time inflation, which is the current accelerated expansion. I should mention here that this is not necessarily the only possibility for dark energy.

Another perceived difficulty with introducing dark energy, is the fact that observations suggest that its contribution to the density budget of the Universe is becoming dominant at present. This is referred to as the *coincidence problem*. In my view, there is no problem with this because it is indeed merely a coincidence that we happen to be around when the dark energy takes over the Universe expansion. In the past, there was also a time when matter dominated radiation, as discussed later. There is no special significance of that either. The only factor crucially determining the time when one component of the Universe content takes over from another is the relative abundance of the different substances. For dynamical dark energy, this is related to the initial conditions mentioned before.

Finally, when considering dark energy that is not the vacuum density, we have to set the vacuum density to zero, based on the *unknown* symmetry conjectured before. This means that such *dark energy does not solve the cosmological constant problem*. .

2.8 Dynamics with spatial flatness

2.8.1 Basic equations

As mentioned earlier, observations suggest that the Universe at present is approximately spatially flat $\Omega_0 \approx 1$. Therefore, we can set $k = 0$ in Eq. (2.10), which means that $\rho = \rho_c$, i.e.,

$$\rho = \frac{3H^2}{8\pi G} ,$$

(2.38)

where $\rho = \Sigma_i \rho_i = \rho_m + \rho_r + \rho_{\rm DE} + \cdots$, where $\rho_{\rm DE}$ is the dark energy density, which could be the vacuum density ρ_Λ. Using Eq. (2.17) and assuming that the Big Bang is the origin of time (so setting $a(t) = 0$ when $t = 0$), Eq. (2.38) suggests

$$a \propto t^{\frac{2}{3(1+w)}} ,$$

(2.39)

where we assumed $w = {\rm constant} > -1$. Then Eq. (2.7) gives

$$H = \frac{2}{3(1 + w)t} ,$$

(2.40)

which, when inserted in Eq. (2.38), results in

$$\rho = \frac{1}{6(1 + w)^2 \pi G t^2} .$$

(2.41)

From the above, we see that, when $w = {\rm constant} > -1$, $H \sim 1/t$ and $\rho \sim 1/Gt^2$ regardless of the value of w. The inverse of the rate of expansion is the typical timescale for the dynamics of the Universe, called the *Hubble time*: $t_H \equiv H^{-1}$. We see that the Hubble time is comparable to the age of the Universe ($t_H \sim t$) when $w = {\rm constant} > -1$. However, even though the rate of

---

*Particles are excitations of the corresponding fields, e.g., photons are excitations of the electromagnetic field.

the Universe expansion H is roughly independent of w, the growth of lengthscales is different for different dominant fluids. Indeed, from Eq. (2.39), we find

$$\text{Radiation Domination}: \quad w = \tfrac{1}{3} \Rightarrow \quad a \propto t^{1/2}, \tag{2.42}$$

$$\text{Matter Domination}: \quad w = 0 \Rightarrow \quad a \propto t^{2/3}. \tag{2.43}$$

Thus, we see that the pressure inhibits the expansion (since the Universe expands faster when $p = 0$). This is because pressure corresponds to extra work.[*]

2.8.2 Matter-radiation equality

The early Universe is radiation dominated. This is because the density of radiation is diluted faster than the density of matter with the Universe expansion, regardless of their late-time abundances. However, when does the switch-over from radiation to matter domination happen? There is a critical time when the dominance of radiation density is challenged by the ever more important matter density; a time when both the contributions of matter and radiation to the density budget of the Universe are equal. This time is called *matter-radiation equality* (see Fig. 2.12). From their abundance at present, we can estimate when this happened.

Since the Universe is approximately spatially flat, its density is roughly the critical density at present $\rho_0 = 3H_0^2/8\pi G = 1.879 h^2 \times 10^{-29}\,\text{g/cm}^3$, where observations of the Hubble constant suggest $h = 0.675$ or so. Thus, we have $\rho_0 \simeq 0.864 \times 10^{-29}\,\text{g/cm}^3$. For matter $\rho_m = \Omega_m \rho$. Since today $\Omega_m \simeq 0.31$, we find $\rho_m^0 \simeq 2.678 \times 10^{-30}\,\text{g/cm}^3$. For radiation, the dominant contribution is coming from the CMB radiation (discussed later) and has density $\rho_r^0 = \epsilon_r^0/c^2 = 4.814 \times 10^{-34}\,\text{g/cm}^3$, where ϵ_r^0 is the energy density of radiation at present. Now, $\rho_m/\rho_r \propto a$ and by definition $(\rho_m/\rho_r)_{\text{eq}} \equiv 1$, where the subscript "eq" denotes matter-radiation equality. This means that

$$\frac{a_0}{a_{\text{eq}}} = \left.\frac{\rho_m}{\rho_r}\right|_0 \simeq \frac{2.678 \times 10^{-30}}{4.814 \times 10^{-34}} = 5563 \tag{2.44}$$

Ignoring late-time dark energy, we assume that the Universe is matter dominated after equality, such that $a \propto t^{2/3}$. Therefore,

$$t_{\text{eq}} = \left(\frac{a_{\text{eq}}}{a_0}\right)^{3/2} t_0 \simeq 33260\,\text{y}, \tag{2.45}$$

where we considered the age of the Universe at present to be $t_0 \simeq 13.8\,\text{Gy}$ (see later). The actual value is more like $t_{\text{eq}} \simeq 4.7 \times 10^4\,\text{y}$, when the calculation properly considers late-time dark energy (redshift $z_{\text{eq}} \simeq 3600$). So, during the first 47,000 years or so, the Universe was radiation dominated. Afterwards, it became matter dominated for about 13 billion years until the last billion years or so, when late dark energy became important.

2.8.3 Phantom dark energy

As we have seen, when $w = \text{constant} > -1$, we have $a \propto t^{\frac{2}{3(1+w)}}$ in a flat Universe (cf. Eq. (2.39)). We have also shown that, when $w = -1$ (vacuum density), we have $a \propto e^{Ht}$, with $H = \text{constant}$ (cf. Eq. (2.33)). What happens when $w < -1$?

---

[*]Note that there is *one* Universe expansion and one scale factor function $a(t)$ and *not* a different expansion of matter $a_m \not\propto t^{2/3}$ and of radiation $a_r \not\propto t^{1/2}$, but a single function $a(t)$ determined by the *dominant* component of the Universe content.

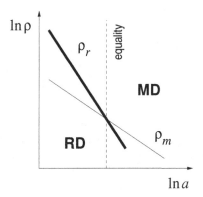

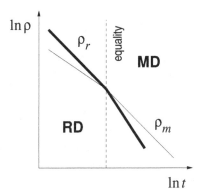

FIGURE 2.12 Log-log plots of the density of matter ρ_m and radiation ρ_r (thin and thick slanted lines, respectively) with respect to the scale factor of the Universe a and the cosmic time t (upper and lower panel, respectively). Because $\rho_m \propto a^{-3}$ and $\rho_r \propto a^{-4}$ the density of radiation is diluted faster than that of matter by the Universe expansion. As a result, regardless of matter being dominant over radiation at late times, there is always a moment in the past when both densities were equal, denoted as "equality" by the dashed lines. Before equality we are in the radiation-dominated (RD) era, while after equality we are in the matter-dominated (MD) era. In the lower panel, it is worth pointing out that the dominant component of the Universe content always scales as $\rho \propto t^{-2}$ (when $w > -1$) according to Eq. (2.41).

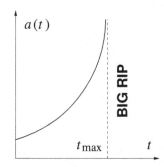

FIGURE 2.13 Evolution of the scale factor $a(t)$ when dominated by phantom dark energy. The curve is convex, meaning curved upwards, because the epansion is accelerating, $\ddot{a} > 0$. The scale factor shoots at infinity at finite time t_{max}, corresponding to the Big Rip singularity.

A substance with $w < -1$ violates the *null energy condition*: $\epsilon + p \geq 0$ (cf. Eq. (9.132)). Such a hypothetical substance is called exotic matter in general relativity and it is necessary for the construction of traversable wormholes. In cosmology, exotic matter is called *phantom dark energy* [Cal02]. If the Universe becomes dominated by phantom dark energy, it undergoes super-accelerated expansion, resulting in all distances (down to atoms) becoming infinite at finite time. This moment, when all distances blow up, is called the Big Rip and signifies a cataclysmic end of the Universe, see Fig. 2.13 (and the discussion in Sec. 9.8.2).

Equation (2.12) suggests that the density of the Universe is increasing with the expansion. Using this in Eq. (2.38) and considering that $a \to \infty$ at the time of the Big Rip $t \to t_{max}$, we find

$$a \propto \left(1 - \frac{t}{t_{max}}\right)^{\frac{2}{3(1+w)}} , \tag{2.46}$$

where $\frac{2}{3(1+w)} < 0$. From its definition in Eq. (2.7), it is straightforward to obtain $H(t)$:

$$H = -\frac{2}{3(1+w)t_{max}} \left(1 - \frac{t}{t_{max}}\right)^{-1} = \frac{2}{3|1+w|} \frac{1}{t_{max} - t} . \tag{2.47}$$

Hence, we find that $\dot{H} > 0$, and the rate of the expansion is increasing (in tandem to the density, cf. Eq. (2.38)) and becomes infinite at the Big Rip. The physical interpretation of the Big Rip is further explained when we revisit the cosmological horizon below. But first we mention briefly the concept of the spacetime metric.

2.9   The metric

Here we briefly discuss a mathematical formulation of the geometry of spacetime. Firstly, we consider spacetime to be such that we could define the notion of distance between points, which when considering spacetime are called *events*. Spaces (and spacetimes) in which the distance between different points (events) can be defined are called *metric* space(time)s.

At the moment, ignore curvature. In his theory of special relativity, Einstein suggested that time is but a fourth dimension, which together with three-dimensional space, comprises four-dimensional spacetime. The distance between two different events in spacetime is called the *spacetime interval* and is given by

$$\Delta s^2 = -(c\Delta t)^2 + (\Delta x)^2 + (\Delta y)^2 + (\Delta z)^2 , \tag{2.48}$$

where c is lightspeed, and we assumed a Cartesian coordinate system. We see that there is a qualitative difference between the three spatial dimensions and time because the contribution of the temporal dimension to the spacetime interval is of opposite sign. Writing this in polar coordinates we have

$$\Delta s^2 = -(c\Delta t)^2 + (\Delta r)^2 + r^2(\Delta \theta)^2 + r^2 \sin^2 \theta (\Delta \phi)^2 \,. \tag{2.49}$$

In the limit of infinitesimal distance $\Delta s \to ds$, the spacetime interval in polar coordinates becomes

$$ds^2 = -(c\,dt)^2 + (dr)^2 + r^2(d\theta)^2 + r^2 \sin^2 \theta (d\phi)^2 \,, \tag{2.50}$$

which is now called the *line element*.

Now, once we switch on curvature the above line element is modified. In particular, for the homogeneous and isotropic spacetime of the Universe, the line element is

$$ds^2 = -(c\,dt)^2 + a^2(t) \left[\frac{(dr)^2}{1 - kr^2} + r^2(d\theta)^2 + r^2 \sin^2 \theta (d\phi)^2 \right] \,, \tag{2.51}$$

where $a(t)$ is the scale factor, and k is the curvature parameter. The above is called the Friedmann-Lemaître-Robertson-Walker (FLRW) metric and was obtained independently by Alexander Friedmann in 1922 and Georges Lemaître in 1927. Howard P. Robertson and Arthur G. Walker in 1935 showed that the above metric is unique for a homogeneous and isotropic spacetime.

As previously discussed, k is a dimensionful constant, which can take a positive, negative, or zero value. If the above metric is spatially flat, then $k = 0$ and Eq. (2.51) becomes

$$ds^2 = -(c\,dt)^2 + a^2(t) \left[(dr)^2 + r^2(d\theta)^2 + r^2 \sin^2 \theta (d\phi)^2 \right] \,. \tag{2.52}$$

Note that this is still different from Eq. (2.50). This is because, even though space is flat when $k = 0$, spacetime is still curved; $a(t)$ and the dynamics of an expanding (or contracting) Universe correspond to curvature in the time direction. Note also that r in Eqs. (2.51) and (2.52) is a comoving coordinate. We can switch back to Cartesian (comoving) coordinates, in which case the metric of a (spatially) flat Universe is

$$ds^2 = -(c\,dt)^2 + a^2(t) \left[(dx)^2 + (dy)^2 + (dz)^2 \right] \,. \tag{2.53}$$

2.10   The cosmological horizon revisited

Assuming a flat Universe, we take a closer look at the concept of the cosmological horizon, which corresponds to the range of causal correlations and defines the observable Universe. In view of Eq. (2.5), the relation between physical distance dr and comoving distance dx is $dr = a\,dx$. The reach of causal correlations is limited by the speed of light, as this is the maximum speed of propagation in space. The physical distance dr travelled by light per time-interval dt is $dr = c\,dt$. Combining these we obtain* $dx = c\,dt/a$. Integrating, we obtain the comoving range x_H of causal correlations, between times t_1 and t_2. The physical range is therefore

$$D_H = a\,x_H = a \int_{t_1}^{t_2} \frac{c\,dt}{a} \,. \tag{2.54}$$

There are the following possibilities for the cosmological horizon, which have to do with the choice of (t_1, t_2).

---

*This is obtained also from Eq. (2.53) considering light propagation along the x direction, so $dy = dz = 0$ and also $ds = 0$ (called null worldline).

- **Particle Horizon:** $(t_1, t_2) = (0, t)$.

$$D_H(t) = a(t) x_H(t) = a(t) \int_0^t \frac{c\,dt'}{a(t')}. \tag{2.55}$$

This corresponds to the extent of causal correlations so far. Instead of zero, the lower limit can be an initial value, for which $t_i \ll t$.

- **Event Horizon:** $(t_1, t_2) = (t, \infty)$.

$$D_H(t) = a(t) x_H(t) = a(t) \int_t^\infty \frac{c\,dt'}{a(t')}. \tag{2.56}$$

This corresponds to the ultimate extent of causal correlations. Instead of infinity, the upper limit can be a final value, such that $t \ll t_f$. In the case of phantom dark energy, t_f cannot be larger than the time of the Big Rip, meaning $(t_1, t_2) = (t, t_{\max})$.

The cosmological horizon is either a particle or an event horizon. D_H is a particle horizon if the barotropic parameter of the Universe w is bigger than $-\frac{1}{3}$. Conversely, D_H is an event horizon if w is smaller than $-\frac{1}{3}$. In each possibility, only one type of horizon is well-defined; the other diverges.[*]

The particle horizon corresponds to the distance travelled by a signal (e.g., light or gravitational waves) since the Big Bang (when $t = 0$), or since a very early time t_i close to the Big Bang. Signals beyond the particle horizon cannot reach us because they need more time than the age of the Universe to do so. This is all fine as long as $w > -\frac{1}{3}$. In the opposite case, however, the particle horizon diverges (i.e., there is no limit to the distance travelled), and D_H is now an event horizon. This corresponds to the maximum distance that a signal will ever travel. This is not infinite (as it would be when $w > -\frac{1}{3}$) because the expansion of space is fast enough to keep the far regions always inaccessible, even if the signal travels with light-speed. In effect, space expands superluminally when $w < -\frac{1}{3}$, or equivalently $\ddot{a} > 0$ (cf. Eq. (2.21)), that is when the Universe is dominated by a dark-energy substance.

One can understand this as follows. The condition $\ddot{a} > 0$ means that $\dot{a}$ grows in time. The Hubble flow is $v_H = \dot{a}x$, where x is constant. Thus, v_H grows in time as well, and this growth is *not bounded* by c, because the Hubble flow is not the movement through space but the expansion of space itself. The expansion is so strong that, if we consider two objects (galaxies say), even when originally causally connected (one could see the other), the Hubble flow pulls them apart so fast that eventually one is inaccessible to the other (they become causally disconnected).[†] The horizon acts like an "inverted" (meaning inside-out) black hole surrounding the observer, in the sense that unbound objects are sucked out by the expansion. This is why D_H is called an event horizon in this case.

If $w > -1$, the horizon grows with the speed of light: $D_H \sim ct$. This is what we would naïvely expect. However, if $w = -1$ (vacuum density) then the horizon remains constant: $D_H = c/H = \text{constant}$. Finally, if $w < -1$ (phantom dark energy) then the horizon shrinks to zero towards the Big Rip: $D_H \to 0$ as $t \to t_{\max}$. This means that the superluminal expansion

---

[*]However, at any given time there exist fossils of previous horizons corresponding to earlier periods of the Universe history with different values of the barotropic parameter.

[†]For individual galaxies, which may be gravitationally bound in a cluster, this would not be so, because they would not follow the Hubble flow. But this is true for the unbound galactic superclusters, now that dark energy dominates the present Universe.

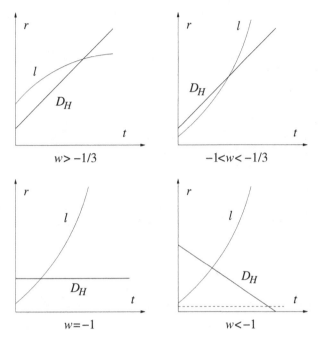

FIGURE 2.14 The evolution of a comoving scale ℓ (curved line) and the cosmological horizon D_H (straight line) for various values of the barotropic parameter. When $w > -\frac{1}{3}$ (upper-left panel), the Universe is undergoing decelerated expansion, and we have a particle horizon, which grows as $D_H \propto t$. A superhorizon comoving scale eventually enters the horizon and becomes subhorizon. When $-1 < w < -\frac{1}{3}$ (upper right panel), the Universe is undergoing accelerated expansion and we have an event horizon, which grows as $D_H \propto t$. A subhorizon comoving scale eventually exits the horizon and becomes superhorizon. When $w = -1$ (lower left panel), the Universe is undergoing accelerated expansion (dominated by vacuum density), and we have an event horizon, which remains constant: $D_H = $ constant. Similarly to the previous case, a subhorizon comoving scale eventually exits the horizon and becomes superhorizon. Finally, when $w < -1$ (lower right panel), the Universe is undergoing accelerated expansion (dominated by phantom dark energy) and we have an event horizon, which shrinks to zero in finite time. Similarly to the previous two cases, a subhorizon comoving scale eventually exits the horizon and becomes superhorizon. However, this time a bound scale, which remains constant (depicted by the dashed line) also eventually becomes superhorizon because the horizon shrinks to zero.

is so intense that even bound objects are pulled apart as the horizon shrinks to ever smaller distances, from intergalactic distances to the size of individual atoms. Hence, Big Rip!*

In general, using the above relations, it is easy to show that the horizon is

$$D_H = \frac{2c/H}{|1 + 3w|} \quad \text{with } w \neq -\frac{1}{3}. \tag{2.57}$$

When $w = -\frac{1}{3}$, then Eq. (2.39) gives $a \propto t$ and $t_H \equiv H^{-1} = t$ (cf. Eq. (2.7)). Hence, Eq. (2.54) suggests that $D_H = (c/H) \ln(t_2/t_1)$. Therefore, in all cases, we have $D_H \sim c/H$, or $D_H \sim R_H$, where $R_H \equiv c\, t_H = c/H$ is called the *Hubble radius*.[†] Pictorially, the evolution of a comoving lenghtscale and of the horizon is shown in Fig. 2.14 for various values of the barotropic parameter.

2.11   The age of the Universe

As we have discussed, Big Bang cosmology suggests that the age of the Universe t_0 is finite. A first estimate is obtained by the characteristic dynamical timescale of the Universe expansion, which is the Hubble time at present: $t_H(t_0) \equiv H_0^{-1} \simeq 14.4\,\text{Gy}$. This is in excellent agreement with astrophysical estimates, which date the oldest objects in the Universe. Several methods exist, the most prominent of which are

- **Nucleocosmochronology:**
  This amounts to estimates based on the radioactive decay of long half-life isotopes, e.g., ^{232}Th. In a sense, it is a similar method to carbon dating, but it is applied to stars.

- **Aging globular clusters:**
  Globular clusters are among the oldest objects in the Cosmos. They can be dated using several methods, the most important of which is the cooling of white dwarf stars and the main sequence turn-off point method.

All these independent methods estimate the age of the Universe to be 12–16 Gy, which seems to agree nicely with the estimate based on the Hubble time. Agreement with observations about the age of the Universe is a major success of Big Bang cosmology. Indeed, imagine if the Big Bang model predicted an age in millions rather than billions of years. In this case, the Universe would be younger than the objects within it! Conversely, imagine if the theory suggested trillions or quadrillions of years for t_0. Then, we would need to explain what happened to really old galaxies. As it stands, the numbers seem to fit perfectly.

However, when looked at more carefully, this agreement seemed to be lost. Before the observation of present dark energy, cosmologists were convinced that the present Universe was matter dominated, with $\Omega = \Omega_m$. Considering also the fact that observations support a flat Universe, it is straightforward to find the age of the Universe as $t_0 = \frac{2}{3} H_0^{-1} \simeq 9.6$ Gy in this scenario, where we used Eq. (2.40) with $w = w_m = 0$. This value does not match the observations and this was the infamous *age problem* of the Universe.

---

*In 2003, R.R. Caldwell, M. Kamionkowski, and N.N. Weinberg [CKW03] calculated the Big Rip scenario when $w = -\frac{3}{2}$. In this case, $t_{\text{max}} = 35$ Gy. They found:

$t_{\text{max}} - 1\,\text{Gy}$:	Galaxy clusters are erased
$t_{\text{max}} - 60\,\text{My}$:	Milky Way is destroyed
$t_{\text{max}} - 3\,\text{months}$:	Solar System is unbound
$t_{\text{max}} - 30\,\text{min}$:	Earth explodes
$t_{\text{max}} - 10^{-19}\,\text{sec}$:	Atoms are dissociated

[†] $R_H = c/H$ is also called the *Hubble horizon* because it corresponds to the distance at which points move at the speed of light away from each other. Thus, points further apart are beyond causal contact. This is reminiscent of Eq. (2.4), which was an estimate of the cosmological horizon at present.

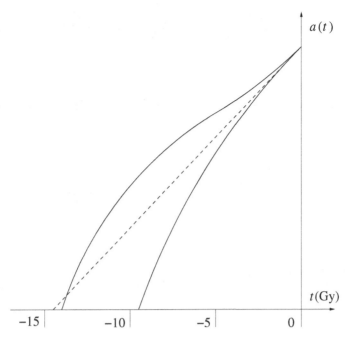

FIGURE 2.15 Plot of the scale factor $a(t)$ with respect to time, taking $t = 0$ at present and counting the time from the present to the past. When considering a matter-dominated Universe, the plot of the scale factor is always curved downwards and we see that the moment of the Big Bang (when $a = 0$) is no more than 10 Gy in the past. In contrast, when considering a Universe that was originally matter-dominated but near the present time becomes dominated by dark energy, the plot of the scale factor becomes curved upwards near the present time. Because in both cases the slope of the curve is the same today (dashed line), corresponding to the measured value of the Hubble constant, we see that in a Universe that was originally matter-dominated but became dark-energy-dominated near the present time, the moment of the Big Bang is moved to the past such that the age of the Universe problem is resolved.

The observations of dark energy have drastically changed this picture. Indeed, dark energy neatly solves the age problem. The reason is easy to understand by considering the behaviour of the scale factor in a standard flat, matter-dominated Universe (one of the three Friedmann universes, in Sec. 2.5.1) and in a flat Universe, where dark energy is taking over at present and results in accelerated expansion. In the latter case, plotting $a(t)$, we get a line that curves downwards when $t < t_0$ but that twists and starts curving upwards when $t \simeq t_0$, signifying that the expansion starts from decelerating and becomes accelerating. In contrast, for a standard matter-dominated Universe, the expansion is always decelerating and the line in the $a(t)$ plot is always curved downwards. In both cases, however, the slope of the line today must be the same, because it corresponds to the Hubble constant $\dot{a}(t_0) = a_0 H_0$ (we can normalise $a_0 = 1$), which is observationally determined. This results in a different duration of the Universe history, as can be clearly seen in Fig. 2.15, with a larger value of t_0 once dark energy is taken into account.

Indeed, for flat Universe with matter and $\Lambda > 0$, the age of the Universe is

$$
\begin{aligned}
t_0 &= \frac{2}{3} \frac{H_0^{-1}}{\sqrt{1 - \Omega_m}} \ln \left(\frac{1 + \sqrt{1 - \Omega_m}}{\sqrt{\Omega_m}} \right) \\
&= \frac{2}{3} \frac{H_0^{-1}}{\sqrt{1 - \Omega_m}} \sinh^{-1} \sqrt{\frac{1 - \Omega_m}{\Omega_m}},
\end{aligned}
\tag{2.58}
$$

t_0/Gy

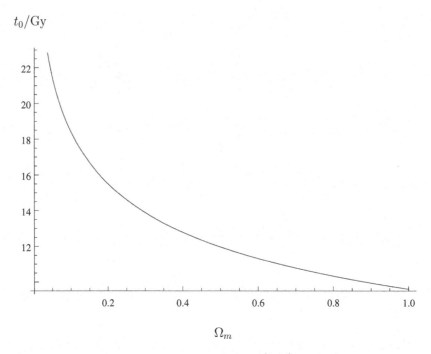

Ω_m

FIGURE 2.16   The age of the Universe as determined by Eq. (2.58), which corresponds to a flat Universe containing matter and vacuum density only. The observations suggest that the density parameter of matter today is $\Omega_m \simeq 0.3$. Using this, the above plot suggests that $t_0 \simeq 14\,\text{Gy}$, which is in excellent agreement with astophysical estimates.

where Ω_m is the matter density parameter at present and we used Eq. (2.28) with $k = 0$. Putting $\Omega_m = 0.31$ in the above, we find $t_0 = 13.8\,\text{Gy}$, which is just about right. A plot of Eq. (2.58) is shown in Fig. 2.16.

Note that, even if the value of t_0 is increased when considering dark energy taking over today, this is not necessarily a solution to the age problem. It depends on the type of dark energy. As we have just seen, vacuum density seems to work. However, as another example, consider a dark-energy substance with constant barotropic parameter $w_{\text{DE}} = -\frac{1}{2}$. Then, using Eqs. (2.12) and (2.38), we find $t_0 = \frac{4}{3}H_0^{-1}(1 + \sqrt{\Omega_m})^{-1}$. Putting $t_0 \simeq 13.8\,\text{Gy}$, we find $\Omega_m \simeq 0.15$, which is significantly lower than the (many) estimates for the matter abundance today.

Thus, the age problem is resolved only with the right kind of dark energy, which includes vacuum density. We should stress here that the resolution of the age problem is additional evidence for dark energy, independent of supernova observations.

2.12   Dark matter

We have reviewed some of the fluids comprising the Universe content. So far, we have discussed matter and radiation which are gases comprised by nonrelativistic and relativistic particles, respectively. We have also briefly introduced dark energy, which is an exotic hypothetical substance with a barotropic parameter negative enough ($w_{\text{DE}} < -\frac{1}{3}$). A particular kind of dark energy is vacuum density, for which the barotropic parameter is $w_\Lambda = -1$. We have also mentioned scalar fields as a possible candidate for dark energy, but we discuss them in much more detail later on.

Here, we briefly return to matter and radiation, which are conventional and not at all exotic, compared to dark energy. However, they may be less conventional than one thinks. Radiation in the Universe is mainly CMB photons and primordial neutrinos (both discussed later). Matter, however, is another matter!

Indeed, it turns out that not all the matter in the Universe is comprised of our familiar planets and stars. "Normal" matter, which makes up stars and planets and us, is called *baryonic matter* in cosmology, because most of it is protons and neutrons (both called baryons), since electrons have masses a thousand times less but are equal in number with protons for neutrality (otherwise the Universe would have a net electric charge, which it does not). The additional mysterious matter constitutes more than five times the amount of baryonic matter and is called *dark matter*. One thing we need to emphasise here is that dark matter should not be confused with dark energy. Dark matter is pressureless as is baryonic matter, while dark energy has pressure negative enough to lead to accelerated expansion, if dominant. In contrast, if dark matter is dominant, the Universe is simply matter-dominated, as it has been in the last 13 billion years. Expansion is decelerated and $a \propto t^{2/3}$ since the Universe is also flat (cf. Eq. (2.39)).

We can define dark matter as follows:

Dark matter: nonluminous matter, whose presence is indirectly inferred from the effect of its gravitational attraction on luminous matter and on the global geometry of the Universe.

Its name is actually misleading, as we are talking about invisible rather than black matter. Its existence and abundance is suggested and supported by numerous uncorrelated observations. The nature, however, of dark matter remains elusive.

2.12.1 Observational evidence for dark matter

Here we list only some of the major independent observations, which point to the existence of dark matter.

- **Star counts:**
  Stellar structure theory provides a clear link between the mass of a star and its surface temperature and absolute luminosity. Hence, observations of a large enough region can estimate the mass density of stellar material. Such observations give $\Omega_{\rm stars} \lesssim 0.01$. However, other observational evidence suggests that the abundance of matter is at least 10 times more.

- **Galaxy rotation curves:**
  This is a dynamical estimate of galactic mass based on the gravitational attraction inferred from the rotation of the galactic disk (in spiral galaxies). Newtonian physics suggests

$$\frac{v_{\rm rot}^2}{R} = \frac{GM}{R^2} \quad \Rightarrow \quad v_{\rm rot}(R) = \sqrt{\frac{GM(R)}{R}}, \tag{2.59}$$

  where $v_{\rm rot}(R)$ is the rotational speed of stars in the galaxy at distance R from the galactic centre and $M(R)$ is the mass within radius R.
  A galaxy rotation curve is a plot of the speed $v_{\rm rot}(R)$ of the rotating matter of the spiral disk as a function of the radius R from the centre. At distances enclosing the visible part of the galaxy, one expects that $M(R) \simeq$ constant since almost all the galaxy is supposedly within radius R. This means that $v_{\rm rot}(R) \propto 1/\sqrt{R}$ (Kepler's law). However, this is not what it is observed. Indeed, observations suggest that $v_{\rm rot}(R)$ approaches a constant for large R, which is a clear indication of a dark halo extending beyond the luminous disk, because it means that $M(R) \neq$ constant near the visible edge of the galaxy. An example of this is shown in Fig. 2.17. From rotation curves, we estimate $\Omega_{\rm halo} \simeq 0.1$.

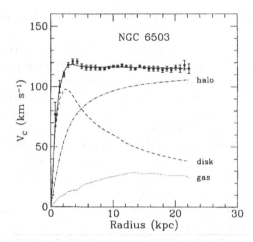

FIGURE 2.17 Observations of the rotation curve of galaxy NGC6503. At radii more than a few kilo-parsecs, the rotation curve becomes flat, suggesting that beyond the contribution of the disk, which follows Kepler's law, there is a dark halo whose density is peaking a few tens of kiloparsecs away from the centre of the galaxy. Reproduced with premission from Ref. [BBS91].

- **Hot gas in galactic clusters:**
  Gas present in intergalactic space is heated when it falls into the potential well of a galactic cluster. The gas temperature then generates pressure, which counter-balances gravitational attraction. The gas temperature is determined by X-ray observations, which allow the mass of the gas to be estimated. However, the inferred mass is much larger than that of the observed gas. This suggests that $\Omega_{\text{cluster}} \simeq 0.3$.

- **Bulk flows:**
  We obtain a dynamical estimate of galactic masses based on the gravitational attraction inferred by the relative peculiar motions of neighbouring galaxies. Basically, we can tell how massive galaxies are when they rotate around each-other. This estimate is roughly $\Omega_m \geq 0.2$.

- **Structure formation:**
  The formation of large scale structure is due to the growth of overdensities, triggered by the presence of initially small irregularities (density perturbations) of primordial origin (more on this later). Structure formation is facilitated through gravitational attraction by the potential wells of the primordial density perturbations. Thus, structure formation probes Ω_m. Observations suggests that the age of the Universe is enough time for the observed structure to form only if the matter abundance is $\Omega_m \geq 0.2$.

- **Big Bang nucleosynthesis:**
  The process of Big Bang nucleosynthesis (BBN) is a major success of Big Bang cosmology. BBN accounts for the observations of the delicate abundance of light elements such as D, ^{7}Li, ^{3}He, and ^{4}He, not produced by stars. We discuss the process in the next chapter (Sec. 3.3.2). Here we note that BBN imposes stringent bounds on baryonic (i.e., regular) matter. Indeed, the latest BBN observations suggest that the density parameter of baryonic matter is $\Omega_B = 0.049 \pm 0.001$. This suggests that most, if not all, dark matter is non-baryonic.

2.12.2   Properties of non-baryonic dark matter

Structure formation requires non-baryonic dark matter to have a number of properties. These requirements set stringent constraints on non-baryonic dark matter candidates. The main properties are:

- **Pressureless:**
  Non-baryonic dark matter must be nonrelativistic (dust), similarly to baryonic matter. Such dark matter is called *cold dark matter* (CDM).

- **Collisionless:**
  The interactions of CDM with itself are weak. Such CDM cannot dissipate its kinetic energy due to the in fall into the gravitational wells of overdensities. As a result, CDM does not form disks nor does it gather at the centre of overdensities. Instead, CDM forms spherical halos around galaxies.

- **Weakly interacting:**
  This is similar to the above but has to do with interactions between CDM and baryonic matter. Indeed, CDM interacts predominantly gravitationally with baryonic matter. This is why CDM is notoriously difficult to detect; it flows through any detector almost unimpeded since the detector is built from baryonic matter.

2.12.3   Candidates for dark matter

Since stars make up about 1% of the total density of the Universe at present while baryonic matter is roughly 5% (says BBN), we also need some baryonic dark matter. The main candidate is:

- **MACHOs:**
  MACHOs stands for massive astrophysical compact halo objects. These are star remnants (neutron stars, burned-out white dwarfs), brown dwarfs (failed stars), Jupiters (giant planets), and so on. MACHOs have been observed by gravitational microlensing events. Their inferred density is bound as $\Omega_{\text{MACHO}} < 0.06$, which agrees nicely with BBN.

For non-baryonic dark matter the main candidates are:

- **WIMPs:**
  WIMPs stands for Weakly Interacting Massive Particles. These are massive particles, with rest energy about a TeV (a thousand times heavier than protons) that are generically produced in many supersymmetric theories. Supersymmetry was originally introduced to unify fermions and bosons and assist, thereby, with the unification of fundamental forces. Nothing to do with dark matter. Yet, supersymmetry naturally produces WIMPs with the right masses and interaction rates to be CDM candidates. This is called the "WIMP miracle." In many supersymmetric theories, the lightest supersymmetric particle (LSP) is stable (does not decay), which makes the LSP a compelling CDM candidate. Prominent examples are the neutralino and the gravitino. Neutralinos interact with ordinary matter through the weak nuclear force (like neutrinos), which result in numerous direct detection experiments. However, at the time of writing, no WIMPs have been observed, and no evidence for supersymmetry has been found yet.

- **Axions:**
  These are ultra light, scalar particles (like the Higgs boson) originally introduced to solve the strong CP-problem in quantum chromodynamics (QCD). Again, no relation to dark matter. Yet, they can be successful CDM candidates. Axions weakly interact

with electromagnetism (axion-photon conversion in strong magnetic fields), which offers some observational possibilities. However, like with WIMPs, no axions have been observed yet.

In a sense, WIMPs and axions are not very exotic. They are both hypothetical particles, which are reasonably motivated by theory, but interact weakly with ordinary matter, so they are very difficult to detect. This is similar to the neutrinos, which were originally also hypothetical, postulated by Wolfgang Pauli in 1930 but were eventually detected in 1956 by Cowan and Reines, who won the Nobel prize in 1995 (with Harrison, Kruse, and McGuire); almost fourty years afterwards! This is why cosmologists tend to believe that CDM is not really a big mystery. It most probably corresponds to a nonrelativistic particle, which will eventually be discovered. Note that WIMPs or axions can account for all necessary CDM.

In recent years, interest has been growing about another CDM candidate, namely:

- **Primordial Black Holes:**
  These are possible to generate in the very early Universe, when there is a spike in the density perturbations spectrum. They are both baryonic and non-baryonic candidates, in the sense that it is irrelevant whether the original overdensities were baryonic or non-baryonic or both. Such primordial black holes need to be massive enough not to evaporate (through emitting Hawking radiation) until today so they can become the dark matter. Additionally, primordial black holes must have low enough abundance not to disturb BBN. They also need to satisfy many astrophysical constraints.[*]

2.12.4 Problems of CDM

There are some generic problems that arise in numerical simulations of the formation of the large scale structure. The main problems are:

- **Cuspy Halos:**
  CDM halos in simulations feature a cusp at galactic centres compared to the mass distribution inferred by rotation curves, which peaks at the centre.

- **Missing Satellites:**
  CDM simulations predict too many small satellite dwarf galaxies compared to observations around spiral galaxies.

Both the above problems can be addressed if the contribution of baryons is taken into account or if there is a mix of CDM with warm dark matter (WDM), corresponding to thermalised particles with rest energy about a keV (a thousand times lighter than electrons). A typical example of a WDM candidate is a sterile neutrino. For scales larger than galaxies WDM behaves like CDM, but for smaller scales structures are washed out.

2.12.5 Direct observational evidence for dark matter

In 2004 a remarkable observation, that of the bullet cluster, provided direct evidence for the existence of dark matter.[†] The bullet cluster (1E 0657-558) consists of two galactic clusters passing through each other. It lies at a distance of a little more than 1 Gpc. X-ray observations of the hot intergalactic gas of the clusters provide the image (see Fig. 2.18) for the location

---

[*]Another related option for CDM are the "Planck particles," which can be hypothetical remnants from primordial black hole evaporations.
[†]Strictly speaking, only one of the two clusters is called "bullet."

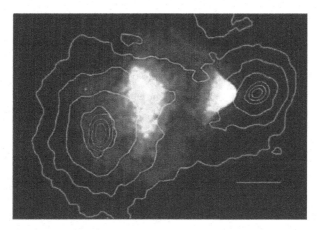

FIGURE 2.18 Observational image of the bullet cluster. X-ray observations of the hot intergalactic gas of the clusters depict the location of baryonic matter. Gravitational lensing observations are used to infer the distribution of the total matter content of the clusters, depicted by contours. It is evident that there is a significant lag of the baryonic matter with respect to the location of the dark matter. © AAS Reproduced with permission from Ref. [CBG+06].

of baryonic matter. Gravitational lensing observations are used to infer the distribution of the total matter content of the clusters. There is a significant lag of the baryonic matter, due to its collisional nature, with respect to the location of the dark matter, which is weakly interacting and collisionless. This is another piece of evidence that dark matter is predominantly non-baryonic. It is also important to point out here that the bullet cluster observation strongly undermines efforts to dispense with dark matter by assuming some modified Newtonian dynamics approach because there is a distinct difference in location of the gravitational pull of visible and dark matter. In my mind, the bullet cluster observation is the final nail on the coffin of efforts to dispense with dark energy by considering modified Newtonian dynamics (MOND).

All the observations agree on the value $\Omega_{\mathrm{CDM}} = 0.264 \pm 0.006$. As we have said, for baryonic matter we have $\Omega_B = 0.049 \pm 0.001$ and for dark energy, observations suggest $\Omega_\Lambda = 0.687 \pm 0.007$. To complete the density budget of the present Universe, we also need to mention radiation (CMB and neutrinos), for which $\Omega_r = (0.910 \pm 0.025) \times 10^{-4}$ [A+18a].

EXERCISES

1. Assume that the Universe content is (nonrelativistic) matter (both baryonic and dark matter) and radiation (relativistic matter) and that the present acceleration of the Universe expansion is due to a nonzero cosmological constant.

   (a) Show that the density of each independent component of the Universe content as a function of redshift is

   $$\rho_i(z) = \Omega_i(t_0)\rho_c(t_0)(z+1)^{3(1+w_i)},$$

   where $\Omega_i(t_0)$ is the density parameter of the given component at present, w_i is the corresponding barotropic parameter (assumed constant) and $\rho_c(t_0)$ is the current value of the critical density.

   (b) Show that the total density as a function of redshift is

   $$\rho(z) = \rho_c(t_0)\left[\Omega_M(1+z)^3 + \Omega_R(1+z)^4 + \Omega_\Lambda\right],$$

   where Ω_M is the density parameter of matter at present, Ω_R is the density parameter of radiation at present, and Ω_Λ is the density parameter of the cosmological constant at present.

   (c) Observations suggest that $\Omega_M \simeq 0.3$ and $\Omega_\Lambda \simeq 0.7$. Find the redshift when $\rho_m = \rho_\Lambda$ and the cosmological constant takes over (starts to dominate) the Universe expansion (from matter).

   (d) Observations suggest that $\Omega_R \sim 10^{-4}$. Find the redshift when $\rho_m = \rho_\gamma$ and matter takes over (starts to dominate) the Universe expansion (from radiation).

   In addition, if the Universe is also spatially flat, then $\rho_c(t_0) = \rho(t_0) \equiv \rho_0$. In this case, this is the ΛCDM model of concordance cosmology.

2. Consider the case in which the Universe has non-negative curvature (i.e., $k \geq 0$) and is dominated by phantom dark energy with equation of state $p < -\rho c^2$.
   Combine the continuity equation with the Friedman equation to show that the Hubble parameter is a growing function of time, $\dot{H} > 0$.

3. Assume that in the early Universe the density is given by

   $$\rho = \frac{3}{8\pi G t_\star^2}\left(\frac{a_c}{a}\right)^2 e^{-2(a/a_c)},$$

   where $a_c \neq 0$ is a constant value of the scale factor $a(t)$ and $t_\star \neq 0$ is the time of the Big Bang, such that $a(t_\star) = 0$. (Note that the Big Bang is *not* the origin of time in this case.)
   Assume also that the Early Universe is spatially flat.

   (a) By solving the Friedmann equation, show that

   $$a(t) = a_c \ln(t/t_\star).$$

   (b) By solving the acceleration equation, show that the barotropic parameter of the Universe is:

   $$w = \frac{1}{3}\left(2\frac{a}{a_c} - 1\right).$$

   (c) Find the time when:
       (i) the barotropic parameter mimics that of a radiation-dominated Universe.
       (ii) the barotropic parameter of the Universe is $w = -\frac{1}{3}$. Determine whether or not the acceleration of the expansion of the Universe is zero at that moment.

4. Suppose that the large-scale dynamics of the Universe in late times could be modelled as if, starting from matter domination, the Universe became abruptly dominated by a single substance with a variable barotropic parameter, such that

$$H = -\dot{w},$$

where H is the Hubble parameter and we assume that w varies continuously from matter domination onward.

(a) Assuming that $w = -1$ at present show that

$$1 + w = \ln(a_0/a),$$

where a_0 is the current value of the scale factor.

(b) Integrating the continuity equation, show that the density is

$$\rho = \rho_0 \exp\left[\frac{3}{2}(1+w)^2\right],$$

where ρ_0 is the density at present.

(c) Find the redshift when:

(i) The Universe stops being dominated by matter.

(ii) The substance dominating the Universe becomes dark energy.

5. Considering a flat Universe with matter and $\Lambda > 0$, show that the age of the Universe is

$$
\begin{aligned}
t_0 &= \frac{2}{3}\frac{H_0^{-1}}{\sqrt{1-\Omega_m}} \ln\left(\frac{1+\sqrt{1-\Omega_m}}{\sqrt{\Omega_m}}\right) \\
&= \frac{2}{3}\frac{H_0^{-1}}{\sqrt{1-\Omega_m}} \sinh^{-1}\sqrt{\frac{1-\Omega_m}{\Omega_m}},
\end{aligned}
$$

where Ω_m is the matter density parameter at present.

3

History of the Universe

3.1   Thermal background   46
     Planck distribution • Black body spectrum •
     Maxwell-Boltzmann distribution • The Hot Big Bang •
     Freeze-out
3.2   The cosmic microwave background radiation   51
     Decoupling and recombination • CMB and its
     anisotropy
3.3   The first few minutes   53
     Neutrino decoupling and electron-positron pair
     annihilation • Big Bang nucleosynthesis
3.4   The very early Universe   57
     Phase transitions • Quark confinement • Electroweak
     transition • Grand unification • The Planck epoch
3.5   Baryon asymmetry   62
3.6   The outline of the Hot Big Bang   64

Now that we have an idea of the dynamics and content of the Universe, we can travel backwards in time and map the history of the Universe. We find that this history can be exotic and violent. Dramatic events occurring in the early Universe leave traces behind which we observe, thereby constraining the theory describing physics at extreme situations. The more we move backwards in time, the more the Universe contracts and the energy density becomes very large. Hence, we need to employ high-energy physics to study phenomena which occur in the early Universe. Also, towards early times, not only distances but also the horizon contracts to minuscule scales. This means that the observable Universe, which now extends over several Gpc, is compressed to minute size, such that particle physics (which governs microscopic scales) becomes relevant. In this sense, cosmology treats the entire Universe as a giant laboratory, to test fundamental physics at energies well beyond Earth-based experiments. Before contemplating this, however, we briefly describe the radiation dominated era, which (as we have seen) constitutes the first 47,000 years of the Universe's evolution. If we time-travel backwards into the radiation era, we find that the Universe is already squeezed to the point that its content is more like the interior of the Sun: hot and dense, fully ionised and in approximate thermal equilibrium.

3.1   Thermal background

Because the density of matter and radiation scales as inverse powers of the scale factor ($\rho_m \propto a^{-3}$ and $\rho_r \propto a^{-4}$ respectively), in the early Universe we expect the density $\rho = \rho_m + \rho_r$ to be huge. Consequently, since particles are substantially squeezed together, the particle interaction rate is much larger than the expansion rate of the Universe. This means that relativistic and nonrelativistic particles are in thermal equilibrium.

To obtain the particle number density for a given particle species in thermal equilibrium we work as follows. Heisenberg's uncertainty relation assigns volume h^3 in momentum space for every quantum state, where h is the Planck constant. This means that the particle number density of a given particle species is

$$n = g \int \mathcal{N} \frac{d^3 P}{(2\pi\hbar)^3} \,, \tag{3.1}$$

where P is the particle momentum, $\hbar \equiv h/2\pi = 1.05 \times 10^{-27} \, \mathrm{g\,cm^2/sec}$ is the reduced Planck constant, g is the particle degrees of freedom (e.g., spin states), and $\mathcal{N}$ is the occupation number per quantum state of energy $E(P)$, which is given by

$$\mathcal{N} = \frac{1}{\exp\left(\frac{E(P)}{k_B T}\right) \pm 1} \,, \tag{3.2}$$

which is also called the Planck function.* In the above, the plus sign corresponds to fermions and the minus sign to bosons. $\mathcal{N}$ is smaller in the case of fermions because of the Pauli exclusion principle. Also, T is the temperature, and $k_B = 8.62 \times 10^{-5} \, \mathrm{eV/K}$ is the Boltzmann constant, which can be thought of as a conversion between temperature and energy. We should stress here that the Planck function is only valid if the species at equilibrium is its own antiparticle (e.g., the photon) or if there are equal number of particles and antiparticles, created and destroyed in pairs (e.g., electrons and positrons).

Thus, Eq. (3.1) becomes

$$n = \frac{g}{(2\pi\hbar)^3} \int_0^\infty \frac{4\pi P^2 dP}{e^{E(P)/k_B T} \pm 1} \,, \tag{3.3}$$

where the particle energy is $E(P) = \sqrt{m^2 c^4 + P^2 c^2}$.

3.1.1 Planck distribution

Concentrate now on relativistic particles, i.e., radiation. In this case, $E \simeq Pc$ and Eq. (3.3) can be written as

$$n_r = \frac{4\pi g (k_B T)^3}{(2\pi\hbar c)^3} \int_0^\infty \frac{y^2 dy}{e^y \pm 1} \,, \tag{3.4}$$

where $y \equiv Pc/k_B T$. The integral above is of order unity; indeed $\int_0^\infty \frac{y^2 dy}{e^y \pm 1} = \zeta(3) \times (1.75 \mp 0.25)$, where $\zeta(3) = 1.2021...$, with $\zeta(x)$ being the Riemann zeta function. Thus, as an order of magnitude estimate, we find the number density of relativistic particles

$$n_r \sim g \left(\frac{k_B T}{\hbar c}\right)^3 \,. \tag{3.5}$$

We see that the number density of radiation is solely determined by the temperature T.

3.1.2 Black body spectrum

We now consider the energy density per momentum interval for radiation, which is given by $(d\epsilon/dP)dP = (dn/dP)E(P)dP$, i.e.,

$$\frac{d\epsilon}{dP} dP = \frac{4\pi g c}{(2\pi\hbar)^3} \frac{P^3 dP}{e^{\frac{Pc}{k_B T}} \pm 1} \quad \Rightarrow \quad \frac{d\epsilon}{dP} = \frac{4\pi g c (k_B T)^3}{(2\pi\hbar c)^3} \frac{y^3}{e^y \pm 1} \,, \tag{3.6}$$

---

*We ignore the chemical potential.

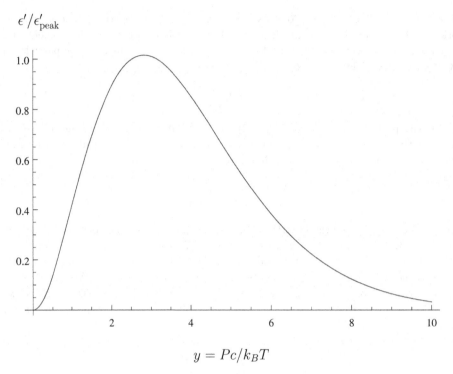

$$y = Pc/k_B T$$

FIGURE 3.1   Black body spectrum. The plot of $\epsilon'(y) \equiv d\epsilon/dP$ peaks roughly at $y \simeq 3$, where $y \equiv Pc/k_B T$.

where $y \equiv Pc/k_B T$ as before. Plotting the above with respect to y, we find that it peaks near $y_{\text{peak}} \approx 3$, i.e., $P_{\text{peak}}c \approx 3k_B T$ (see Fig. 3.1). This means that most radiation particles have energy $E_r \simeq 3k_B T$. This is called the black body spectrum of radiation.

We obtain the total energy density of radiation, by integrating Eq. (3.6) over all momenta. We find

$$\epsilon_r = \int_0^\infty \frac{d\epsilon}{dP} dP = \frac{\pi^2 g}{30(c\hbar)^3} (k_B T)^4 \int_0^\infty \frac{y^3}{e^y \pm 1} dy \,, \tag{3.7}$$

where we used $dP = (k_B T/c)dy$. For bosons we have $\int_0^\infty \frac{y^3}{e^y-1} dy = \frac{\pi^4}{15}$, while for fermions, we need to multiply with a factor of $\frac{7}{8}$ since $\int_0^\infty \frac{y^3}{e^y+1} dy = \frac{7\pi^4}{120}$. Thus, a collection of relativistic species of both fermions and bosons has energy density given by

$$\epsilon_r = \frac{\pi^2 g_*}{30(c\hbar)^3} (k_B T)^4 \,, \tag{3.8}$$

where $g_* \equiv \sum_{\text{bosons}}^i g_i + \frac{7}{8} \sum_{\text{fermions}}^j g_j$ is the number of *effective relativistic degrees of freedom*. If we consider only light then Eq. (3.8) is the Stefan–Boltzmann law with $g_* = 2$, corresponding to the two polarisation states of light. In general, in the early Universe we have $g_* = \mathcal{O}(10 - 100)$.

A collection of fluids in thermal equilibrium is called the *thermal bath*. For relativistic fluids, the typical energy of a particle of the thermal bath is $E_r \simeq 3k_B T$.

3.1.3   Maxwell-Boltzmann distribution

We now investigate the case of nonrelativistic particles, i.e., matter. In this case $mc^2 \gg Pc$, which means $E(P) \simeq mc^2 + P^2/2m$. Now, if the nonrelativistic particles are in thermal equilibrium with relativistic particles at temperature T, then the momentum of all particles in the thermal bath is the same and is given by $Pc \sim k_B T$. Using this, Eq. (3.3) becomes

$$n_m = \frac{4\pi g (2m k_B T)^{3/2}}{(2\pi\hbar)^3} \, e^{-mc^2/k_B T} \int_0^\infty \hat{y}^2 e^{-\hat{y}^2} d\hat{y}, \qquad (3.9)$$

where $\hat{y} \equiv P/\sqrt{2m\,k_B T}$. Again, the integral above is of order unity; indeed $\int_0^\infty \hat{y}^2 e^{-\hat{y}^2} d\hat{y} = \frac{\sqrt{\pi}}{4}$. Thus, as an order of magnitude estimate, we find the number density of nonrelativistic particles

$$n_m \sim g \left(\frac{m k_B T}{\hbar^2} \right)^{3/2} e^{-\frac{mc^2}{k_B T}}. \qquad (3.10)$$

The above shows that, in the case of nonrelativistic particles, apart from T, the particle number density is also determined by the particle mass m.

3.1.4   The Hot Big Bang

We see that when $k_B T \ll mc^2$, the number density of nonrelativistic particles in thermal equilibrium is exponentially suppressed (called "Boltzmann suppressed"). This is because the thermal bath can create particles and antiparticles of mass m, only if the typical energy $\sim k_B T$ of the relativistic particles is larger than the rest energy mc^2 of the particles of mass m. If $k_B T \ll mc^2$, no pair production of particles of mass m can take place. However, the existing particles of mass m find the corresponding antiparticles and annihilate. This is why their number density is exponentially suppressed. In contrast, when the energy of the typical relativistic particle of the thermal bath is larger than the rest energy of the particles of mass m, annihilations are replenished by new pair production. Schematically, we can write

$$q\,\bar{q} \longleftrightarrow \gamma\,\gamma \qquad \text{when } k_B T \gg mc^2$$
$$q\,\bar{q} \longrightarrow \gamma\,\gamma \qquad \text{when } k_B T \ll mc^2,$$

where the rest mass of particle q is mc^2, $\bar{q}$ stands for its antiparticle and γ stands for a photon—a product of pair annihilation. The energy mc^2 is called the *threshold energy* for particle q. It is straightforward to see that Eqs. (3.5) and (3.10) meet each other at the threshold energy.

Thus, when $k_B T \gg mc^2$ the number of particles in a comoving volume remains constant, and the fluid can be considered independent as it experiences no net decay. The condition $Pc \sim k_B T \gg mc^2$ suggests that the particles are relativistic. Therefore, we conclude that radiation in thermal equilibrium is an independent fluid. As such, we expect $n_r \propto a^{-3}$ and $\rho_r \propto a^{-4}$ as discussed in the previous chapter. From Eqs. (3.5) and (3.8), we see that $n_r \propto T^3$ and $\rho_r = \epsilon_r/c^2 \propto T^4$. Thus, we obtain

$$T \propto \frac{1}{a}, \qquad (3.11)$$

which suggests that the temperature of the thermal bath filling the Universe decreases with the expansion. Therefore, the Universe cools down as it expands. As a result, we expect the early Universe to be not only very dense but also very hot. This is why the thermal history of the Universe is called the *Hot Big Bang*, which should not be confused with the term Big Bang, which denotes the original explosion which started it all. The Hot Big Bang is the history of the Universe, more or less until the present, when the Universe content is dominated first by radiation and then by matter. Note, however, that dark-energy domination is not considered a part of the Hot Big Bang, either at early or at late times.

For matter in thermal equilibrium things are different from radiation. Because pair annihilations are not replenished in this case, n_m is exponentially suppressed and the fluid is *not* independent.

Therefore, for a given species in thermal equilibrium the story is as follows. At early times, the temperature is high such that $Pc \sim k_B T \gg mc^2$, and the species is relativistic (radiation) with density $\propto a^{-4}$. As the Universe expands and cools, T decreases and after some time $Pc \sim k_B T < mc^2$. Then the species becomes nonrelativistic (matter) with density mn_m, which is exponentially suppressed so that the species practically disappears. All this, of course, provided that the species in question remains in thermal equilibrium.

The picture that emerges is thus the following. At very early times the content of the Universe includes a large number of species, which are all relativistic and are generated by pair creation. This is called the *primordial soup*. As time passes and the Universe cools down, the characteristic energy of the thermal bath $k_B T$ reduces below the threshold energy of various species, which become Boltzmann suppressed and disappear, leaving behind only the lightest particles, such as photons and neutrinos. However, a particle species does not always stay in thermal equilibrium to the thermal bath. If it doesn't, then we say it "freezes out" off thermal equilibrium. We discuss this freeze-out next.

3.1.5   Freeze-out

Whether a given fluid in the expanding Universe remains in thermal equilibrium depends on the relative magnitude of the expansion rate $H \equiv \dot{a}/a$ and the interaction rate $\Gamma_{\text{int}} \equiv \langle \sigma v \rangle n$ between the particles of the fluid with the rest of the thermal bath, where σ is the cross section of particle interactions and v is the average particle speed. The fluid in question remains in thermal equilibrium if $\Gamma_{\text{int}} \gg H$. If, however, $\Gamma_{\text{int}} < H$ then the fluid decouples from the thermal bath and the corresponding particle species is said to freeze-out. The above can be understood better by considering that

$$\Gamma_{\text{int}} < H \quad \Rightarrow \quad t_{\text{int}} \sim \Gamma_{\text{int}}^{-1} > H^{-1} \sim t \,,$$

where t_{int} is the typical interaction time-scale. The above means that if $\Gamma_{\text{int}} < H$, then the typical interaction time is larger than the age of Universe, so interactions cease.

What happens after freeze-out? Well, it depends on whether the species in question is stable or not. If it is stable, then the particle number is conserved in a comoving volume, which means that $n \propto a^{-3}$ because the comoving volume grows as $V \propto a^3$, or equivalently $\dot{n} + 3Hn = 0$. If the particles decay however, the equation is modified to

$$\dot{n} + 3Hn = -\Gamma_{\text{dec}} n \,, \tag{3.12}$$

where Γ_{dec} is the decay rate. Similarly to interactions, if $\Gamma_{\text{dec}} \ll H$ then the particle is effectively stable and $n \propto a^{-3}$. In the opposite case, when $\Gamma_{\text{dec}} \gg H$, then Eq. (3.12) suggests $n \propto e^{-\Gamma_{\text{dec}} t}$ for $\Gamma_{\text{dec}} = $ constant, meaning that the particle species becomes exponentially suppressed. This should not be confused with Boltzmann suppression discussed in the previous section, because it may well occur for a relativistic species, which is not part of the thermal bath. The general solution to Eq. (3.12) is

$$n = n_{\text{fr}} \left(\frac{a_{\text{fr}}}{a} \right)^3 e^{-\Gamma_{\text{dec}}(t - t_{\text{fr}})} \quad \text{for } \Gamma_{\text{dec}} = \text{constant} \,, \tag{3.13}$$

where the subscript "fr" denotes the moment of freeze-out of the particle species off the thermal bath.

Our discussion in this section allows one to trace the fate of a constituent of the thermal bath, throughout the Hot Big Bang. A prominent example is the case of thermal CDM. This is a CDM species (e.g., a WIMP), which is part of the thermal bath in the early Universe, such that its abundance is determined, first by Eq. (3.5) and then by Eq. (3.10), once it becomes nonrelativistic and Boltzmann suppressed. The exponential decrease of its density is halted when the CDM freezes-out of the thermal bath. From then on, its number density scales as $n \propto a^{-3}$ until the present since it is a stable particle species. We see that the only input we require in order to estimate the abundance of this kind of CDM candidate is the mass and the interaction rate with the thermal bath, which determines when the freeze-out occurs. It so happens that, for WIMPs, these values are exactly right (hence "WIMP miracle," cf. Sec 2.12.3). For axions, however, things are more complicated because they are never part of the thermal bath. Hence, they remain "cold" (that is nonrelativistic) despite being ultralight.

3.2 The cosmic microwave background radiation

3.2.1 Decoupling and recombination

In the latest stages of the Universe evolution, the thermal bath is dominated by photons; as the neutrinos have decoupled some time ago (see Sec. 3.3.1). The typical energy of photons of the thermal bath is $E_\gamma \sim \epsilon_\gamma / n_\gamma \sim k_B T$, where we considered Eqs. (3.5) and (3.8). This agrees with the fact that most photons correspond to the peak of the black body distribution, which has energy $E_\gamma \simeq 3 \, k_B T$. In the early Universe, the temperature is very high, which means that E_γ is large. If $E_\gamma > 13.6$ eV, which is the binding energy of the hydrogen atom, this means that most photons would be energetic enough to ionise hydrogen. The same is true for other, heavier atoms, if they are around. Thus, there can be no atoms in the early Universe. Instead, the Universe content is comprised of radiation and fully ionised plasma.

Plasma can be thought of as a superposition of two electrically charged fluids: the electron fluid and the nuclei fluid. The interaction rate between free electrons and photons (Thomson scattering) is very high, because we have to remember that photons are the quanta of the electromagnetic field, which are naturally strongly affected by the presence of electrically charged particles, like the electrons. As a consequence of this strong interaction, the photon mean-free-path is very small, meaning that the photons cannot travel unimpeded into the plasma; they keep bumping into electrons. Therefore, the early Universe is opaque, very similar to the interior of our Sun, which is also made of opaque plasma.

As time passes, the temperature decreases and eventually photons become unable to ionise hydrogen (or any other heavier atom). This means that the free electrons can combine with the protons (or other nuclei) to form neutral atoms without the latter being destroyed by energetic photons (photo-disintegrated). This process is called recombination and took place when the Universe was about 378,000 years old (redshift about 1090). Recombination is actually a mis-nomer because electrons and nuclei had never been combined before. Note that recombination occurs after $t_{\rm eq} \simeq 47,000$ years, that is after the end of the radiation-dominated era and the onset of the matter-dominated era.

The scattering of photons on atoms is drastically reduced exactly because atoms are electri-cally neutral. This means that, after recombination, the photon mean-free-path grows rapidly up to horizon size. Light decouples from matter, and the Universe becomes transparent. The process is called *decoupling*. It is distinct to recombination, even though it follows right after it.

This is exactly what happens at the photosphere of the Sun. The temperature of the solar material decreases enough that neutral atoms can form and light can travel unimpeded. There is more of the Sun beyond the photosphere (chromosphere, corona). However, this part of the Sun is transparent, and this is why the visible surface of the Sun is the photosphere.

Decoupling in the early Universe and in the Sun are directly analogous; only in the case of the Sun, the temperature decreases as one moves in space away from the Sun's centre, while in the case of the Universe, the temperature decreases as one moves in time along the Universe expansion. But seeing far away in astronomy is also seeing into the past, as it takes time for light to reach the observer. This means that if we can see very far in the Universe, we may look at the time that light scattered for the last time on the free electrons before becoming free. This *last scattering surface*, therefore, is a surface in time, because decoupling happened everywhere. Yet, for the observer, it is a firewall, similar to the Sun's photosphere, surrounding the field of vision. This is because, if anyone looked far enough, they would see the moment when light decoupled from matter, regardless of direction. The radius of the last scattering surface is not too different from that of the horizon itself, because decoupling happened relatively soon after the Big Bang; 378,000 years is very little compared to 14 billion years.[*]

3.2.2 CMB and its anisotropy

As we have seen, after decoupling, the released radiation travels freely, filling the Universe with a uniform background. Today, this background radiation is observed at microwave wavelengths and is called the cosmic microwave background radiation (CMB). It was accidentally discovered in 1964 by Arno Penzias and Robert Woodrow Wilson, who won the Nobel prize in 1978. The CMB is one of the major pillars of support for the Big Bang model, since it demonstrates that the early Universe was hot and dense, disproving thereby the rival theory of a steady-state universe.[†]

Even after decoupling, radiation retains its black body spectrum. The reason for this is the following. For a photon $Pc = E_\gamma = hc/\lambda \propto 1/a \propto T$. This means that $y \equiv Pc/k_B T =$ constant. Hence, the functional form of the energy density per momentum interval, as given by Eq. (3.6), remains invariant apart from an overall decrease in amplitude, which corresponds to a lowering of the frequency as $f = E_\gamma/h \propto 1/a$. This is why the peak of the CMB photons corresponds to microwave frequency, even though when emitted it was close to the visible light, like the Sun (a bit redder though). Thus, despite the absence of interactions (scattering), photons retain their original thermal distribution. This fossil black body radiation is characterised by a value of the temperature, which is observed to be $T_{\rm CMB} = 2.7255 \pm 0.0006$ K (equivalently $k_B T_{\rm CMB} = 0.2348 \pm 0.0001$ meV). In view of Eq. (3.11), the temperature at decoupling is $T_{\rm dec} = (a_0/a_{\rm dec})T_{\rm CMB} = 0.26$ eV$/k_B \approx 3000$ K, which is slightly lower than the Sun's photosphere (between 4600 K and 6000 K).[‡] The CMB radiation has been called (by Martin White) "the most perfect black body ever measured in nature." Its extreme isotropy is very strong support for the cosmological principle. Yet, there are anisotropies in the CMB, which are very revealing.

Firstly, we discuss the *dipole anisotropy*. The CMB radiation has a dipole anisotropy at the level of 10^{-3} due to Doppler redshift from Earth's motion. The CMB photons have energy $Pc = E_\gamma = hf$. Accounting for the Doppler shift, the corresponding Planck function is

$$\mathcal{N}_\gamma = \frac{1}{e^{\frac{hf}{k_B T}} - 1} \quad \rightarrow \quad \frac{1}{e^{\frac{h(f/\mathcal{D})}{k_B T}} - 1} = \frac{1}{e^{\frac{hf}{k_B \mathcal{D} T}} - 1}, \tag{3.14}$$

where the Doppler factor is $\mathcal{D} \simeq 1 + \vec{v} \cdot \hat{r}/c$, for $v \equiv |\vec{v}| \ll c$, with $\hat{r}$ being the unit vector along the line of sight and $\vec{v}$ being the velocity of the Earth in the frame where the CMB

---

[*]As we have seen, at present, the expansion begins to accelerate so that we have switch-over from particle to event horizon, both of which are currently comparable.

[†]The steady-state theory proposed that the dilution of matter due to the Universe expansion is replenished by a continuous creation of matter, so that the the Universe is in a stationary situation, remaining always the same. Among others, it was advocated by Herman Bondi, Thomas Gold, and Fred Hoyle (who also coined the term "Big Bang," remember?).

[‡]Because of the difference in abundance of photons versus free electrons at last scattering.

is isotropic. Thus, the Doppler-boosted thermal radiation appears exactly thermal with temperature $T_{obs} = \mathcal{D}T_{CMB}$. Therefore, the observed temperature features a sinusoidal modulation, such that $T_{obs} = T_{CMB}\left\{1 + \frac{v}{c}\cos\theta + \mathcal{O}[(v/c)^2]\right\}$, where θ is the angle between the line of sight and $\vec{v}$. We conclude that the CMB radiation has a dipole anisotropy with magnitude $\delta T_{CMB} \equiv (v/c)T_{CMB} = 3.36 \times 10^{-3}$ K. The observed value of δT_{CMB} allows the calculation of v. Indeed, we find that the Milky Way (which carries along the Sun and the Earth) is moving with speed 627 ± 22 km/sec relative to the reference frame of the CMB.

Apart from the dipole anisotropy, the CMB radiation also features a primordial anisotropy of magnitude $\Delta T_{CMB} \simeq 10^{-5}\,T_{CMB}$. The primordial anisotropy is due to density perturbations, which are local variations of the density of the Universe of primordial origin. They affect the CMB because gravitationally collapsing overdensities redshift light in their potential wells. These primordial density perturbations are the seeds for the formation of structure in the Universe, such as galaxies and galactic clusters. The Hot Big Bang does not account for the origin of these primordial perturbations, but inflation does. Thus, the CMB primordial anisotropy reveals crucial information about cosmic inflation. Hence, we talk about it in much more detail in Chapter 5.

In 2006, the Nobel prize was given to John C. Mather, NASA Goddard Space Flight Center, and George F. Smoot from UC Berkley "for their discovery of the black body form and anisotropy of the cosmic microwave background radiation." The discovery was made by the spacecraft COBE (Cosmic Background Exporer). To this day, the constraint imposed on inflation from the amplitude of the primordial density perturbations, which are behind the CMB primordial anisotropy, is called the "COBE constraint."

3.3 The first few minutes

We have already discussed two milestones of the early history of the Universe: (matter-radiation) equality and decoupling (and recombination), which take place a few tens and a few hundreds of thousands of years after the Big Bang, respectively. Before equality, we are in the radiation-dominated era. The density of the Universe is given by Eqs. (2.41) and (3.8), which suggest

$$\frac{3}{32\pi G t^2} = \rho = \frac{\epsilon_r}{c^2} = \frac{\pi^2 g_*}{30}\frac{(k_B T)^4}{c^2(c\hbar)^3} \quad \Rightarrow \quad \left(\frac{t}{1\text{ sec}}\right) = \frac{2.4}{\sqrt{g_*}}\left(\frac{1\text{ MeV}}{k_B T}\right)^2, \qquad (3.15)$$

where we considered that the barotropic parameter of the Universe is $w = \frac{1}{3}$ and that the early Universe is approximately spatially flat. A number of important events occur when the energy of the thermal bath is $k_B T \sim 1$ MeV. Equation (3.15) suggests that this corresponds to a time from a few seconds to a few minutes after the Big Bang. In this section, we take a look at this period. We are now deep in the radiation era, and the Universe is so compressed that conditions are similar to the centre of the Sun. Firstly, we look into the whereabouts of neutrinos and electrons.

3.3.1 Neutrino decoupling and electron-positron pair annihilation

Neutrinos are ultralight, electrically neutral relativistic particles, weakly interacting with the thermal bath. Neutrinos decouple from the thermal bath when its energy drops to $k_B T \simeq 2.3 - 3.5$ MeV. After the decoupling of neutrinos, their distribution retains its black body form exactly as is the case for photon decoupling and the CMB.

Since the rest energy of electrons (and positrons, of course) is $m_e c^2 = 0.511$ MeV, the threshold energy of the electron fluid is around 1 MeV or so. This means that, once the energy of the thermal bath drops below this value, electron-positron pair production ceases (by most photons), and the electrons become nonrelativistic and Boltzmann-suppressed, because of not-replenished electron-positron ($e^\pm$) pair annihilation. Since the exothermic interaction $e\,e^+ \longrightarrow \gamma\,\gamma$ becomes one-way, the $e^\pm$ pair annihilations heat up the photon fluid. Photons are heated by $e^\pm$ annihilation but neutrinos are not because the latter have already decoupled from electrons. Hence, $e^\pm$ annihilation results in a boost of photon over neutrino black body temperatures.

Today, a neutrino black body background is expected to exist with temperature: $T_\nu(t_0) \approx 1.945\,^\circ\mathrm{K}$. The low-frequency end of this black body spectrum is affected by the fact that neutrinos are expected to have a nonzero rest energy with $m_\nu c^2 \sim 0.1\,\mathrm{eV}$. Their contribution to the density budget of the Universe at present is given by

$$\Omega_\nu(t_0) = \frac{\sum m_\nu c^2}{93.14 h^2\,\mathrm{eV}}\,, \tag{3.16}$$

where the sum considers all three flavours of neutrinos and $h \approx 0.6752$.

The exponential decrease of the number density of the electron fluid, even though Boltzmann-suppressed, is halted by the fact that there is a tiny imbalance between matter and antimatter in the Universe. Indeed, during $e^\pm$ pair annihilation, every positron e^+ annihilates with an electron e leaving only one electron per proton for neutrality, i.e., the excess of electrons over positrons is matched by an excess of protons over antiprotons. At that time, there is about 1 proton per 10^{10} photons, which means that the residual electron number density is $n_e \sim 10^{-10} n_\gamma$, where n_γ is the photon number density. After all positrons disappear, the surviving electrons are stable particles, with their number density decreasing as $n_e \propto a^{-3}$, even though the electron fluid is still in thermal equilibrium. Because radiation is an independent fluid, we also have $n_\gamma \propto a^{-3}$, which suggests that the ratio $n_e/n_\gamma \sim 10^{-10}$ remains constant. Equation (3.10) holds no more (despite that electrons are in thermal equilibrium) because it is based on the Planck function in Eq. (3.2), which is valid only when there are equal numbers of particles and antiparticles. The matter–antimatter asymmetry is also called *baryon asymmetry* and is not accounted for by the standard model of particle physics. Similarly to the primordial density perturbations, the Hot Big Bang model does not explain the origin of the baryon asymmetry but can accommodate it once a mechanism for its generation is provided. Generating the baryon asymmetry is called *baryogenesis* (cf. Sec. 3.5).*

3.3.2   Big Bang nucleosynthesis

The decoupling of neutrinos and the virtual disappearance of electrons have profound consequences because they affect a seminal process; *Big Bang nucleosynthesis* (BBN). This is defined as the process of the formation of nuclei of the light elements through nuclear fusion processes occurring in the early Universe. The typical nuclear binding energy is of the order of 1 MeV or so. This means that, for very large temperatures $k_B T \gg 1$ MeV, nuclei with multiple nucleons (i.e., protons and neutrons) are destroyed (photo-disintegrated) by the energetic photons of the thermal bath. Consequently, formation of heavy nuclei can happen only after $t_{\mathrm{BBN}} \sim 1$ sec, when the energy of the thermal bath decreases below 1 MeV. Similarly to photon decoupling, which leaves a relic radiation behind (the CMB), BBN also leaves a distinct relic behind, which is observable today. This is the primordial abundance of light elements, which are not generated in stars. These observations open a window to the early Universe, as they offer a glimpse of the state of affairs a few seconds after the Big Bang. BBN is a huge success and a major support of Big Bang cosmology.

---

*CDM can also feature a matter antimatter imbalance, in which case it is called *asymmetric* CDM.

3.3.2.1 A sketch of BBN

BBN is a delicate and complicated process. Here we only sketch a broad outline. We concentrate on the formation of ^{4}He, which is the main product of BBN. To this end, we follow the fate of neutrons, because they largely determine the abundance of ^{4}He. There are three stages for the evolution of BBN: nucleon freeze-out, neutron decay, and nuclear fusion.

Nucleon freeze-out

We are still at an energy of about 1 MeV, a few seconds after the Big Bang. The rest energies of protons and neutrons are $m_p c^2 = 938.3$ MeV and $m_n c^2 = 939.6$ MeV, respectively. They are both in thermal equilibrium. Because the energy of the thermal bath is lower than the threshold energy of both species $k_B T \ll 1$ GeV nucleons are Boltzmann-suppressed and their number densities are given by Eq. (3.10). Applying this equation, the ratio of number densities is

$$\frac{n_n}{n_p} = \left(\frac{m_n}{m_p}\right)^{3/2} \exp\left[-\frac{(m_n - m_p)c^2}{k_B T}\right] \approx \exp\left(-\frac{1.3 \text{ MeV}}{k_B T}\right), \tag{3.17}$$

where we considered that $m_n/m_p \approx 1$ in the last equation. Thus, when the energy of the thermal bath is higher that 1.3 MeV, the number densities of neutrons and protons are roughly the same $n_n \approx n_p$. However, for energy lower that 1.3 MeV we have $n_n < n_p$. This is so, as long as thermal equilibrium still applies.

Equilibrium is maintained due to nuclear reactions of the form $n\,\nu_e \longleftrightarrow p\,e$ or $n\,e^+ \longleftrightarrow p\,\bar{\nu}_e$. However, after the neutrinos decouple from the thermal bath and the electrons virtually disappear through $e^\pm$ annihilation, such nuclear reactions cannot occur any more. Indeed, below the energy of 0.8 MeV the reaction rates drop below the Hubble scale H, and the number density ratio freezes at the value

$$\left.\frac{n_n}{n_p}\right|_{\text{fr}} \approx e^{-\frac{1.3}{0.8}} \approx \frac{1}{5}. \tag{3.18}$$

Neutron decay

After freeze-out, the ratio n_n/n_p decreases further. This is because, in contrast to protons, neutrons are not stable particles, but decay through β-decay: $n \longrightarrow p\,e\,\bar{\nu}_e$. The evolution of the number densities for both neutrons and protons is determined by Eq. (3.13), with $\Gamma_{\text{dec}} = 0$ for the stable protons and $\Gamma_{\text{dec}} = (882\,\text{sec})^{-1}$ for the neutrons. Thus, after freeze-out the ratio of number densities is

$$\frac{n_n}{n_p} = \left.\frac{n_n}{n_p}\right|_{\text{fr}} \times e^{-\Gamma_{\text{dec}}(t - t_{\text{fr}})} \approx \frac{1}{5} \times e^{-\Gamma_{\text{dec}} t}. \tag{3.19}$$

Neutrons cease to decay once locked inside nuclei. This is because inside a nucleus the emitted electron from a β-decaying neutron is absorbed by a proton in the nucleus through the inverse β-decay (which turns the proton into a new neutron). Therefore, the formation of heavy nuclei in BBN is the only way to salvage primordial neutrons and allow them to survive until the present. Now, let's see how nuclei are formed.

Nuclear fusion

As the neutrons begin to decay away, the energy of the thermal bath decreases enough for nuclear fusion to operate, without the produced nuclei being destroyed by energetic photons. A typical chain of nuclear fusion reactions is

$$\begin{array}{ccc}
p\ n & \longrightarrow & D\ \gamma \\
D\ p & \longrightarrow & ^3\text{He}\ \gamma \\
n\ ^3\text{He} & \longrightarrow & ^4\text{He}\ \gamma\,,
\end{array}$$

where $D \equiv {}^2\text{H}$ denotes deuterium, an isotope of hydrogen, with a proton and a neutron in its nucleus. Other similar reactions also involve tritium, which is another isotope of hydrogen, with a proton and two neutrons in its nucleus. At high temperatures, reactions also occur in reverse because photons are energetic enough to photo-disintegrate the light nuclei.

The binding energy of deuterium is 2.22 MeV. This means that, when $k_B T < 2.22$ MeV most photons cannot destroy deuterium. However, the number density of deuterium n_D is much smaller than that of photons, because $n_D < n_p = n_e \sim 10^{-10} n_\gamma$. Consequently, even a tiny fraction of energetic photons is enough to photo-disintegrate all the deuterium. As a result, the production of deuterium begins only after the energy of the thermal bath falls below 0.07 MeV. Deuterium production is a necessary step towards the production of stable ${}^4\text{He}$ because deuterium is involved in all possible chains of nuclear reactions leading to ${}^4\text{He}$ formation. This is called the *deuterium bottleneck*. Once deuterium is formed, all other steps are much more efficient and occur rapidly. Hence, the locking of all the neutrons into ${}^4\text{He}$ takes place when the energy of the thermal bath decreases to 0.07 MeV. Using Eq. (3.15), it is straightforward to find that the time of ${}^4\text{He}$ formation is about 350 sec. Inputting this in Eq. (3.19), we find

$$\frac{n_n}{n_p}\bigg|_{\text{final}} \approx \frac{1}{5} \times e^{-\frac{350}{882}} \approx \frac{1}{7.4}. \tag{3.20}$$

Now we are in a position to calculate the abundance Y of ${}^4\text{He}$. We find

$$Y \equiv \frac{\rho_{{}^4\text{He}}}{\rho_B} = \frac{2n_n}{n_n + n_p} \simeq 0.24, \tag{3.21}$$

where $\rho_{{}^4\text{He}} = m_{{}^4\text{He}} n_{{}^4\text{He}} \approx 2m n_n$ is the density of ${}^4\text{He}$, $\rho_B = m(n_n + n_p)$ is the density of baryons (protons and neutrons), $m \equiv m_n \approx m_p$, the mass of a ${}^4\text{He}$ nucleus is $m_{{}^4\text{He}} = 4m$ and where all the neutrons are assumed to be absorbed inside ${}^4\text{He}$ nuclei, with each nucleus containing two neutrons (so $n_{{}^4\text{He}} = \frac{1}{2} n_n$). Therefore, we find that 24% of the baryonic density is in ${}^4\text{He}$ nuclei.

Why doesn't ${}^4\text{He}$ fuse to produce even heavier elements, as in stars? The reason is that the Universe expansion soon decreases the temperature such that further fusion reactions cannot happen. The reaction rates drop below the Hubble scale very soon after the formation of ${}^4\text{He}$. In fact, all nuclear reactions stop. Consequently, we get a snap-shot of intermediate products like deuterium, tritium, and ${}^3\text{He}$, as well as minute traces of ${}^7\text{Li}$ and ${}^7\text{Be}$.[*] Thus, there is a narrow window of energy for the cooling thermal bath when nuclei heavier than protons are not photo-disintegrated but also when fusion reactions can still occur. The conditions are exactly as in the core of our Sun, where similar fusion reactions take place. But the Universe crosses this stage quickly so only a few light nuclei have a chance to form. Yet, about a quarter of the baryonic density is now in ${}^4\text{He}$ nuclei. They correspond to almost all the surviving neutrons. BBN completes about 350 sec $\approx$ 6 min after the Big Bang.

3.3.2.2 BBN and observations

The fusion reaction rates are measured accurately in the lab. This means that the number density of each species is known as a function of time. Consequently, the expected abundance of ${}^4\text{He}$ and the other light elements (D, ${}^3\text{He}$ and ${}^7\text{Li}$) is well determined. However, comparing these with observations is not straightforward because these light elements are also produced astrophysically: ${}^3\text{He}$ and ${}^4\text{He}$ are continuously generated in stellar interiors and released in space by supernovae explosions, while some ${}^7\text{Li}$ is due to cosmic ray spallation, i.e., it is produced by extremely rare collisions between travelling nuclei in interstellar space. An estimate of the

---

[*]Tritium and ${}^7\text{Be}$ are unstable isotopes and decay radioactively to ${}^3\text{He}$ and ${}^7\text{Li}$, respectively.

primordial abundances is obtained by extrapolating observations at progressively earlier times. The primordial abundances are determined by two crucial factors: the number of neutrino species and the baryon density ρ_B of the Universe.

The number of neutrino species is important because an extra neutrino species increases the number of effective relativistic degrees of freedom g_*, which in turn increases the radiation density of the Universe $\rho_r = \epsilon_r/c^2$ for the same temperature (cf. Eq. (3.8)). This increases the Hubble rate by virtue of the Friedmann equation (2.38) resulting in an earlier freeze-out for the neutrons, which generates a greater abundance of ^{4}He, because the neutrons are not as much Boltzmann-suppressed before freeze-out. The difference in abundance is about 1% per extra neutrino species. Observations suggest that the number of neutrino species is very close to three, which confirms the existence of a neutrino background and, more importantly, agrees nicely with the standard model of particle physics that considers three flavours of neutrinos corresponding to the three families of elementary particles.

Now we discuss the baryon density $\rho_B = mn_B$, where $m = m_n \approx m_p$ as before and $n_B = n_n + n_p$. A larger value of n_B increases the ratio n_D/n_γ, which controls the deuterium bottleneck, and results in a larger value of $(n_n/n_p)_{\text{final}}$. This leads to a greater abundance of ^{4}He. The value of ρ_B also affects the abundances of all the other light elements, which differ between them exponentially, being about 0.01% for D and ^{3}He, and $\sim 10^{-10}$ for ^{7}Li. The theoretical abundance of light elements changes with ρ_B in different and uncorrelated ways, as can be seen in Fig. 3.2; in some cases growing and in other cases diminishing with ρ_B. It is remarkable that, in a narrow range of values for the baryon density (which corresponds to the density parameter at present $\Omega_B = 0.049 \pm 0.001$), observations are indeed simultaneously satisfied. This is why BBN is strong support for the Hot Big Bang.* It also confirms the existence of non-baryonic dark matter, since the present density parameter for matter is $\Omega_m = 0.313 \pm 0.007$ [A+18a], which means that baryonic matter is by far not enough.

BBN is the earliest confirmation of the Hot Big Bang, which is therefore an accurate description of the Universe history since $t_{\text{BBN}} \sim 1$ sec. Indeed, BBN is regarded as "the holy cow" of the standard model of cosmology, meaning that the latter can admit radical modifications to the Hot Big Bang as long as they do not disturb BBN. It is imperative that the Universe be radiation-dominated with no exotic relics playing an important role by the time of BBN. What happens earlier, though, is a different story. Before contemplating various possibilities for the Universe's pre-BBN history, we take a look at what the Hot Big Bang model itself suggests.

3.4 The very early Universe

3.4.1 Phase transitions

At earlier times (and higher energy density) than BBN the Hot Big Bang is reduced to a guideline corresponding to what is expected under the crucial assumption that there is no new and exotic physics. The history of the very early Universe is thus merely an extrapolation of the Hot Big Bang towards early times. At early times the temperature and the typical particle energy of the thermal bath $\sim k_B T$ can become extremely large. Indeed, the flat Friedmann equation (2.38) suggests that $\rho \to \infty$ when we approach the Big Bang, $a \to 0$. As a result, the physical properties of matter may undergo abrupt changes, which are called *phase transitions*.

Phase transitions correspond to the spontaneous breaking of underlying symmetries in Nature, which are manifest at high energy but not in the vacuum. A useful analogy to understand phase transitions and the breaking of symmetries is to consider the phase transition of liquid

---

*Even though there is a minor tension regarding the abundance of ^{7}Li (by a factor of ~ 3).

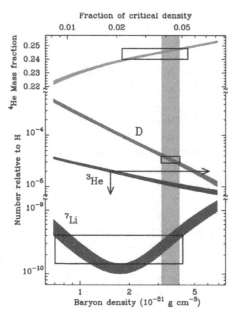

FIGURE 3.2 The theoretical abundances of light elements, such as ^{4}He, ^{3}He, D, and ^{7}Li, compared to the observations depicted by the horizontal bands. We find that there is a narrow band of the baryon density parameter, which satisfies all the theoretical predictions (albeit with some tension for ^{7}Li), even though the abundances are exponentially different. Reproduced with permission from Ref. [BNT00].

water turning to ice as the temperature decreases below zero degrees Celsius. At temperatures higher than zero, water is in liquid form and is characterised by spherical symmetry, that is, it is isotropic. The temperature can change by tens of degrees, but as long as water remains in liquid form, its properties largely remain the same. However, if the temperature is reduced from $+1°$C to $-1°$C a dramatic change takes place. Water freezes, which means it ceases to be liquid and becomes a solid, whose properties are radically altered. Ice is a hexagonal crystal, which means that the isotropy of the liquid water has been reduced to the hexagonal symmetry of the ice lattice. This breaking of the original spherical symmetry with the decrease in temperature, is very much like the breakdown of symmetries in the nature of the material filling the Universe as the latter expands and cools. Phase transitions are potentially violent phenomena especially if they occur through bubble nucleation, similarly to water on the boil, which is another phase transition, between the liquid and the gaseous form of water.

Below we briefly explore the history of the very early Universe based on what known physics suggests for the nature of the Universe content, as it goes through a series of phase transitions.

3.4.2 Quark confinement

At temperatures larger than $k_B T_{\rm QCD} \simeq 200$ MeV nucleons and other hadrons are destroyed by energetic photons into their constituent quarks. The Universe is filled with *quark-gluon plasma*, the gluons being the massless mediator particles of the strong nuclear force. The density of the Universe during this quark epoch is larger than nuclear density. Quark-gluon plasma was artificially generated in 2010 in the Relativistic Heavy Ion Collider of Brookhaven National Laboratory and later on (in 2015) in the Large Hadron Collider in CERN. In a tiny fraction of a second, the produced quark-gluon plasma hadronized (meaning that the free quarks were confined) into stable composite hadrons, which are baryons (made of three quarks like the nucleons) and mesons (made of a quark and an antiquark). This is what happens as the Universe expands and cools. Hence, the quark confinement phase transition is also known as quark-hadron transition (or QCD transition*). Equation (3.15) suggests that it takes place about a few microseconds ($\sim 10^{-6}$ sec) after the Big Bang.

3.4.3 Electroweak transition

At temperatures larger than $k_B T_{\mathrm{EW}} \simeq 250$ GeV, electromagnetism mixes with the weak nuclear force into the *electroweak* force. The crucial quantity is the expectation value of the (electroweak) Higgs field, whose associated particle is the Higgs boson, discovered at CERN in 2012. At high temperatures, this expectation value is zero and electromagnetism is unified with the weak nuclear force. When the temperature decreases with the expansion below T_{EW} ($\sim 10^{15}$ K), the Higgs field assumes a nonzero expectation value and the electroweak force is split into electromagnetism and the weak nuclear force. Most of the particles of the standard model (meaning the particles of usual everyday matter) have masses proportional to the expectation value of the Higgs field, which means that after the electroweak phase transition, they become massive. This is true for all fermions of matter, such as electrons and quarks (although the mass of composite particles like protons and neutrons also includes their binding potential energy). The same is true for some of the particles which mediate forces, such as the $W^{\pm}$ and Z bosons, that mediate the weak nuclear force. The photon, which mediates electromagnetism though, is left massless. At temperatures higher than T_{EW}, all these particles are massless. The photon, in particular, mixes with one of the W-bosons and gives rise to the *hypercharge* field. Thus, in the very early Universe there are no electric or magnetic fields, but hyperelectric and hypermagnetic fields, which are configurations of the hypercharge field.

The standard model of particle physics is really the theory of electroweak unification. In 1979, Sheldon Glashow, Abdus Salam, and Steven Weinberg were awarded the Nobel prize for developing it. In 1983 the W and Z bosons were discovered at CERN, for which Rubbia and van der Meer won the Nobel prize in 1984. Thus, electroweak unification is an experimentally verified fact. As mentioned, the last piece of the standard model, the Higgs boson, was found in 2012.

It is still an open question whether the electroweak phase transition is second- or first-order, meaning whether it proceeds smoothly or violently (with bubble nucleation, as in boiling water), respectively.[†] A first-order phase transition is desired for electroweak baryogenesis. Indeed, a violent phase transition means that the Universe temporarily exits from thermal equilibrium. This allows processes that favour the creation of matter over antimatter to occur. In contrast, if thermal equilibrium is maintained, then the reverse processes are also allowed, so any excess of matter over antimatter is wiped out and baryogenesis cannot happen.

Equation (3.15) suggests that the electroweak phase transition occurs just 10^{-11} sec (a few tens of pico-seconds) after the Big Bang.

3.4.4 Grand unification

Earth-based experiments cannot venture towards energies much higher than the electroweak energy scale. At the moment, colliders can explore energies about 10 TeV or so. They have found no new physics so far. Therefore, considering earlier times than the electroweak phase-transition is mere speculation. The existing knowledge in particle physics does not feature any important

---

*QCD stands for quantum chromodynamics, which is the theory that describes the strong nuclear force and governs the behaviour of quarks.

[†]At the moment, the prevailing view is that the transition is a smooth crossover.

energy scales for an exponentially large region in energy scale, called the *particle desert*. However, there is one other milestone at the other end of the particle desert, at extremely high energies.

At temperatures larger than $k_B T_{\rm GUT} \sim 10^{15-16}$ GeV, the electroweak interaction unifies with the strong nuclear force in the context of a *grand unified theory* (GUT) corresponding to a single GUT-force. At the moment there are many candidate GUTs, but none has experimental confirmation. Cosmology, however, has ruled out several classes of GUTs because they produce dangerous relics (called *topological defects*), which are harmful to the observationally confirmed Hot Big Bang. In particular, many GUTs produce a specific class of topological defects (called *cosmic strings*), which would generate density perturbations with characteristics that would be in conflict with observations.* As such, they are excluded. Above the GUT energy scale, all particles are massless, and there is no net baryon number. If the breaking of grand unification is a first-order phase transition, the phase transition is violent and proceeds with bubble nucleation. Similarly to the electroweak case, a first-order GUT transition can support baryogenesis mechanisms.

Grand unification is facilitated by the GUT Higgs field, which is the GUT equivalent of the electroweak Higgs field. In an analogous process to the electroweak phase-transition, as the temperature decreases and the Universe cools down below the GUT energy scale, the GUT Higgs assumes a nonzero expectation value. This renders the X and Y GUT-bosons (the equivalent to the W and Z weak bosons) massive, and the GUT symmetry is broken to the electroweak and the strong nuclear force.

How do we know that the GUT energy is $\sim 10^{15-16}$ GeV? The GUT energy scale is determined by the weak variance (running) of the coupling constants which parameterise the strength of the fundamental interactions. The values of these coupling constants have a logarithmic dependence on energy (due to renormalisation effects), which means they slowly change as we consider higher and higher energies, towards the Big Bang. Their change is such that they approach each other at high energy. As a result, the three fundamental interactions: electromagnetism and the weak and strong nuclear forces, are almost equally strong at energies near the GUT scale. However, the running couplings do not meet exactly (see Fig. 3.3). This is one of the main motivations behind supersymmetric GUTs, in which the three couplings exactly merge at the supersymmetric GUT-scale $\sim 2 \times 10^{16}$ GeV.

Eq. (3.15) suggests that the GUT transition occurs at time $\sim 10^{-(37-38)}$ sec, which is extremely close to the Big Bang. Yet, there is another milestone earlier than this. You see, we have left out one of the fundamental interactions; gravity. Gravity is theorised to merge with the other fundamental interactions, which are already joined in the context of a GUT, in a *theory of everything* (ToE). The breakdown of the ToE marks the edge of the Hot Big Bang and takes place at the *Planck time*.

3.4.5   The Planck epoch

At temperatures comparable to $k_B T_P \sim 10^{19}$ GeV quantum gravity effects become important. The quantum gravity scale is set by the Planck energy $M_P c^2 = 1.22 \times 10^{19}$ GeV. Over the Planck energy, the classical description of spacetime breaks down. This means that the Friedmann equation (2.38) is valid no more. In fact, close to the Planck-scale we expect the Friedmann

---

*GUT cosmic strings used to be a prominent alternative to cosmic inflation for the formation of the density perturbations, which are needed to seed the large scale structure in the Universe. However, CMB observations in the '90s demonstrated that the density perturbations cannot be sourced by cosmic strings and the cosmic string paradigm has collapsed. This is an example of how drastic an effect observation can have on theory in Cosmology and how tightly, in fact, the theory is connected to the real world.

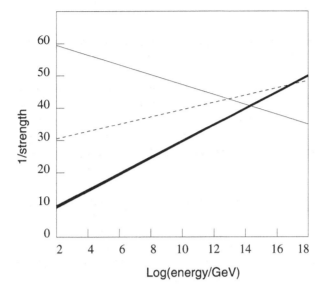

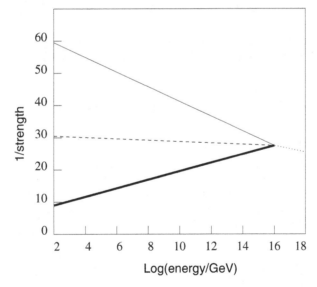

FIGURE 3.3 Logarithmic running of the coupling constants, which parameterise the strength of the interactions, suggests that there may be new physics connecting the fundamental forces: electromagnetism (thin solid line), weak and strong nuclear force (dashed and thick solid line respectively) at large energy scales, in the context of a Grand Unified Theory (upper panel). In supersymmetric theories, the couplings can exactly merge at the energy scale $\sim 2 \times 10^{16}$ GeV, implying that, above such energy there is a single force with a single coupling constant (lower panel).

equation to feature corrections of the form

$$H^2 = \frac{8\pi G}{3}\rho \left[1 + \sum_{n\geq 1} C_n \left(\frac{\rho}{\rho_P}\right)^n\right],\tag{3.22}$$

where the C_n are coefficients of order unity and $\rho_P \equiv M_P/\ell_P^3$ is the Planck density, with $\ell_P = \hbar/M_P c$ being the Planck length (see Sec. 2.6.2). For example, to first-order approximation, loop quantum gravity suggests that $C_1 \simeq -2.44$ (and similarly for other ToEs, e.g., string theory). According to Eq. (3.15), the Planck time, when the density becomes ρ_P, is $t_P \sim 10^{-43}$ sec.

For densities above ρ_P, general relativity (GR) breaks down and spacetime is envisaged in fluctuating form, which is called *spacetime foam*. Spacetime foam is characterised by quantum nucleation of primordial black holes and wormholes, even baby universes, where a region of spacetime closes into itself (like a closed universe) and detaches, possibly connected only through a wormhole. Arguably, the most disturbing characteristic of spacetime foam is the continuous appearance and disappearance of closed timelike curves, which are a violation of causality since an event can cause itself. GR admits closed timelike curves (e.g., near the ring singularity of a Kerr black hole) but they are typically shielded by event horizons.* Therefore, in spacetime foam the causal structure of spacetime is fluctuating and constantly changing. This means that there is no causality or arrow of time. Consequently, the question "what happens before t_P?" is not well-defined, as time ceases to be linear. Thus, we see that it is more accurate to state that the Universe does not begin with a singularity (the Big Bang) at $t = 0$, but "emerges" at $t_P \sim 10^{-43}$ sec from spacetime foam. What is the meaning of t_P in this case? Well, it can be defined as the time that would have passed were the Universe classical and started with a Big Bang at $t = 0$, for a given value of ρ_P and $H(\rho_P)$.

The above have some interesting philosophical implications. They refute the famous argument of *prima causa* (first cause) of Thomas Aquinas for the existence of God. Roughly speaking, the argument goes as follows: everything that happens has a cause. Tracing all causes backwards one should arrive at the first cause, which is identified with God. We have seen that, going backwards in time we encounter an epoch where causality is no more and the argument is not applicable.

A schematic history of phase transitions in the early Universe is shown in Fig. 3.4.

3.5   Baryon asymmetry

Before concluding this chapter it is instructive to briefly discuss the baryon asymmetry. As we have mentioned, the ratio of the number densities of protons to photons is very small; there is roughly a proton every few billion photons $n_p \sim 10^{-10} n_\gamma$. Considering all the baryons (so including also the neutrons) the number density ratio is

$$\eta = (6.10 \pm 0.05) \times 10^{-10},\tag{3.23}$$

where $\eta \equiv n_B/n_\gamma$. Can this number be explained in the same way as the ^{4}He abundance?

At temperature higher than the QCD transition $T > T_{\text{QCD}}$, the thermal bath can create pairs of quarks and antiquarks, which are the constituents of baryons (and mesons). Thus we expect $n_B \sim n_\gamma$ and $\eta \sim 1$ (even though hadronisation has not happened yet). However, at lower temperature $T < T_{\text{QCD}}$, quark creation cannot happen but the existing quarks continue to annihilate with antiquarks so that the corresponding number density n_B becomes Boltzmann

---

*Naked Kerr black holes are not surrounded by event horizons, but they are not allowed by the cosmic censorship hypothesis.

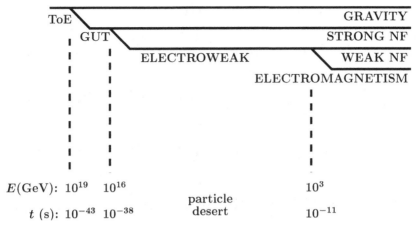

FIGURE 3.4 Schematic history of phase transitions in the early Universe, assuming all four fundamental forces unify above a certain energy scale. Electromagnetism and the weak nuclear force unify in the electroweak force according to the experimentally verified standard model of particle physics at energy about 1 TeV, which in the early Universe is realised 10^{-11} sec after the Big Bang. At higher energies, not much is expected (the particle desert) until the energy $\sim 10^{16}$ GeV, when the electroweak force is unified with the strong nuclear force in the context of a grand unified theory (GUT). This occurs only $\sim 10^{-38}$ sec after the Big Bang. At an energy a thousand times larger, the GUT force is expected to become unified with gravity, in the context of a theory of everything (ToE), such as string theory or loop quantum gravity. This is at the time 10^{-43} sec after a theorised Big Bang, which really marks the beginning of linear time.

suppressed. Then we have

$$\eta \sim \left(\frac{m_N c^2}{k_B T} \right)^{3/2} \exp \left(-\frac{m_N c^2}{k_B T} \right), \tag{3.24}$$

where $m_N \simeq 0.94$ GeV is the nucleon mass ($m_N \approx 3 m_q$, with m_q being the quark mass). Freeze-out occurs when $k_B T \simeq 0.8$ GeV. At this temperature, the above gives $\eta \sim 10^{-500}$! Obviously, this is much much smaller than the required $\eta \sim 10^{-10}$. What does this mean?

It means that annihilation between quarks or baryons with their antiparticles ceases at some point because there is an imbalance in their numbers. In short, the annihilation stops when we run out of antibaryons. In effect, the remaining baryons, even though heavy and in thermal equilibrium with the photons, act as stable particles and the ratio η is kept constant because $n_B, n_\gamma \propto a^{-3}$. Therefore, the original imbalance is

$$n_B - n_{\bar{B}} = \eta n_\gamma \tag{3.25}$$

where $\bar{B}$ denotes the antibaryons. Thus, the baryon asymmetry means that all the antiquarks (later antibaryons) annihilate with quarks (later baryons) leaving behind only a tiny amount of matter (quarks or later baryons). Therefore, we expect no significant amounts of antimatter to be present today, in agreement with cosmic ray observations.* The standard model of particle physics does feature an imbalance between matter and antimatter, but it is not enough to explain the observed imbalance in the Universe. We, therefore, need a new mechanism that enhances the excess of baryons over antibaryons.

---

*If abundant antimatter existed, annihilation with regular matter would generate cosmic rays, which are not observed.

Mechanisms for the generation of the original imbalance between matter and antimatter are called *baryogenesis* mechanisms. They have to satisfy the Sakharov conditions, which are:

- **Baryon number violation** To obtain an excess of baryons over antibaryons when originally there is none. This is possible in GUTs.

- **CP-violation** So that there is only generation of an excess of baryons over antibaryons and not the other way around.

- **Off thermal equilibrium** So that baryon asymmetry generation is irreversible.

The latter means that baryogenesis can take place either in phase transitions or during inflation.

3.6   The outline of the Hot Big Bang

Based on the (observationally supported) cosmological principle the Hot Big Bang provides a successful account of the history of the Universe as far back as a few seconds (at the latest) after the Big Bang explosion. It utilises standard and well-tested physics, such as general relativity, statistical physics (thermodynamics), and nuclear physics. When the latest advances in particle physics are also considered, we can push back the validity of the Hot Big Bang to times of a few microseconds after the Big Bang. However, it is but a guideline to even earlier times.

The Hot Big Bang is strongly supported by several independent observations. The main pillars supporting the Hot Big Bang are:

- **Universe expansion**
  As evidenced by the observed self-similar Hubble flow.

- **Age of the Universe**
  The latest estimate of which ($t_0 \simeq 14\,\text{Gyrs}$) is in agreement with an array of astrophysical observations of the oldest objects in the Universe.

- **CMB radiation**
  This is a kind-of afterglow of the Big Bang. A perfect black body radiation is observed with temperature $T_{\text{CMB}} \simeq 2.73\,\text{K}$, which is almost perfectly isotropic.

- **Big Bang nucleosynthesis**
  As observationally supported by the primordial abundance of light elements.

It is fair to say that the Hot Big Bang provides the backbone to our understanding of the Universe history. However, as usual in science, any answers introduce new open questions. Indeed, the Hot Big Bang suffers from a number of problems, the most important of which are:

- **Baryon asymmetry**
  This is the observed imbalance of matter over antimatter. If this imbalance did not exist, almost all matter (including us) would have annihilated with antimatter and the main constituent of the Universe at present would be the CMB radiation.

- **Dark matter and dark energy**
  For the Hot Big Bang to work, more matter than baryons is required; this is the dark matter (CDM). Dark energy domination is not really part of the Hot Big Bang, but it is still an issue that the Hot Big Bang does not address. In general, even though there is no information about their nature, the Hot Big Bang can accommodate dark matter (and dark energy), once the latter are given.

- **Origin of structures**
  Were the cosmological principle exactly obeyed, there would be no galaxies and the Universe today would be filled with a thinned-out gas at temperature T_{CMB}. The

break-away from the cosmological principle is facilitated with the primordial density perturbations, whose origin is not explained by the Hot Big Bang.

- **CMB primordial anisotropy**
  This is at the level of 10 parts in a million ($\sim 10^{-5}$) and is not explained by the Hot Big Bang. It is associated with the primordial density perturbations.

- **Horizon and flatness problems**
  These are both very important problems. They have to do with the Universe being too uniform and also being so close to spatial flatness, even though the latter is a repeller. They will be discussed in the next chapter.

- **Origin of the Universe expansion**
  Why is the Universe expanding in the first place? Since asking why presupposes causality, one may rephrase this more accurately by asking: Once it exits spacetime foam, why is the Universe expanding?*

Most (if not all) of the above problems can be overcome by the theory of cosmic inflation discussed in the following chapters. Note that the Hot Big Bang can accommodate the formation of structures (such as galaxies) and the observed baryon asymmetry provided their origins are given (for example a baryogenesis mechanism).

TABLE 3.1 Milestones of the Hot Big Bang history of the Universe.

	t (sec)	$k_B T$ (GeV)
BIG BANG	0	∞
End of quantum gravity	10^{-43}	10^{19}
GUT-transition	10^{-37}	10^{16}
Electroweak transition	10^{-11}	10^{2}
QCD transition	10^{-6}	0.1
$e^{\pm}$ pair annihilation	1	1
Big Bang Nucleosynthesis	10	0.1
Matter-radiation equality	10^{12}	10^{-9}
Decoupling and recombination	10^{13}	10^{-9}
Galaxy formation	10^{15}	10^{-11}
TODAY	10^{17}	10^{-13}

---

*Note that the Universe does not need a cause to exit spacetime foam.

EXERCISES

1. A weakly interacting massive particle (WIMP) has mass $m = 28.8\,\mathrm{TeV}/c^2$. Just before the electroweak phase transition, at $k_B T_{\mathrm{fr}} = 800\,\mathrm{GeV}$, its rate of interactions with the thermal bath dramatically decreases and the WIMP freezes out off thermal equilibrium.

    (a) Find the time of the WIMP freeze-out. Find also if the WIMP is relativistic or not at this time.

    (b) As an order of magnitude estimate, find the ratio n_W/n of the WIMP number density over the number density of the Universe at the time of the WIMP freeze-out.

    (c) Suppose that the WIMP is a stable particle. Find the time t_{dom} when the Universe becomes dominated by this WIMP. Thus, explain whether this WIMP is a valid dark matter candidate.

    [Hint: At matter-radiation equality $t_{\mathrm{eq}} = 4.7 \times 10^4\,\mathrm{y} = 1.5 \times 10^{12}\,\mathrm{sec}.$]
    [Hint: You may assume that, roughly $\sqrt{g_*} \sim 1.$]

2. Consider two particle species A and B. Species B has particle mass $m = 100\,\mathrm{MeV}/c^2$ and is part of the thermal bath. Species A is relativistic but freezes out from the thermal bath at a very early time, much earlier than the time of Big Bang nucleosynthesis (BBN) $t_{\mathrm{BBN}} \sim 1\,\mathrm{sec}.$

    (a) Species A is unstable and decays with decay rate Γ_{dec}. Show that, well after its freeze-out, the number density of species A is given by

    $$n_A \sim \left(\frac{k_B T}{\hbar c}\right)^3 \exp\left(-\frac{\Gamma_{\mathrm{dec}}}{2H}\right).$$

    (b) Below its threshold energy, species B becomes Boltzmann-suppressed. After this time, the number densities of the two species become comparable, $n_A \sim n_B$, at temperature T_X. If $T_X = T_{\mathrm{BBN}}$, where $T_{\mathrm{BBN}} \sim 1\,\mathrm{MeV}/k_B$ is the temperature at BBN, show that $\Gamma_{\mathrm{dec}} \simeq 186\,H_{\mathrm{BBN}}.$

    Using this, calculate the half-life of species A, $t_{\mathrm{half}} = \ln 2/\Gamma_{\mathrm{dec}}$. Also calculate the threshold time for species B, which corresponds to the moment species B became nonrelativistic.

4

Inflation Basics

4.1	Problems of the Hot Big Bang	67
	Horizon problem • Flatness problem • Relic problem • Density perturbations • Origin of the expansion	
4.2	Inflation's basic idea	72
	Definition • Superluminal expansion • Reheating	
4.3	Resolution of the Hot Big Bang problems	73
	Solution of the horizon problem • Solution of the flatness problem • Solution of the relic problem • Generating the density perturbations • Inflation and the origin of the expansion	
4.4	Quasi-de Sitter inflation	86
	Explosive exponential expansion • Inflationary e-folds • Particle production in quasi-de Sitter inflation	

4.1 Problems of the Hot Big Bang

The previous chapter ended with a list of successes of the Hot Big Bang accompanied by a list of problems. In this chapter, we discuss these problems and then address them with a silver bullet, the theory of cosmic inflation. We begin with two generic and fundamental challenges of the Hot Big Bang model, namely the horizon and flatness problems.

4.1.1 Horizon problem

Despite the observational support for the Hot Big Bang, the fact is that, fundamentally, the cosmological principle is incompatible with a finite age for the Universe. This paradox is the so-called *horizon problem* and has to do with the apparent uniformity of the Universe over distances, which are causally unconnected. For example, the CMB appears to be correlated at regions beyond the causal correlation scale (the horizon); it appears to be in thermal equilibrium and at a preferred reference frame. How can material in a region of space, which has never interacted with material in another region of space, well beyond causal contact, be nevertheless in thermal equilibrium with the latter at the same temperature? To make this even clearer, imagine observing the CMB in one direction and then observing it in the opposite direction. In both cases, the CMB light has been travelling for 14 billion years from the last scattering surface (LSS) to reach you, the observer, sitting at the centre of the LSS. It will take at least this amount of time (in fact, much more considering the Universe expansion) for the two opposite sides to send light to each other (i.e., at least another 14 billion years). Yet, the temperature of the CMB is the same in both directions (see Fig. 4.1). Moreover, relativity does not single out a preferred reference frame. This implies that Doppler shifts in the frequency of the CMB over causally unconnected regions (with different centre of mass frames) would result in $\delta T/T \sim 1$ even if the temperature of the material were the same in both regions. Yet, observations suggest

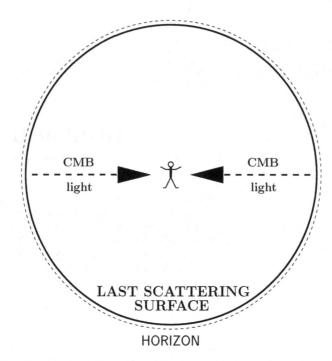

FIGURE 4.1   The observer sits at the centre of the last scattering surface, where the CMB is emitted, depicted by the solid circle. The radius of the last scattering surface is slightly smaller than the cosmological horizon depicted by the dashed circle. CMB light received from two antipodal directions, shown here by the dashed arrows, has been travelling for 14 billion years to reach the observer. The patches of the last scattering surface where the CMB light originated did not have the time to interact with each other (causally unconnected), yet the black body spectrum of the CMB is at the same temperature in both cases.

that matter in these regions does lie at the same reference frame when the CMB is emitted and the fractional perturbation of the CMB temperature is much smaller.

The definition of the horizon problem is:

Horizon Problem: The CMB appears to be correlated over scales beyond causal contact (superhorizon scales) at the time of its emission (decoupling)

One can understand the problem schematically by considering the spacetime diagram in Fig. 4.2. The slanted lines depict signals that travel with the speed of light, which is the fastest possible propagation through space. Strictly speaking, these lines should be curved because of the expansion of the Universe. However, for simplicity we ignore the Universe expansion here.* Along the time axis, the present time $t_0 \simeq 1.4 \times 10^{10}$ y is depicted at the tip of our past light-cone, that is the region of spacetime from which we can receive signals, and lies below the large slanted lines in the diagram. The horizontal line corresponds to the time of the decoupling of radiation from matter, when the CMB was emitted at $t_{\mathrm{dec}} \simeq 3.8 \times 10^5$ y. The small triangles correspond to the light-cones at the time of decoupling and designate the regions, which are in causal contact when the CMB was emitted; at t_{dec} an observer cannot be affected by events occurring beyond their light-cone. Because $t_0 \gg t_{\mathrm{dec}}$ our field of vision today includes a large

---

*Formally, the Universe expansion can be factored out by considering comoving spatial distances (for which $dr = a(t)dx$) and *conformal time* τ, related with cosmic time as $dt = a(t)d\tau$.

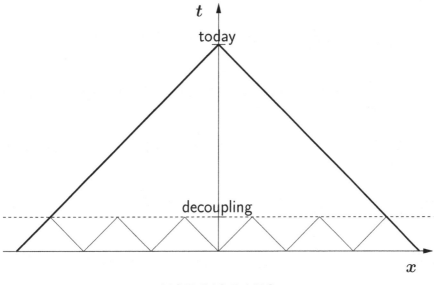

t

today

decoupling

x

HOT BIG BANG

FIGURE 4.2 Spacetime diagram showing the horizon problem. In the Hot Big Bang, our past light cone, depicted by the slanted thick solid lines, encapsulates many areas of the last scattering surface, depicted by the horizontal dashed line, that corresponds to decoupling, when the CMB was emitted. Indeed, it is shown that within our past light cone lie many nonoverlapping light cones at the time of decoupling, depicted by the slanted thin solid lines. The expansion of the Universe is ignored; the horizon problem has to do with the finite age of the Universe and the fact that space expands subluminally in the Hot Big Bang.

number of regions (about 40,000 in fact), which were beyond causal contact at the time of the CMB emission,* and yet these uncorrelated regions appear to be in thermal equilibrium at the same temperature.

The scale of the horizon problem can be better felt with an analogy. Imagine you go to a party wearing your purple sweater, only to discover that another person there is wearing another purple sweater, similar to yours. Most likely this is no more than a coincidence. Even if a third person appears with a purple sweater, this may be dismissed as a weird and unlikely coincidence too. Now, imagine you go to a rock concert with your favourite purple sweater on, only to discover that every single fan of the 40,000 filling the stadium is wearing a purple sweater like yours. You cannot dismiss this as a coincidence. It's more like "the great purple sweater conspiracy"!

Analytically the horizon problem can be easily understood as follows. First, we write the size of the observable Universe as $R_{\mathrm{obs}}(t_0)$. By definition, $R_{\mathrm{obs}}(t_0)$ is given by the size of the cosmological horizon at the present time t_0, i.e.,

$$R_{\mathrm{obs}}(t_0) \equiv D_H(t_0). \tag{4.1}$$

The comoving scale of the observable Universe $R_{\mathrm{obs}}(t)$ scales with the expansion as any other lengthscale, so we have $R_{\mathrm{obs}}(t) \propto a(t)$. For the cosmological horizon, we have shown that $D_H(t) \simeq c/H^{-1}$ (cf. Eq. (2.57)). Combining these and using Eq. (4.1), we can compare the

---

*Which reflects how much the range of causal correlations has grown since then.

two lengthscales (the observable Universe and the horizon) in the past $t < t_0$ to find

$$R_{\rm obs}(t) = R_{\rm obs}(t_0) \frac{a(t)}{a(t_0)} \Rightarrow \frac{R_{\rm obs}(t)}{D_H(t)} = \frac{D_H(t_0)}{D_H(t)} \frac{a(t)}{a(t_0)} \simeq \frac{H(t)}{H(t_0)} \frac{a(t)}{a(t_0)} \Rightarrow \frac{R_{\rm obs}(t)}{D_H(t)} \simeq \frac{\dot{a}(t)}{\dot{a}(t_0)} , \quad (4.2)$$

where we used that $\dot{a} = aH$. In the Hot Big Bang, $\ddot{a} < 0$ and, therefore, $\dot{a}(t)$ is a decreasing function of time. Consequently, $R_{\rm obs}(t) > D_H(t)$ for all $t < t_0$. Hence, the present horizon has always been beyond causal contact. In fact, the situation is more acute the earlier the time we consider $t \ll t_0$, which suggests that the Universe towards early times becomes increasingly causally disconnected. Since the CMB is isotropic over distances much larger than the causal horizon at decoupling, the question is how causally disconnected patches of the Universe have come into thermal equilibrium and in a single centre of mass reference frame.[*]

4.1.2   Flatness problem

As we have shown, the Friedmann equation can be written as (cf. Eq. (2.25))

$$|\Omega - 1| = \frac{|k|c^2}{(aH)^2} , \quad (4.3)$$

where $\Omega(t) \equiv \rho/\rho_c$ is the density parameter for the Universe, with $\rho_c(t)$ being the critical density. For a flat Universe, $\rho = \rho_c$ and $\Omega = 1$ (and $k = 0$). Thus, we may call the quantity $|\Omega - 1|$ "deviation from flatness". The above equation may be used to trace the behaviour of the deviation from flatness in the past. Indeed, it is straightforward to find

$$\frac{|\Omega_0 - 1|}{|\Omega - 1|} = \left[\frac{a(t)H(t)}{a(t_0)H(t_0)} \right]^2 = \left[\frac{\dot{a}(t)}{\dot{a}(t_0)} \right]^2 , \quad (4.4)$$

where $\Omega_0 \equiv \Omega(t_0)$ is the deviation from flatness today. In the Hot Big Bang $\ddot{a} < 0$ and, therefore, $\dot{a}(t)$ is a decreasing function of time. Consequently, $|\Omega_0 - 1| > |\Omega - 1|$ for all $t < t_0$, i.e., the deviation from flatness grows with time. Hence, as time passes the Universe increasingly deviates away from spatial flatness, meaning that a flat Universe is a repeller (see Fig. 4.3).

However, the latest observations suggest that $\Omega_0 = 1.000 \pm 0.002$ [A+18a], that is the Universe at present is approximately spatially flat, which as we have seen is an unstable condition. This requires severe fine-tuning of initial conditions, for if initially the Universe deviated from flatness in the slightest, it would clearly be closed or open today. The situation is not too different from balancing a pencil on its tip. A slight deviation brings the pencil down. This is called the *flatness problem*.

The definition of the flatness problem is:

Flatness Problem: The Universe at present appears to be spatially flat despite the fact that a flat Universe is unstable.

To quantify the amount of fine-tuning required we can estimate the magnitude of the deviation from flatness at the time of Big Bang nucleosynthesis (BBN), which is the earliest time when we know for sure that the Hot Big Bang is valid. A simple calculation, suggests

$$\frac{|\Omega_{\rm BBN} - 1|}{|\Omega_0 - 1|} = \left[\frac{\dot{a}(t_0)}{\dot{a}(t_{\rm BBN})} \right]^2 \approx \left(\frac{t_{\rm eq}}{t_0} \right)^{2/3} \left(\frac{t_{\rm BBN}}{t_{\rm eq}} \right) \sim 10^{-16} , \quad (4.5)$$

---

[*]Another aspect of the horizon problem is the observed homogeneity of the Universe, which also points to a preferred reference frame. Indeed, a relativistic boost in any direction would render the Universe inhomogeneous, as remote locations with the same average density will not be simultaneous anymore.

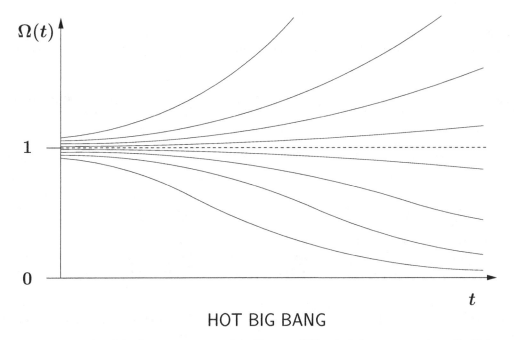

HOT BIG BANG

FIGURE 4.3 Plot of the density parameter of the Universe $\Omega(t)$, which demonstrates that a flat Universe where $\Omega = 1$ is a repeller in the Hot Big Bang. If initially Ω is slightly displaced from unity, then soon the Universe becomes strongly curved, either open or closed, with $|\Omega| > 1$. The flatness problem is that today $\Omega(t_0) \approx 1$, which requires substantial fine-tuning as $\Omega(t)$ needs to be extremely close to unity at the onset of the Hot Big Bang to remain very near unity today.

where $\Omega_{\rm BBN} \equiv \Omega(t_{\rm BBN})$ is the density parameter of the Universe at the time of BBN: $t_{\rm BBN} \sim 1\,{\rm sec}$, $t_{\rm eq} = 4.7 \times 10^4\,{\rm y} = 1.5 \times 10^{12}\,{\rm sec}$ is the time of equality, when the radiation era ends and the matter era begins and the present time is $t_0 = 1.4 \times 10^{10}\,{\rm y}$. We have considered that in a flat Universe $a \propto t^{1/2}$ in the radiation era and $a \propto t^{2/3}$ in the matter era (cf. Eqs. (2.42) and (2.43) respectively) and we have neglected the recent dark energy period. Thus, we see that at BBN the deviation from flatness must be tiny, at most $|\Omega_{\rm BBN} - 1| \sim 10^{-19}$. This is an extremely small number (it involves 19 zeros!). Considering times earlier than BBN intensifies the fine-tuning. For example, if the Hot Big Bang is taken to be valid as early as the time of the breakdown of grand unification (GUT phase transition) at $t_{\rm GUT} \sim 10^{-38}\,{\rm s}$ then the corresponding deviation from flatness cannot be larger than $|\Omega_{\rm GUT} - 1| \sim 10^{-57}$, which is ridiculously small.*

4.1.3 Relic problem

The flatness and horizon problems of the Hot Big Bang are generic and inescapable. However, on many occasions, when considering high energy physics theories, one has additional issues arising from the appearance of unwanted by-products, which may be harmless in the context of the theory but can be disastrous if this high energy theory were realised in the early Universe.

---

*Pushing further back to the Planck time $t_P \sim 10^{-43}\,{\rm s}$ intensifies the upper bound on the deviation from flatness down to 10^{-63} or so. This is comparable to the fine-tuning required for the cosmological constant, at least in supersymmetric theories.

This is because these objects tend to influence the Universe evolution in a way which is not compatible with the observationally supported Hot Big Bang. As such, cosmology can impose severe constraints on theory. Sometimes, however, the theoretical prejudice is so strong that the problem appears to be with the Hot Big Bang instead.

This was the case of the magnetic monopoles. They were originally thought to be a generic product of the breaking of grand unification. They would have a nonzero magnetic charge, so that they would act as sources of the magnetic field, like isolated magnetic poles, in the same way that electrons are sources of the electric field. They would be very heavy (nonrelativistic) objects, amply generated at the GUT phase transition. Consequently, they were expected to dominate the Universe soon after the GUT transition, because their density would decrease as $\rho_{\text{mon}} \propto a^{-3}$, which is less rapidly than the density of the background radiation $\rho \propto a^{-4}$. As a result, monopole domination would give rise to a prolonged matter era, which would be in contradiction with the Hot Big Bang story.

These days, however, the thinking regarding grand unification has changed, with monopoles not necessarily produced at the GUT transition. So magnetic monopoles are not viewed as much of a problem for cosmology anymore. However, many other similar relics are threatening the Hot Big Bang. For example, in many supergravity theories, gravitino particles (when stable) are overabundant if the Hot Big Bang is valid at energy higher than $k_B T > 10^9$ GeV, which, even if huge, is still a million times smaller than the GUT energy. Other examples are moduli particles (associated with the moduli fields in string theory) or primordial black holes (remnants from spacetime foam). All these examples act as heavy matter (like the monopoles), which would dominate the Universe and would not allow the radiation era to begin in time for BBN.

One can argue that, if theory does not agree with cosmological observations, this is tough for theory; it has to change! However, theoretical physicists think that the cosmology must change instead. To overcome this conflict, it would be very convenient to find a mechanism that sweeps all such unwanted relics aside keeping both cosmologists and theorists happy. As we discuss below, such a mechanism is cosmic inflation.

4.1.4   Density perturbations

Why is the cosmological principle not exact? Of course, if it were exact I would not be here writing this book, and neither would you be reading it. The Universe would be filled with the CMB radiation and a thinned-out gas at temperature T_{CMB}, evenly distributed throughout space without galaxies, stars, planets, or life. Therefore, it is absolutely essential that the uniformity in the Universe is not pefrect and never was perfect (for otherwise it would have remained so forever). We know this to be true, but the Hot Big Bang cannot explain why. Cosmic inflation can!

4.1.5   Origin of the expansion

Why is the Universe expanding in the first place? The Hot Big Bang assumes the Universe expansion but does not account for it. As we have discussed, the Hot Big Bang is likely to be preceded by the Planck era, where causality does not apply, so there is no need to explain why. However, can the classical expanding Universe emerge from spacetime foam? Or as Barbara Rayden put it, "What put the Bang in the Big Bang?" Cosmic inflation might have something to do with this as well.

4.2   Inflation's basic idea

4.2.1   Definition

Cosmic inflation is defined as:

Cosmic inflation: a period of accelerated expansion of
space in the early Universe, preceding the Hot Big Bang

As we have shown in Sec. 2.3.5, the condition for accelerated expansion is

$$\ddot{a} > 0 \quad \Leftrightarrow \quad w < -\frac{1}{3}. \tag{4.6}$$

Thus, during inflation the Universe traverses an epoch when the pressure is negative enough to lead to accelerated expansion. This means that inflation is a period of dark energy domination.

4.2.2 Superluminal expansion

The condition $\ddot{a} > 0$ means that $\dot{a}(t)$ is a *growing* function of time. The Hubble flow is $v_H = \dot{a}x$, where x is the comoving distance (between two comoving observers). The growth of $\dot{a}$ results in a growth of v_H, which becomes larger than lightspeed. Thus, in a nutshell, cosmic inflation can also be defined as

---
Cosmic inflation: a period of superluminal expansion of
space in the early Universe, preceding the Hot Big Bang

General relativity allows inflation because the latter does not correspond to displacement of matter or energy through space with velocity faster than lightspeed (which is forbidden). Instead, it is space itself which is expanding faster than light can travel. Indeed, inflationary expansion is readily obtained by general relativity if certain conditions are satisfied.[*]

4.2.3 Reheating

As we have discussed $T \propto 1/a$ (cf. Eq. (3.11)). This means that the rapid growth of the scale factor during inflation drastically diminishes the temperature of any preexisting thermal bath. The effect is called supercooling.[†] This disappearance of the thermal bath implies that the Universe is out of thermal equilibrium during inflation, which offers possibilities for baryogenesis. However, after the end of inflation we should recover the Hot Big Bang, which has a huge temperature at early times (before BBN). This means that the thermal bath must be (re)created for the Hot Big Bang to begin. The process is called *reheating*. Therefore, reheating amounts to entropy production after the end of inflation, which transfers the energy of inflation to the matter and radiation of the newly created thermal bath of the Hot Big Bang. Of course, reheating has to occur before BBN if the successes of the Hot Big Bang cosmology are to be retained. The term reheating is actually a misnomer, because there is no need to have a thermal bath before inflation; however, this was the thinking in the early days of inflation and the name has stuck. Reheating is a complex process and is important not only for the Universe history after the end of inflation, but also in determining the observational signatures of inflation, which are used to constrain and test the inflation theory. This is why we return to it later on in more detail (see Chapter 6).

4.3 Resolution of the Hot Big Bang problems

In the late '70s a number of authors came up with embryonic versions of the idea of inflation (e.g., D. Kazanas (1980) [Kaz80], K. Sato (1981) [Sat81]). The most striking proposals though,

---

[*]And this is another reason why the cosmic egg fallacy is wrong, hence fallacy.
[†]See, however, Sec. 7.6.

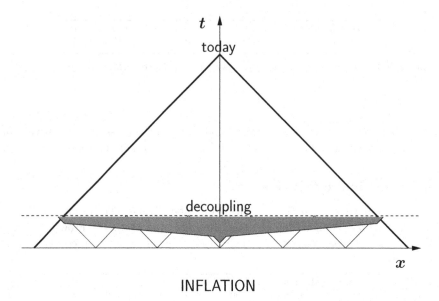

INFLATION

FIGURE 4.4 Spacetime diagram showing the resolution of the horizon problem by inflation, following directly from Fig. 4.2. Our past light cone, depicted by the slanted thick solid lines, allows a vision of the last scattering surface, depicted by the horizontal dashed line. The latter corresponds to decoupling, when the CMB was emitted. Because of inflationary superluminal expansion, the entire field of vision originated in a causally connected patch, which used to be within the light cone before inflation. This is depicted by the gray shaded area. Having the entire visible last scattering surface causally connected accounts for the observed isotropy in the CMB.

were put forward by Alexei A. Starobinsky (1979) [Sta80], whose inflation model is still one of the most successful ones at the time of writing, and independently by Alan H. Guth (1980) [Gut81], who was the first to address comprehensively the problems of the Hot Big Bang through inflation. He also coined the name "inflation," which has the unfortunate consequence of never coming up in internet seach engines in contexts other than economics (but these engines did not exist in 1980), hence adding the "cosmic" before it. Here we briefly show how the idea of accelerated expansion in the early Universe overcomes all the problems of the Hot Big Bang we discussed in the beginning of this chapter.

4.3.1 Solution of the horizon problem

Superluminal expansion during inflation results in superhorizon correlations. This means that inflation produces correlations beyond the causal horizon by enlarging an initially causally connected region to a size large enough to encompass the observable Universe. The region which corresponds to the observable Universe, lies within the light-cone before the onset of inflation. Being initially causally self-correlated, it can become uniform by interacting with itself. After it is inflated to super-horizon distance scales, this original uniformity sets up uniform initial conditions after inflation, even if the original region now corresponds to many causally-disconnected regions. This is depicted schematically in Fig. 4.4, where during inflation the growth of an initially causally correlated region is shown to be faster than lightspeed so as to engulf the entire field of vision today. After the end of inflation, uniformity persists because of the causal nature of the Universe evolution. Hence, we see that, provided it lasts long enough, cosmic inflation homogenises the observable Universe.*

Analytically, the solution of the horizon problem is straightforward. In order to have the comoving scale of the observable Universe R_{obs} causally connected, we need to make sure that $R_{\text{obs}}(t_i) \leq D_H(t_i)$, where t_i is the initial time of inflation. In view of Eq. (4.2) (which is always valid), we see that the condition for the solution of the horizon problem is

$$\dot{a}_i \leq \dot{a}_0 \,, \tag{4.7}$$

where $\dot{a}_i \equiv \dot{a}(t_i)$ and $\dot{a}_0 \equiv \dot{a}(t_0)$ with t_0 being the present time. The physical meaning of this is more apparent if we rewrite the above condition as

$$\frac{\dot{a}_i}{\dot{a}_{\text{end}}} \leq \frac{\dot{a}_0}{\dot{a}_{\text{end}}} \,, \tag{4.8}$$

where "end" denotes the end of inflation and $\dot{a}_{\text{end}} \equiv \dot{a}(t_{\text{end}})$.[†] Since after the end of inflation we have $\ddot{a} < 0$, the fraction $\dot{a}_0/\dot{a}_{\text{end}}$ is smaller than unity. Equation (4.8) simply demands that the fraction $\dot{a}_i/\dot{a}_{\text{end}}$ corresponding to the inflationary period is even smaller. Because $\dot{a}$ is growing during inflation, we see that the condition in Eq. (4.8) is satisfied only if inflation lasts long enough. Thus, we conclude that the horizon problem is solved if inflation lasts long enough.

Without resorting to inflation the horizon problem is very difficult to overcome. Most proposals do so by postulating a pre-Big Bang period, such that correlations have the time to grow large enough to encompass the observable Universe. To retain causal correlations, this implies that the evolution of the Universe through the Big Bang era remains causal, so the energy density never rises up to the Planck density, where as we discussed causality breaks down.

4.3.2 Solution of the flatness problem

From Eq. (4.4), we see that $|\Omega(t) - 1| \propto \dot{a}^{-2}$, where $\dot{a} = aH$. As we have discussed, during the Hot Big Bang $\dot{a}(t)$ is diminishing so the deviation from flatness grows, and a flat Universe (with $\Omega = 1$) is a repeller. During inflation, however, $\dot{a}(t)$ is growing (because $\ddot{a} > 0$ by definition), which means that the deviation from flatness diminishes, and a flat Universe becomes an attractor instead (see Fig. 4.5). In view of Eq. (4.4), the analytic criterion is

$$\frac{\dot{a}_i}{\dot{a}_0} = \sqrt{\frac{|\Omega_0 - 1|}{|\Omega_i - 1|}} < 0.1 \,, \tag{4.9}$$

where we used that the observations suggest $|\Omega_0 - 1| < 0.01$ and we assumed that the original deviation from flatness at the onset of inflation is $|\Omega_i - 1| \sim 1$ on dimensional grounds, which is to say that there is no reason for it to be either huge or minuscule. The above bound is similar but a little tighter to the bound in Eq. (4.7), which corresponds to the solution of the horizon problem. Following the discussion in Sec. 4.3.1 we see that the condition in Eq. (4.9) implies that the flatness problem is solved provided inflation lasts long enough. The longer inflation lasts, the smaller the deviation from flatness becomes by the end of inflation and the flatter the observable Universe is. This may well account for the tiny values required for the deviation from flatness at the onset of the Hot Big Bang.

---

*Among other things, inflation homogenises the expansion rate, which of course is tied with the density through the Friedmann equation.

[†]$\dot{a} > 0$ because the Universe is expanding.

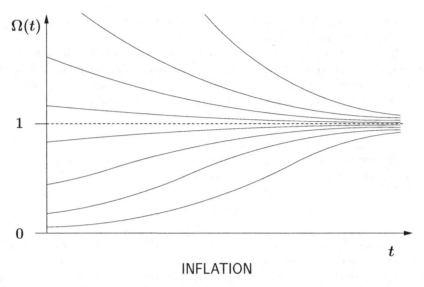

INFLATION

FIGURE 4.5   Plot of the density parameter of the Universe $\Omega(t)$, which demonstrates that a flat Universe where $\Omega = 1$ is an attractor in inflation. Even if initially Ω is substantially displaced from unity, soon it approaches a flat Universe $\Omega = 1$ thereby accounting for the flatness problem because it explains that today $\Omega(t_0) \approx 1$ without requiring substantial fine-tuning of initial conditions.

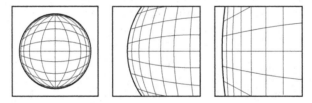

FIGURE 4.6   Cosmic inflation reduces the spatial curvature similarly to the surface curvature of an inflated balloon, which approaches flatness the more the balloon is inflated.

A useful analogy to understand this, which considers curved space as an elastic rubber sheet (a common analogy in general relativity), involves the curvature of the surface of an inflated balloon. Indeed consider an area of fixed physical radius, which can be thought of as the size of the horizon (whose growth is negligible compared to the superluminal expansion). The more the balloon is inflated, the flatter the area becomes (see Fig. 4.6). This is why we say that, during inflation, the rapid expansion inflates the curvature away. Hence, the choice of the term "inflation" is spot on.

In a similar manner, any anisotropy in the original curvature is also inflated away, and the patch of space which corresponds to the observable Universe becomes highly isotropic. Thus, by solving the horizon and flatness problems, inflation imposes homogeneous and isotropic initial conditions to the Hot Big Bang cosmology. This means that inflation fixes the cosmological principle as an initial condition for the Hot Big Bang. In a sense, the observational support for the cosmological principle is also support for the theory of inflation. It must be said that most, if not all, the proposed alternatives to inflation do not address the flatness problem. Instead, they are forced to accept the fine-tuning of the deviation from flatness at the onset of the Hot Big Bang in a similar way to how they regard the cosmological constant problem.

4.3.3 Solution of the relic problem

The resolution of the relic problem is straightforward. The density of the relics is drastically reduced because of the rapid expansion. For example, for monopoles (or gravitinos, or string moduli, or primordial black holes) whose density is diluted as matter, we have at the end of inflation

$$\rho_{\rm mon}^{\rm end} = \rho_{\rm mon}^{i} \left(\frac{a_i}{a_{\rm end}}\right)^3, \tag{4.10}$$

where $\rho_{\rm mon}^{\rm end} \equiv \rho_{\rm mon}(t_{\rm end})$ and $\rho_{\rm mon}^{i} \equiv \rho_{\rm mon}(t_i)$. At the end of inflation, reheating converts the inflation energy into radiation. For simplicity, let us assume that this happens promptly. This means that at the end of inflation the radiation density is

$$\rho_{r}^{\rm end} \simeq \rho_{\rm inf}^{\rm end} = \rho_{\rm inf}^{i} \left(\frac{a_i}{a_{\rm end}}\right)^{3(1+w_{\rm inf})}, \tag{4.11}$$

where $\rho_{\rm inf}$ is the density of the dark energy substance responsible for inflation (which decays into radiation at reheating), and $w_{\rm inf}$ is its barotropic parameter. According to Eq. (4.6), $w_{\rm inf} < -\frac{1}{3}$. From the above two equations, we see that the ratio of the monopole to the radiation density at the end of inflation is

$$\left.\frac{\rho_{\rm mon}}{\rho_r}\right|_{\rm end} = \left.\frac{\rho_{\rm mon}}{\rho_{\rm inf}}\right|_{i} \left(\frac{a_i}{a_{\rm end}}\right)^{-3w_{\rm inf}} \sim \left(\frac{a_i}{a_{\rm end}}\right)^{-3w_{\rm inf}}, \tag{4.12}$$

where we have considered that $\rho_{\rm mon}^{i} \sim \rho_{\rm inf}^{i}$ because the onset of inflation occurs when the density of the inflationary dark energy takes over ("i" stands for "initial" here). In the radiation era, $\rho_{\rm mon}/\rho_r \propto a$. Thus, at the time of BBN we find

$$\left.\frac{\rho_{\rm mon}}{\rho_r}\right|_{\rm BBN} = \left.\frac{\rho_{\rm mon}}{\rho_r}\right|_{\rm end} \frac{a_{\rm BBN}}{a_{\rm end}}, \tag{4.13}$$

where $a_{\rm BBN} \equiv a(t_{\rm BBN})$. Demanding that $\rho_{\rm mon}^{\rm BBN} \ll \rho_r^{\rm BBN}$ and the above, we obtain the constraint

$$\frac{a_{\rm BBN}}{a_{\rm end}} \ll \left(\frac{a_{\rm end}}{a_i}\right)^{-3w_{\rm inf}}. \tag{4.14}$$

The above can be understood as follows. The ratio $a_{\rm BBN}/a_{\rm end}$ can be very large the earlier inflation happens. This means that the ratio $a_{\rm end}/a_i$, raised to the power $-3w_{\rm inf} > 1$, has to be even larger. The latter ratio is further enlarged the longer inflation lasts, which means that the relic problem is resolved if inflation lasts long enough.

In the above, we made the crucial assumption that the dangerous relic appears before inflation begins (or that there is a lot of inflation after it appears). This may be so for magnetic monopoles if the energy of inflation is slightly below the GUT energy scale, which is likely as we discuss later. However, there are examples where this is not so, e.g., in hybrid inflation, it is the GUT phase transition that terminates the inflationary phase (see Sec. 7.2.4). If the GUT transition produced magnetic monopoles then such inflation could not inflate them away. Similarly, for primordial black holes, if they are produced at (or near) the end of inflation (so not by spacetime foam). For gravitinos, the story is similar; the temperature of the thermal bath of the Hot Big Bang cannot be too large (below 10^9 GeV$/k_B$ or so), so reheating cannot be very efficient.

However, the situation is not so grim. As mentioned already, GUT theories do not necessarily produce monopoles, even if supergravity is true (which is a big if) the gravitino particles do not have to be stable and most inflationary models do not give rise to primordial black holes. With these examples, I mean to say that the dangerous relics may not be there in the first place, so even if inflation does not deal with them, too bad for the theories that generate them!

4.3.4   Generating the density perturbations

As we have shown, provided it lasts long enough, cosmic inflation can impose the cosmological principle as an initial condition for the Hot Big Bang. However, if the cosmological principle were exact then there would be no structure in the Universe; no galaxies, no stars with planets harbouring intelligent life. All these structures require the uniformity of the Universe to be violated. Thus, our existence demands that the cosmological principle is only approximately respected.

4.3.4.1 Inflation and the primordial density perturbations

It so happens that cosmic inflation can also provide a small violation of the cosmological principle. This is due to quantum effects during inflation, which, as explained below, can introduce a tiny variation in the density ρ of the Universe, called the *primordial density perturbation $\delta\rho/\rho$*. This density perturbation reflects itself onto the CMB through the Sachs-Wolfe effect (see next chapter), which describes how the CMB photons become redshifted when crossing regions of higher density than average (called overdensities) because they lose energy while struggling to exit from the gravitational potential wells of these growing overdensities. Hence, variations in the density of the Universe cause perturbations in the apparent temperature of the CMB radiation:

$$\frac{\Delta T}{T}\bigg|_{\text{CMB}} \simeq \frac{1}{2}\frac{\delta\rho}{\rho} \simeq 10^{-5}. \tag{4.15}$$

Indeed the CMB observations suggest that $\delta\rho/\rho = (1.832 \pm 0.013) \times 10^{-5}$, which albeit tiny turns out to be enough to explain the formation of structures in the Universe, such as galaxies and galaxy clusters.

Starting from an initial density perturbation, structure formation proceeds through the process of *gravitational instability*. This amounts to intensifying, in a runaway manner, the contrast between overdense and underdense regions and is based on the fact that overdensities grow more massive by attracting matter from surrounding underdensities, depleting them even further.[*] Indeed, numerical simulations have shown that initial density perturbations of magnitude $\delta\rho/\rho \sim 10^{-5}$ suffice to generate the observed structures given 14 billion years of growth. Therefore, it seems that inflation can successfully impose both the cosmological principle onto the Universe and the deviations from it, which are necessary for structure formation.

The horizon and flatness problems are overcome by all inflationary models, because their resolution is due to the idea of inflation itself. In contrast, the production of the density perturbations, even though also a generic prediction of inflation, results in perturbations (and the associated CMB temperature primordial anisotropy) with characteristics that are dependent on the inflation model considered. Hence, observations of $(\Delta T/T)_{\text{CMB}}$ can distinguish between inflationary models. In the last two decades, the increasing precision of CMB observations has enabled cosmologists to discard classes of inflationary models and severely constrain the surviving ones shedding light on the background theory. This is why the generation of the density perturbations by inflation is of utmost importance, and we talk about it in some detail in Chapter 6. Here, however, we are only interested in the model independent aspect of it, which we discuss below in a descriptive and heuristic manner, leaving a more rigorous treatment for later.

---

[*]The overdensities' typical amplitude is $\delta\rho/\rho \sim 10^{-5}$ before it is intensified by gravitational instability which enlarges it to order unity, when the growth of the overdensity becomes nonlinear.

4.3.4.2 Particle Production and the Quantum Vacuum

Inflation produces the primordial density perturbations through a process called *particle production* that arises from considering the superluminal expansion of space in conjunction with the notion of the quantum vacuum.

We are all familiar with the concept of the classical vacuum. A box filled with vacuum is a box which is totally empty of matter and energy, i.e., the energy in the box is $E = 0$. However, in quantum theory, this cannot be true, and the box has to have $E \neq 0$. This is due to *Heisenberg's uncertainty principle*, which, for energy, can be expressed mathematically as*

$$\Delta E \cdot \Delta t \sim \hbar. \tag{4.16}$$

The meaning of the above expression is that the energy of a closed system (such as our box) cannot be precisely determined, but it has to fluctuate by an amount ΔE, which, in a given time period Δt, has to satisfy the above constraint. The crucial point is that the reduced Planck constant† $\hbar$ featured above, is small but positive, which means that, for any finite time interval, ΔE has to be nonzero. In fact, the smaller the period of time considered the larger the fluctuation of the energy has to be.

Hence, the uncertainty principle amounts to a controlled violation of energy conservation. This violation manifests itself as the brief appearance of pairs of particles and antiparticles, the reason being that other quantities, such as the electric charge or other quantum numbers, are indeed conserved. Consequently, the quantum vacuum is not empty, but instead, it is filled with constantly appearing and disappearing pairs of *virtual particles* and antiparticles (see Fig. 4.7). The energy of these virtual particles is called vacuum (or zero-point) energy. The above may sound like a bunch of flowery ideas, but it so happens that there is experimental proof of the existence of virtual particles. This is the famous Casimir experiment, first realised by Casimir and Polder in 1948.

Before combining the notion of the quantum vacuum with cosmic inflation we need to briefly discuss black holes. Black holes are predicted by general relativity and are thought to exist astrophysically, for example in the centres of galaxies. Recently, gravitational wave observations verified the existence of coalescing binary black holes. A black hole is an extremely compact object with a gravitational field which is locally so intense that anything approaching close enough can never escape. Indeed, a black hole is surrounded by an event horizon, which is the surface on which the escape velocity from the black hole is equal to the speed of light. This means that one would need to travel with the speed of light away from the black hole simply to remain on the event horizon. Within the horizon the gravitational attraction is so strong that, even when travelling at lightspeed, one cannot avoid being pulled inwards. Since no matter or energy can travel faster than the speed of light, the event horizon forms a boundary on the causal structure of spacetime in the sense that events within the event horizon cannot affect events outside it. Thus, black holes act as giant vacuum cleaners, sucking up the surrounding space and all material it contains. When the inflow of space becomes superluminal we have the appearance of the event horizon. One can think of the event horizon as a surface which is permeable from one direction only (from the outside). A classical black hole, being enclosed within an event horizon, can only absorb matter and energy which is captured by its pull from

---

*In quantum mechanics the uncertainty relation is expressed between momentum and position rather than energy and time, because time cannot be promoted into a quantum-mechanical operator. However, quantum field theory generalises quantum mechanics in four-dimensions by incorporating special relativity. The four-dimensional equivalent of the uncertainty relation does include energy and time as the fourth component of momentum and position, respectively.

†$\hbar \equiv h/2\pi$, where h is Planck constant.

$$E = 0 \qquad\qquad E \neq 0$$

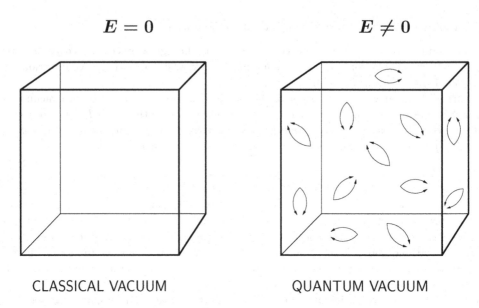

CLASSICAL VACUUM   QUANTUM VACUUM

FIGURE 4.7   A box filled with nothing is an empty box in the classical vacuum, with zero energy $E = 0$. In contrast, in the quantum vacuum a box filled with nothing cannot be empty, but it is filled with virtual particles and antiparticles, which appear (and disappear) in pairs, corresponding to nonzero energy $E \neq 0$, called vacuum energy.

its environment. Infalling matter increases the gravitational field of the black hole, which in turn increases the radius of its event horizon as the gravitational pull further extends its effect in the surrounding space. Hence, a classical black hole can only increase in mass and size.

Such was the understanding of black holes until Stephen Hawking studied them in conjunction with the quantum vacuum. Hawking considered the appearance of virtual particle pairs in the vicinity of the event horizon. Suppose that one of the particles of the pair falls within the event horizon as shown in Fig. 4.8. The other part of the pair may follow it, but since it is still outside the event horizon, it has a nonzero (albeit tiny) probability of escaping from the black hole, whereas, by definition of the event horizon, the particle within it has no chance of escaping. Thus, since pairs of virtual particles are constantly appearing from nothing (through the controlled violation of energy conservation by the uncertainty principle) in the vicinity of the event horizon, the net effect of the above possibility is that a tiny fraction of virtual particles does escape from the black hole. Such particles cannot meet their counterparts of the original pair because the latter have fallen within the event horizon and can never escape. Therefore, these particles cannot annihilate with their partners; their conserved quantum numbers cannot disappear. Consequently, the virtual particles which avoided falling within the event horizon (while their partners did not) survive and become real particles, which can be detected by distant observers.*

Stable real particles can survive indefinitely. For stable particles that used to be virtual particles that escaped from the event horizon of a black hole, this means that Δt in the uncertainty relation (that allowed their existence) can become very large. Thus, we find $\Delta E \sim \hbar/\Delta t \to 0$, i.e., over large time periods, energy conservation has to be restored. Since the particles that escaped from the black hole can be detected by an observer to have a positive energy E_{out}, with

---

*How exactly this happens is still debated, but it has a cool name: *quantum decoherence*.

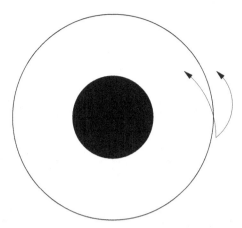

FIGURE 4.8 Virtual particles appearing near the event horizon of a black hole. There is a small possibility that, while one of the virtual particles in a pair falls within the event horizon, the other one manages to escape the black hole, thereby becoming a real particle. In such a way, the black hole event horizon appears to radiate particles.

respect to this distant observer, the pair partners of these particles (the ones which did fall into the black hole) must have energy $E_{in} \approx -E_{out} < 0$, where $\Delta E = E_{in} + E_{out}$. Hence, from the viewpoint of the distant observer, the black hole appears to be absorbing particles of negative energy that equals in magnitude the positive energy of the particles radiated away having escaped from the event horizon. Receiving negative energy the black hole reduces in mass and its event horizon reduces in size, while emitting energy (equivalent to its mass reduction) in the form of *Hawking radiation*. In 1974, Stephen W. Hawking found that the emitted radiation has a black body spectrum corresponding to a characteristic temperature, called the *Hawking temperature* T_H [Haw74]. Surprisingly, T_H was found to increase the smaller the black hole becomes, i.e., the emission of Hawking radiation intensifies as the black hole shrinks. This runaway behaviour ends up in the evaporation of the black hole. Turning this around, the emission of Hawking radiation from supermassive astrophysical black holes, such as the ones in the centres of galaxies, is negligible and is overwhelmed by absorption of positive energy from their environment, such as surrounding matter or even the CMB radiation.

What does black hole evaporation have to do with cosmic inflation? Well, as we have discussed, the cosmological horizon during dark energy domination (i.e., in accelerated expansion) is an event horizon. Indeed, during inflation, the cosmological horizon corresponds to an event horizon of an "inverted" black hole; a black hole "inside-out," centred at the observer. This can be understood by considering that, during inflation, matter inside the horizon is being "sucked out" by the superluminal expansion of space, in analogy with the fact that nearby matter is being "sucked in" by a black hole. In this manner, the virtual particle pairs of the quantum vacuum can be pulled outside the horizon before they have a chance to annihilate as shown in Fig. 4.9. In this case, being over superhorizon distances apart, they are beyond causal contact and can never find each other to annihilate. Thus, they cease being virtual particles and become real particles instead. This is why this process is called *particle production*.* Therefore, during

---

*Note that the existence of an event horizon suffices to have particle production. It is the division of the causal structure of spacetime, enforced by the event horizon, which gives rise to the particle production process; the separation of the virtual particle pairs. Thus, the event horizon, once it exists, can be seen to emit Hawking radiation without the need of a black hole or any concentration of mass.

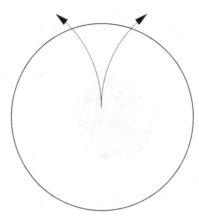

FIGURE 4.9 Virtual particles appearing during inflation may travel beyond the event horizon, in which case they are pulled apart to superhorizon scales by the superluminal expansion, thereby becoming causally disconnected and turning into real particles. Thus, the event horizon during inflation appears to radiate particles. Because there is an event horizon around every location, inflating space is filled with (light and nonconformally invariant) particles.

inflation, the event horizon is filled with Hawking radiation (since it corresponds to the outside of the "inverted" black hole). Since the horizon is centred at the observer and we can put an observer anywhere in space, this means that all space is filled with Hawking radiation, i.e., particle production occurs everywhere once we have superluminal expansion.

Now, in quantum theory, particles correspond to waves with an associated de Broglie wavelength. For example, a massive particle of mass m is characterised by a wavelength $\lambda = 2\pi\hbar/mc$. This dual nature of particles is hard to visualise, although we are all familiar with many "particle-like" properties of waves, such as the reflection of an incident sea wave on a concrete wall. It is also easy to accept that waves, like particles, carry energy and momentum, e.g., laser beams or tsunami waves. Particles correspond to waves travelling on an otherwise smooth sea, which we call a field (the particle is part of the field as the sea wave is part of the sea), for example photons travel through the sea of the electromagnetic field. Virtual particles of the quantum vacuum, therefore, correspond to spontaneously arising ripples on the calm sea of their corresponding fields. This is why they are also referred to as quantum fluctuations of these fields.

Based on the above, the particle production process during inflation can be also understood as the stretching of quantum fluctuations by the superluminal expansion of space, to superhorizon distances. This stretching transforms the quantum fluctuations into classical perturbations of the fields. One can imagine them as mountains and valleys, corresponding to classical variations of the field in question over superhorizon distances. A variation of the values of fields gives rise to a variation of their energy density. Therefore, this is roughly the way that particle production during inflation produces variations of the density of the Universe, which generate the primordial density perturbations, that source the formation of structures in the Universe. The specific ways that field perturbations can source the density perturbations is explained in some detail later on. For the moment, it is important to understand the gist of the idea. The main aspect is that the field perturbations must affect the expansion of the Universe, which is therefore perturbed too because the rate of expansion is directly linked with density in the Friedmann equation (2.10).*

---

*A related picture is through the *Unruh effect*, which takes its name from William George Unruh, who proposed it in 1976 [Unr76]. According to this effect an accelerated observer in empty spacetime perceives an apparent event horizon. This is because there are virtual particles in the quantum vacuum, which

The density perturbations, once produced, do not remain superhorizon forever. Indeed, after the end of inflation, the Universe continues its expansion in the subluminal way of the Hot Big Bang. The cosmological horizon, however, ceases to be an event horizon and becomes a particle horizon, which grows with the speed of light, i.e., faster than the enlargement of the lengthscales of perturbations, which always follow the Universe expansion. As a result, the growing horizon reaches the sizes of previously superhorizon overdensities, which thereby reenter the horizon and become subhorizon. When superhorizon, an overdensity is "frozen," meaning it does not collapse gravitationally. The reason is that the overdensity is causally self-disconnected, that is one end of it cannot attract gravitationally the other end, and any collapse would need superluminal motion (which is forbidden). This changes when the overdensity reenters the horizon sometime after inflation, because then it becomes causally connected and can collapse through gravitational instability. Thus, the story of a galaxy begins as a quantum fluctuation during inflation, which becomes a superhorizon overdensity that collapses after horizon reentry in time to form a galaxy through gravitational instability.[*]

Schematically, we can write

$$[\Delta E \cdot \Delta t \sim \hbar] + \text{INFLATION} \;\to\; \frac{\delta\phi}{\phi} \;\to\; \frac{\delta\rho}{\rho} \;\to\; \text{GALAXIES}\,,$$

where ϕ denotes one of the fields that undergo particle production, whose typical variation during inflation is given by the Hawking temperature $\delta\phi \simeq k_B T_H$. Thus, according to this scenario, all structures in the Universe, such as galaxies, stars, planets, and ultimately ourselves, originated as quantum fluctuations (like the ones we observe in the lab through the Casimir experiment) during a period of superluminal expansion of space.

Not all the fields undergo particle production during inflation; only "light" fields can do so. We call a field light when its mass is small enough such that its Compton wavelength is larger than the horizon size, so that its virtual particles (corresponding to its quantum fluctuations) are able to reach and exit the horizon being caught by the superluminal expansion. The Compton wavelength ℓ_C of a particle species with mass m is determined again by the uncertainty relation as[†]

$$\Delta x \cdot \Delta P \sim \hbar \Rightarrow \Delta x \sim \frac{\hbar}{\Delta P} \;\Rightarrow\; \ell_C \equiv \frac{\hbar}{mc}\,, \tag{4.17}$$

where x and P are the position and momentum of the particle, respectively. Virtual particles can travel over horizon distance before annihilation if $\ell_C > D_H \sim c/H$, which is equivalent to the condition

$$E_V = mc^2 < \hbar H\,, \tag{4.18}$$

where E_V is the energy of the virtual particles. From the uncertainty relation again we find

$$\Delta E \cdot \Delta t \sim \hbar \;\Rightarrow\; \Delta t \sim \frac{\hbar}{\Delta E} = \frac{\hbar}{E_V} \;\Rightarrow\; \Delta t > \frac{1}{H}\,. \tag{4.19}$$

---

cannot reach the observer (they move in the opposite direction to the observer's acceleration), while their virtual partners can. As a result, the accelerating observer is bathed in a warm gas of thermalised particles. As Unruh himself put it in 1990, "you could cook your steak by accelerating it." Similarity with Hawking radiation is not accidental because gravity and acceleration are linked through the strong *equivalence principle*, which states that "the local effects of gravity and acceleration are identical." Thus, in a sealed blind room one cannot tell if they are standing on Earth's surface or being accelerated inside some rocket in deep space. The equivalence principle is a cornerstone of the theory of general relativity.

[*]Of course, if inflation lasts a long time, there are overdensities which remain superhorizon today, because their dimensions were blown up by inflation to such a huge size that there has not been enough time for the growth of the particle horizon after the end of inflation to reach their size in 14 billion years. Such overdensities are bigger than the observable Universe and therefore cannot be probed, so their existence is really academic.

[†]The Compton wavelength corresponds to the range of interactions mediated by the (virtual) particles with mass m.

This suggests that light virtual particles survive longer than expansion timescale, which implies that they can be affected by the expansion. Therefore, the latter can pull the virtual particles out of the horizon and beyond causal contact so that they turn from virtual into real particles, because they can no longer annihilate over superhorizon distances. Thus, we can say that, during inflation, the wavelength of quantum fluctuations of light fields with mass $m < (\hbar/c^2)H$ can be stretched to superhorizon size, so that they become superhorizon classical perturbations.

However, not all light fields undergo particle production during inflation either. There is, in fact, another more subtle requirement to particle production apart from the mass of the field. Indeed, to undergo particle production a light field must not be conformally invariant. A conformally invariant field is unaffected by the expansion of the Universe, because it perceives this expansion as a conformal transformation to which it is insensitive. In terms of its virtual particles, this means that they are not pulled outside the horizon during inflation by the superluminal expansion. Hence, the quantum fluctuations of a conformally invariant field (even if light) are not stretched by the expansion to become classical perturbations. An example of a conformally invariant field is the photon. Thus, inflation does not amplify the quantum fluctuations of the electromagnetic field, even though photons are massless.[*]

Therefore, in a nutshell cosmic inflation amplifies the quantum fluctuations of suitable light fields and turns them into classical perturbations, of superhorizon size. They, in turn, may source perturbations in the density of the Universe, which after horizon reentry are responsible for the formation of structures, such as galaxies and galaxy clusters.

4.3.4.3 Characteristics of the primordial density perturbations

Is there any observational support for this amazing scenario? Very much so. High precision observations of the CMB temperature perturbations, which are due to the primordial density perturbations, have revealed a remarkable agreement with the predictions of inflation. In Chapter 5, we review this evidence. Here it suffices to say that the main characteristics of the primordial density perturbations as generated by inflation and confirmed by observations are:

- **Adiabatic:** The observed density perturbations are predominantly adiabatic because the density contrast is common to all constituents of the Universe content before the overdensities begin to evolve (collapse due to gravitational instability). In simple inflation models, the generated density perturbations are indeed adiabatic because their density contrast is controlled by a single degree of freedom, the field called inflaton (see later), which determines the dynamics of inflation.

- **Gaussian:** Another feature of the observed CMB primordial temperature anisotropy, which stems from the primordial density perturbations and agrees well with inflation is that it appears to be predominantly Gaussian. The nature of quantum fluctuations is stochastic (that is random), which suggests that the perturbation of the field(s) $\delta\phi$, which undergo particle production during inflation, has a Gaussian bell-shaped distribution centred at $k_B T_H$. Thus, inflation naturally predicts Gaussian perturbations, as is indeed observed, reflecting the randomness of quantum fluctuations.

- **Scale-invariant:** The simplest realisations of inflation suggest that the Hawking temperature T_H remains roughly constant during the inflationary phase of the Universe. This means that the variations of the fields, produced by the particle production process, are of the same amplitude even though they may correspond to much different lengthscales, because the quantum fluctuations, which exit the horizon early, are

---

[*]This means that to generate a primordial magnetic field during inflation one needs to break the conformality of electromagnetism.

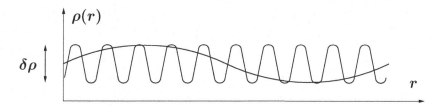

FIGURE 4.10 A scale-invariant spectrum of density perturbations corresponds to a Fourier sum (integral) of perturbation modes with the same amplitude $\delta\rho$ but different wavelengths (corresponding to different lengthscales). Two such modes are depicted with different wavelengths but the same amplitude.

stretched to much larger distances than those that exit the horizon later on. As a result we expect that the spectrum of the density perturbations is almost scale-invariant, i.e., their amplitude is independent of the lengthscale considered (see Fig. 4.10). A small deviation from exact scale invariance is also expected, and has to do with the fact that the inflationary expansion has to end (for the Hot Big Bang to begin). The observations confirm that the density perturbations are indeed almost scale invariant with a subdominant tilt in the spectrum exactly as inflation predicts.

- **Acausal:** The most important feature of the observations, which supports the inflationary scenario for the generation of the primordial density perturbations, is that the density perturbations show correlations beyond the causal horizon (at the time of the CMB emission), a feature that only the superluminal expansion of an inflationary phase can explain.

Alternative explanations of the density perturbations and the resulting CMB primordial anisotropy typically require the existence of a pre-Big Bang era, which is also used to overcome the horizon problem as explained before. In this case, the perturbations are not acausal as in inflation, but still manage to stretch over "superhorizon" scales. There is an important difference though, in that in the case of inflation the superhorizon correlations were predicted over a decade before they were observed in the 1990s. In contrast, alternatives to inflation, which aim to explain the observed cosmological perturbations, offer what Robert Brandenberger calls "postdictions" (instead of predictions) since the observations are known in advance. Needless to say that predictions (and not postdictions) confirmed by observational evidence are a strong support for a proposed theory.

4.3.5 Inflation and the origin of the expansion

Naïvely, if one starts with a static Universe $\dot{a} = 0$ and ends up with an expanding one $\dot{a} > 0$, this means that, at some point, $\ddot{a} > 0$, inflation is inevitable and needed to provide the initial boost for the Universe expansion. However, we suspect that the classical Universe emerges from spacetime foam. One can argue then that, once an arrow of time is fixed, expansion or contraction are equiprobable, while a static Universe is unlikely. The parts of space that undergo contraction, soon end up in a Big Crunch, form black holes and disappear, embedded in the parts which expand and soon dominate classical space. The question is, does the expanding space also inflate? Recent work on the emergence of classical spacetime from spacetime foam seems to indicate that this is possible only when the classical spacetime is dominated by an effective cosmological constant. A comological constant, however, as we have discussed, is a special case of dark energy. Thus, it seems that the emergent classical spacetime is dominated by dark energy and therefore undergoes inflation; a specific type of inflation called quasi-de Sitter (see next section). In quasi-de Sitter inflation, the growth of the scale factor is exponential, that is, space is expanding in an explossive way. This is why the prominent cosmologist Yakov B. Zel'dovich once argued that inflation *is* in fact the Big Bang, the original explosion that started it all. On similar lines, Alan Guth said, "The inflationary Universe is a theory of the 'bang'; of the Big Bang!" (1997).

Here we should also mention the "no-hair" theorem, which states that extracting information from inside the event horizon of a black hole is impossible; one can only know of the black hole's global properties such as its mass, angular momentum, and electric charge. Similarly, for inflation, the no-hair theorem demands that information about superhorizon scales is inaccessible because they correspond to the "inside" of the "inverted" black hole surrounding the observer. Because the lengthscale of the observable Universe is entering the horizon today (by definition) and because we need inflation to begin before the scale of the observable Universe leaves the horizon so as to solve the horizon and flatness problems, the lengthscale, which exits the horizon right after the onset of inflation, is still superhorizon at present. This means that any pre-inflation era is utterly inaccessible to us. Hence, not only does inflation impose the cosmological principle as an initial condition of the Hot Big Bang, not only does it create the small deviations from the cosmological principle needed to form the galaxies and galactic clusters, but it also erases any memory of its own initial conditions. This suggests that inflation is the onset of the Universe history as far as we are concerned.

4.4 Quasi-de Sitter inflation

4.4.1 Explosive exponential expansion

As we discuss in Chapter 6, most particle-physics models of inflation correspond to (quasi)-de Sitter inflation, where the Universe is dominated by an effective cosmological constant Λ_{eff}. The solution of the Einstein equations in general relativity with zero matter and positive cosmological constant is called de Sitter spacetime.* De Sitter spacetime is maximally symmetric both in space and in time ("perfect" cosmological principle). The case of inflation due to Λ_{eff} is called quasi-de Sitter because inflation has to end and the Hot Big Bang has to begin, so spacetime is not really respecting the maximally symmetric de Sitter solution of the Einstein equations. In fact, there is a boundary to the de Sitter patch corresponding to the end of inflation.

In the case of quasi-de Sitter inflation, the density of the Universe is also constant, determined by Λ_{eff} as

$$\rho = \frac{\Lambda_{\text{eff}} c^2}{8\pi G} , \tag{4.20}$$

while for the barotropic parameter we have

$$\rho \propto a^{-3(1+w)} = \text{constant} \ \Rightarrow \ w = -1 . \tag{4.21}$$

By solving the flatness problem, inflation renders the Universe spatially flat which means

$$H^2 = \frac{8\pi G}{3}\rho = \frac{1}{3}\Lambda_{\text{eff}} c^2 = \text{constant} . \tag{4.22}$$

This means that the physical dimensions of the event horizon remain constant $D_H = c/H$ (cf. Sec. 2.10). Then, by integrating the relation $\dot{a} = aH$, it is straightforward to obtain (cf. Sec. 2.6.1)

$$a(t) \propto e^{Ht} . \tag{4.23}$$

---

*With negative cosmological constant it is called anti-de Sitter spacetime.

Thus, the Universe expands exponentially fast. Such inflation leads to the most rapid expansion of space, which still respects the null energy condition, such that $-1 \leq w < -\frac{1}{3}$. This can be understood as follows. As we have discussed, in the case of a flat Universe dominated by a substance whose barotropic parameter remains constant $w > -1$, the scale factor grows as $a \propto t^{\frac{2}{3(1+w)}}$ (cf. Eq (2.39)). Thus, the closer the value of w is to -1, the more intense the expansion. Hence, the limit of quasi-de Sitter inflation $w \to -1$ corresponds to the most rapid expansion.

4.4.2 Inflationary e-folds

The usefulness of the cosmic time is undermined by the exponential growth of the scale factor in quasi-de Sitter inflation. This is why a new "time" coordinate is introduced, corresponding to the logarithmic change of $a(t)$ and is defined as

$$dN \equiv -da/a = -H\,dt\,. \tag{4.24}$$

N is called the number of e-folds. Integrating the above and counting from the end of inflation we find

$$N = \ln\left(\frac{a_{\text{end}}}{a}\right)\,, \tag{4.25}$$

where $a < a_{\text{end}}$ corresponds to some moment during inflation. Note that as time progresses N decreases to zero.

How many e-folds does inflation last? Well, to have a feeling about this we employ the requirement for the solution of the horizon problem in Eq. (4.7). Most successful inflationary models operate at the energy of grand unification (GUT-scale inflation). Considering this and assuming, for simplicity, that reheating is prompt and the Hot Big Bang follows immediately after the end of inflation we find

$$1 \geq \frac{\dot{a}_i}{\dot{a}_0} = \frac{\dot{a}_i}{\dot{a}_{\text{end}}}\frac{\dot{a}_{\text{end}}}{\dot{a}_0} \quad \Rightarrow \quad \frac{a_{\text{end}}}{a_i} \approx \frac{\dot{a}_{\text{end}}}{\dot{a}_i} \geq \frac{\dot{a}_{\text{end}}}{\dot{a}_{\text{eq}}}\frac{\dot{a}_{\text{eq}}}{\dot{a}_0} \sim \left(\frac{t_{\text{GUT}}}{t_{\text{eq}}}\right)^{-1/2}\left(\frac{t_{\text{eq}}}{t_0}\right)^{-1/3}$$

$$\Rightarrow \quad N_{\text{tot}} \equiv \ln\left(\frac{a_{\text{end}}}{a_i}\right) \quad \geq \quad 62\,, \tag{4.26}$$

where we considered that in quasi-de Sitter inflation H remains roughly constant during inflation (so $H_i \approx H_{\text{end}}$) and that the end of inflation occurs at $t_{\text{GUT}} \sim 10^{-38}$ s. We also used that in the radiation era $\dot{a} \propto t^{-1/2}$ and in the matter era $\dot{a} \propto t^{-1/3}$, for a flat Universe.* As we have discussed, the solution of the flatness problem is a bit more demanding, roughly $N_{\text{tot}} \geq 65$ or so. In most cases, however, inflation may last for hundreds, even thousands of e-folds, so the horizon and flatness problems are well overcome.

We should mention a subtlety here. As evident from Eq. (4.23), the limit $a \to 0$ is approached only when $t \to -\infty$. This implies that the moment of the Big Bang is transposed indefinitely in the past.† So, what does it mean to say that $t_{\text{GUT}} \sim 10^{-38}$ s ? Well, this would be the time after the Big Bang when the GUT transition happens, had the radiation era been extended backwards in time until the Big Bang itself $a(t = 0) = 0$ (see Fig. 4.11). Realistically though, quasi-de Sitter inflation cannot be extended infinitely in the past. As we have seen, while inflation progresses, the magnitude and the anisotropy in the curvature of space is inflated away. Therefore, going backwards in time, the curvature becomes more prominent and anisotropic, and the simple de Sitter solution ceases to apply.

---

*Flat because of inflation.

†For this reason, the particle horizon in quasi-de Sitter inflation is $D_H = a \int_{-\infty}^{t} c\,dt/a$, which diverges.

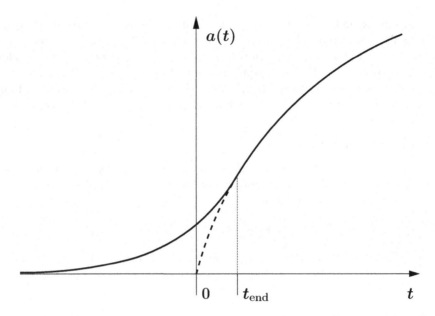

FIGURE 4.11 A period of quasi-de Sitter inflation transposes the moment of Big Bang ($a = 0$) infinitely to the past. The behaviour of the scale factor of the Universe is shown by the solid line. The meaning of the time when inflation ends t_{end} is the time after the Big Bang, had the radiation era been extended backwards in time until the Big Bang itself, in which case the time of the Big Bang would have been $t = 0$. This case would correspnd to the dashed line for the scale factor.

4.4.3   Particle production in quasi-de Sitter inflation

The Hawking temperature of de Sitter space is [HM83]

$$T_H = \frac{\hbar}{k_B} \frac{H}{2\pi}. \tag{4.27}$$

Because $H = $ constant, we see that the typical amplitude $\delta\phi = k_B T_H$ of the perturbations of the fields which undergo particle production remains constant. As we have discussed, this results in a scale-invariant spectrum of superhorizon density perturbations.

In fact, in quasi-de Sitter inflation H is only approximately constant, which means that the produced perturbations are only approximately scale-invariant, as they are characterised by a subdominant tilt in their spectrum. This is exactly what is observed, suggesting that cosmic inflation is quasi-de Sitter. Fig. 4.12 depicts pictorially the evolution of two quantum fluctuations, which exit the horizon during quasi-de Sitter inflation one after the other, become superhorizon perturbations and eventually reenter the horizon some time during the Hot Big Bang. Such can be the true history of a galaxy and the galaxy cluster it finds itself in.

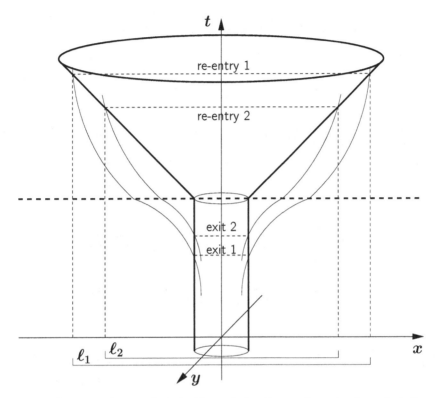

FIGURE 4.12 Spacetime diagram depicting the evolution history of two density perturbations of different ultimate size generated by quasi-de Sitter inflation. The perturbations begin as quantum fluctuations, which exit the horizon one after the other during a period of quasi-de Sitter inflation, thereby becoming classical perturbations. The physical size of the perturbations when they exit the horizon is (roughly) the same because the horizon remains (roughly) constant in quasi-de Sitter inflation. Once superhorizon, the sizes of the perturbations grow with the expansion (i.e., they follow the behaviour of the scale factor $\ell \propto a$). The perturbation, which exited the horizon earlier, becomes larger in size. The amplitude of the perturbations however, does not change (they are "frozen" when superhorizon). Because in quasi-de Sitter inflation the perturbation amplitude is determined by the Hawking temperature Eq. (4.27), which remains (roughly) constant during inflation, both perturbations retain (roughly) the same amplitude despite the fact that they have different sizes. After inflation, the horizon grows with the speed of light, while the growth of the perturbations is subluminal. As a result, the horizon catches up with the perturbations, which one by one reenter the horizon, at different times, because they have different sizes. Note that the perturbation, which exited the horizon earlier, reenters the horizon later. The sizes of the perturbations at horizon reentry can be significantly different $\ell_1 \gg \ell_2$, but their amplitude is (roughly) the same, which is why the spectrum of perturbations over different lengthscales is (almost) scale-invariant in excellent agreement with the observations.

EXERCISES

1. Consider that, just below the Planck scale, copious production of string moduli particles takes place. The moduli are massive objects, which do not thermalise and remain always nonrelativistic. These moduli are so numerous that they dominate the density of the Universe immediately after their formation. The Universe remains moduli dominated until the time when a period of quasi-de Sitter inflation takes place.

   (a) Assume that quasi-de Sitter inflation lasts for N e-folds, where

   $$N \equiv \ln\left(\frac{a_{\text{end}}}{a_{\text{beg}}}\right),$$

   where the subscripts "beg" and "end" denote the beginning and the end of inflation, respectively.

   Assuming that reheating is prompt and the thermal bath of the Hot Big Bang is immediately formed at the end of inflation, show that, at that time

   $$(\rho_M/\rho)_{\text{end}} = e^{-3N},$$

   where ρ and ρ_M denote the density of the Universe and the moduli, respectively.

   (b) Show that, after the end of inflation, the dentiy ratio ρ_M/ρ grows. This means that the moduli eventually dominate the Universe again at some time t_{dom}. Assuming that this moment lies in the radiation era, show that the domination time is

   $$t_{\text{dom}} = e^{6N} t_{\text{end}}.$$

   (c) In a log-log plot, depict the behaviour of the densitiy of the Universe ρ and the density of the moduli ρ_M as functions of the scale factor $a(t)$ from the time of the moduli formation until the time t_{dom}, corresponding to $a_{\text{dom}} = a(t_{\text{dom}})$.

   (d) Suppose that the period of quasi-de Sitter inflation occurs at the electroweak phase transition, when $t_{\text{ew}} \sim 10^{-11}$ sec (i.e., $t_{\text{end}} = t_{\text{ew}}$). Considering that most of the matter in the Universe is dark matter, calculate N such that the moduli can become the dark matter today.
   [Hint: At matter-radiation equality $t_{\text{eq}} \sim 10^{11}$ sec.]

   (e) Briefly discuss what happens if N is much larger that the value obtained in (d).

2. Suppose that the early Universe underwent power-law inflation with $a \propto t^q$, where $q > 1$, between the time of grand unification $t_{\text{gut}} \sim 10^{-38}$ sec and that of electroweak unification $t_{\text{ew}} \sim 10^{-11}$ sec .
   Assuming that $|\Omega_{\text{gut}} - 1| \sim 0.1$ at the onset of inflation, show that the following lower bound on q must be satisfied:

   $$q > 1.5$$

   in order that $|\Omega_0 - 1| < 0.01$, according to observations. You may assume that the Universe is always approximately spatially flat since $|\Omega - 1| \ll 1$ always.
   Hence, find an upper bound on the value of the barotropic parameter w during inflation.
   [Hint: The present time is $t_0 \sim 10^{10}$ yrs, and the time of equal matter-radiation densities is $t_{\text{eq}} \sim 10^4$ yrs $\sim 10^{12}$ sec.]
   [Hint: Late dark energy may be ignored.]

5

CMB Primordial Anisotropy and Structure in the Universe

5.1	Connecting the CMB with density perturbations	91
	CMB spectrum • Sachs-Wolfe effect • Acoustic peaks • Polarisation	
5.2	Baryon acoustic oscillations........................	99
5.3	The curvature perturbation........................	100
	The two-point correlator • Curvature perturbation and observables • Tensors	
5.4	Structure formation................................	104
	The Jeans length • Linear growth of overdensities • Turnaround and virialization • The cosmic web	

In this chapter, we take a closer look at the observations of the CMB primordial anisotropy. This is a powerful tool to explore the characteristics of the primordial density perturbations, which are responsible for the formation of structure in the Universe, such as galaxies and galactic clusters. As discussed in the last chapter, inflation is the most compelling mechanism to explain these primordial density perturbations, which reflect themselves on the CMB primordial anisotropy. Thus, the CMB anisotropy provides insights to the mechanism of inflation and important constraints on inflation models, which are becoming ever tighter as the precision of the observations increases. In contrast to the horizon and flatness problems of the Hot Big Bang, which are solved by all inflation models provided inflation lasts long enough, the constraints due to the CMB primordial anisotropy differentiate between inflation models. As such, they have already ruled out classes of inflation models. At the same time they set stringent constraints on the surviving ones, thereby providing insights on the background theory; the theory behind the inflation model in question. Before moving on to inflation model building, it is instructive to review how we can extract all this information for inflation from the CMB primordial anisotropy.

5.1 Connecting the CMB with density perturbations

5.1.1 CMB spectrum

The fractional difference of the CMB temperature in the vicinity of a region of the last scattering surface with position unit vector $\hat{n}$ is defined as

$$\frac{\delta T}{T}(\hat{n}) \equiv \frac{T(\hat{n}) - \langle T \rangle}{\langle T \rangle}, \tag{5.1}$$

where

$$\langle T \rangle = \frac{1}{4\pi} \int T(\hat{n}) d\Omega = T_{\text{CMB}}, \tag{5.2}$$

with $d\Omega = d(\cos\theta)d\phi$ (θ, ϕ being polar coordinates) and $T_{\rm CMB} = 2.7255 \pm 0.0006\,{\rm K}$. After subtracting the dipole, which as we mentioned previously is due to the peculiar motion of our Solar System, the observations suggest [A$^+$18a]

$$\left.\frac{\Delta T}{T}\right|_{\rm CMB} \equiv \left.\frac{\delta T}{T}\right|_{\rm rms} = (0.916 \pm 0.006) \times 10^{-5}\,, \tag{5.3}$$

where with "rms" we mean root-mean-square. The rms value is $x_{\rm rms} \equiv \sqrt{\langle x^2 \rangle}$, where $\langle \cdots \rangle$ denotes an *ensemble average* (see below). The above is called the COBE constraint, which is used to determine the energy scale of inflation (see later). Let us see how this estimate is obtained.

Because it is defined on the celestial sphere (the last scattering surface), the CMB fractional temperature perturbation is analysed in spherical harmonics as

$$\frac{\delta T}{T}(\hat{\boldsymbol{n}}) = \sum_{\ell=0}^{\infty} \sum_{m=-\ell}^{\ell} a_{\ell m} Y_{\ell m}(\hat{\boldsymbol{n}})\,, \tag{5.4}$$

where for the spherical harmonics* we have $Y_{\ell m}^* = (-1)^m Y_{\ell m}$ and

$$\int Y_{\ell m}(\hat{\boldsymbol{n}}) Y_{\ell' m'}^*(\hat{\boldsymbol{n}}') d\Omega = \delta_{\ell\ell'} \delta_{mm'}\,, \tag{5.5}$$

where $d\Omega$ is the angle element and the Kronecker's delta δ_{ij} is zero when $i \neq j$ and unity when $i = j$. This process is the analogue of the Fourier decomposition but on a sphere instead of an infinite plane (flat volume). This is why the spectrum of modes is not continuous as with Fourier decomposition. It is discrete, because the modes (called multipoles) of oscillation of a sphere are specific, as the circumference of the sphere has to equal a discrete number of half-wavelengths so that the wave consistently wraps around the sphere meeting with itself. Indeed, $\ell = 0$ corresponds to the monopole mode, where the entire sphere is shrinking or swelling at the same time, $\ell = 1$ corresponds to the dipole mode, where the entire sphere is squeezed in one direction and flattened in the perpendicular one, $\ell = 2$ corresponds to the quadrupole and so on (see Fig. 5.1). Thus, small wavelengths probing small scales on the last scattering surface, correspond to large multipoles. Note that the primordial CMB anisotropy corresponds to standing waves (i.e., which are in phase everywhere) because, although they have energy (and momentum), there is no preferred direction so this energy is not travelling.

As with Fourier decomposition, the information on the signal is included in the amplitude of the oscillations $a_{\ell m}$. So, the *spectrum of the CMB anisotropy* C_ℓ is given by

$$\langle a_{\ell m}, a_{\ell' m'}^* \rangle = \delta_{\ell\ell'} \delta_{mm'} C_\ell \tag{5.6}$$

where the $\langle \cdots \rangle$ symbol denotes an ensemble average, meaning an average over all possible realisations of the last scattering surface seen by all possible vantage points, or corresponding to all possible universes, which have the same macroscopic identities (such as curvature or age)

---

*Defined as

$$Y_{\ell m}(\theta, \phi) \equiv \left[\frac{2\ell+1}{4\pi} \frac{(\ell-m)!}{(\ell+m)!}\right]^{1/2} P_\ell^m(\cos\theta)\, e^{im\phi}\,,$$

with P_ℓ^m being the associated Legendre polynomials

$$P_\ell^m(x) = (-1)^m \frac{(1-x^2)^{m/2}}{2^\ell \ell!} \frac{d^{\ell+m}}{dx^{\ell+m}} (x^2-1)^\ell\,,$$

with $d^{\ell+m}/dx^{\ell+m}$ denoting the $(\ell+m)$-th derivative.

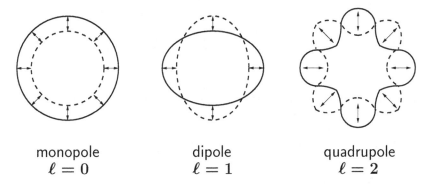

monopole dipole quadrupole
$\ell = 0$ $\ell = 1$ $\ell = 2$

FIGURE 5.1 The multipoles correspond to modes of oscillation of a sphere. The lowest multipoles are the monopole with $\ell = 0$, the dipole with $\ell = 1$, the quadrupole with $\ell = 2$, the octupole with $\ell = 3$ and so on. The higher the multipole the smaller the wavelength of the oscillation it corresponds to. As a result, high multipoles probe small-scale features on the sphere.

to ours. Of course, because we cannot experience all these universes or even all possible vantage points within our own Universe (we have a single vantage point, that of Earth) it would seem that the ensemble average above is impossible to calculate. Fortunately, there is a way out of this dead end, under the name of the *ergodic theorem*, which states that the ensemble average of a statistically homogeneous distribution equals the position average. Thus, the spectrum of the CMB anisotropy C_ℓ (more accurately $\ell(\ell+1)C_\ell$) can be (roughly) thought of as the rms-squared value of $\delta T/T$ within a disk of angular radius $\theta = \pi/\ell$ randomly placed on the sky.

From Eq. (5.6), averaging over m gives

$$C_\ell = \frac{1}{2\ell+1} \sum_{m=-\ell}^{\ell} |a_{\ell m}|^2 = \langle |a_{\ell m}|^2 \rangle. \tag{5.7}$$

This is all fine as long as ℓ is large and the statistics are robust. For small ℓ, however, meaning oscillations with large wavelength covering much of the celestial sphere, the statistics are not good as we do not have many such disks covering the last scattering surface. Thus, the reliability of the measurements is undermined for small ℓ. This is called *cosmic variance*, which can be defined as the rms difference between the local measurement minus the mean over many vantage points, i.e., it is the rms deviation of $|a_{\ell m}|^2$ from $C_\ell = \langle |a_{\ell m}|^2 \rangle$. Cosmic variance is quantified as

$$(\Delta C_\ell)^2 \equiv \langle (|a_{\ell m}|^2 - C_\ell)^2 \rangle = \langle |a_{\ell m}|^4 - C_\ell^2 \rangle, \tag{5.8}$$

where in the last equality the signal is assumed to be Gaussian.* From the above we can obtain that

$$(\Delta C_\ell)^2 = \frac{C_\ell^2}{2\ell+1}, \tag{5.9}$$

which demonstrates that cosmic variance is insignificant for $\ell \gg 1$. However, it is about 50% for the quadrupole ($\ell = 2$).

The CMB correlation function is

$$\left\langle \frac{\delta T}{T}(\hat{n}), \frac{\delta T}{T}(\hat{n}') \right\rangle = \frac{1}{4\pi} \sum_{\ell=2}^{\infty} (2\ell+1) C_\ell P_\ell(\cos\theta), \tag{5.10}$$

---

*Statistical isotropy suggests $\langle a_{\ell m} \rangle = 0$.

where we considered $\hat{\boldsymbol{n}} \cdot \hat{\boldsymbol{n}}' = \cos\theta$ and we used the expansion of the spherical harmonics. P_ℓ are the Legendre polynomials.* In the above, we excluded the monopole $\ell = 0$ (which vanishes) and the dipole $\ell = 1$ $(m = 0)$ from the sum. Inverting Eq. (5.10) we obtain

$$C_\ell = \frac{1}{4\pi} \int d\Omega \, d\Omega' P_\ell(\cos\theta) \left\langle \frac{\delta T}{T}(\theta,\phi), \frac{\delta T}{T}(\theta',\phi') \right\rangle , \qquad (5.11)$$

with $d\Omega' = d(\cos\theta')d\phi'$. This demonstrates that C_ℓ encodes the information of the CMB fractional temperature perturbations. As we said, $\ell(\ell+1)C_\ell(\theta)$ is the typical squared CMB temperature perturbation on an angular scale $\theta = \pi/\ell$. C_ℓ is real and positive.

5.1.2   Sachs-Wolfe effect

We can relate the temperature variations on the last scattering surface with the gravitational potential of the overdensities at the time of decoupling. These overdensities must be time dependent so that a CMB photon falling into a potential well does not gain energy equal to the amount it loses when exiting. Otherwise, any effect on the photon energy cancels out. Fortunately, the Universe expansion makes sure of this time dependence, which is intensified for subhorizon scales when the overdensities are collapsing because of gravitational instability. A full relativistic treatment, originally performed by Reiner K. Sachs and Arthur M. Wolfe in 1967 [SW67], yields the result

$$\frac{\delta T}{T}(\hat{\boldsymbol{n}}) = \frac{1}{3}\frac{\Phi}{c^2} , \qquad (5.12)$$

where Φ is the gravitational potential of the overdensity. This is linked to the density perturbation through the Poisson equation

$$\nabla^2 \Phi = 4\pi G \delta\rho . \qquad (5.13)$$

Now we perform a Fourier transformation following the formula[†]

$$f(\boldsymbol{x},t) = (2\pi)^{-3} \int d^3k \, f(\boldsymbol{k},t)e^{i\boldsymbol{k}\cdot\boldsymbol{x}} , \qquad (5.14)$$

where $k \equiv |\boldsymbol{k}|$. Thus, using the linear independence of the Fourier modes, Eq. (5.13) becomes[‡]

$$(k/a)^2 \Phi_{\mathbf{k}} = 4\pi G \, \delta\rho_{\mathbf{k}} , \qquad (5.15)$$

where we defined the subscript "$\mathbf{k}$" such that $f(\boldsymbol{k},t) \equiv f_{\mathbf{k}}(t)$, and we considered that Eq. (5.13) features derivatives with respect to the physical coordinates, whereas k is the modulus of the comoving momentum, so that k/a is the physical momentum.[§] Dividing the above with the factor $3H^2$ and using the flat Friedmann equation $3H^2 = 8\pi G\rho$ (cf. Eq. (2.38)) we find

$$\frac{1}{3}\left(\frac{k}{aH}\right)^2 \Phi_{\mathbf{k}} = \frac{1}{2}\frac{\delta\rho}{\rho}\bigg|_{\mathbf{k}} . \qquad (5.16)$$

---

*Defined as

$$P_\ell(x) = P_\ell^{m=0}(x) = \frac{1}{2^\ell \ell!}\frac{d^\ell}{dx^\ell}(x^2 - 1)^\ell ,$$

where d^ℓ/dx^ℓ denotes the ℓ-th derivative. So, $P_0 = 1$, $P_1 = x$, $P_2 = \frac{1}{2}(3x^2 - 1)$ and so on. Note that the zeroth derivative is the function itself.

[†]The Fourier theorem states that any periodic signal $f(x)$ can be represented as a sum of harmonic signals of different amplitudes $f(k)$ and frequency k. If the signal is not periodic, the spectrum of frequencies is not discrete but continuous and the sum becomes an integral over all values of k

[‡]Recall that $\nabla e^{i\boldsymbol{k}\cdot\boldsymbol{x}} = i(k/a)e^{i\boldsymbol{k}\cdot\boldsymbol{x}}$, and the perturbation $\delta\rho_{\mathbf{k}}$ is always positive.

[§]Thus we have $\boldsymbol{k}\cdot\boldsymbol{x} = (k/a)\cdot\boldsymbol{r}$ with $\boldsymbol{r} = a\boldsymbol{x}$ being the physical position and $\boldsymbol{x}$ being the comoving one.

Evaluating this at the moment a given superhorizon overdensity reenters the horizon we obtain

$$\frac{1}{3}\frac{\Phi}{c^2} \simeq \frac{1}{2}\frac{\delta\rho}{\rho}\bigg|_H \,, \tag{5.17}$$

where we considered that at horizon reentry (denoted by the subscript H) the physical dimensions of the overdensity are comparable to the size of the horizon $\lambda \sim a/k \sim c/H$ (i.e., $k \simeq aH/c$). With this in mind, Eqs. (5.12) and (5.17) suggest

$$\frac{1}{2}\frac{\delta\rho}{\rho}\bigg|_H \simeq \frac{1}{3}\frac{\Phi_{\rm rms}}{c^2} = \frac{\Delta T}{T}\bigg|_{\rm CMB} \simeq 10^{-5} \,, \tag{5.18}$$

which is Eq. (4.15). The above expression features the root-mean-square (rms) values of the quantities over all of the sky. This reflection of the density perturbations onto the CMB primordial anisotropy is called the *Sachs-Wolfe effect*.

5.1.3 Acoustic peaks

Now that we have established a link between the fractional perturbation of the CMB temperature and the amplitude of the overdensities when they enter the horizon we attempt a brief explanation of the form of the CMB spectrum, suggested by the observations. The form of the observed spectrum depends on whether a given multipole corresponds to a superhorizon or a subhorizon lengthscale on the last scattering surface.

The multipole that corresponds to the horizon at the time of decoupling, when the CMB is emitted, is estimated as follows. The angle subtended by the horizon at decoupling θ_H is $\theta_H \simeq R_{\rm dec}/D_H(t_0)$, where we considered that the radius of the last scattering surface is roughly the size of the present horizon $D_H(t_0)$, and $R_{\rm dec}$ stands for the lengthscale of the decoupling horizon as it has been enlarged by the Universe expansion until today $R_{\rm dec} = (a_0/a_{\rm dec})D_H(t_{\rm dec})$ (see Fig. 5.3). Thus, we obtain

$$\theta_H \simeq \frac{R_{\rm dec}}{D_H(t_0)} = \frac{a(t_0)}{a(t_{\rm dec})}\frac{D_H(t_{\rm dec})}{D_H(t_0)} \simeq (1 + z_{\rm dec})\frac{t_{\rm dec}}{t_0} \simeq 0.024 \,{\rm rad} = 1.3° \,, \tag{5.19}$$

where $z_{\rm dec} = 1090$, we have used Eq. (2.9) and that the horizon is roughly $D_H \sim ct$.[*] The corresponding multiple $\ell_H = \pi/\theta_H$ is about $\ell_H \simeq 130$.

For superhorizon scales $\theta > 3\theta_H$ the CMB spectrum levels off. Indeed, for multipoles $2 \le \ell \le 40$ the spectrum is approximately flat in the *Sachs-Wolfe plateau*. Now, the Sachs-Wolfe effect is the red/blue-shift of radiation due to perturbations in the gravitational potential (due to overdensities) on the last scattering surface. A related effect, called the *integrated* Sachs-Wolfe effect has to do with the red/blue-shift of radiation due to perturbations in the gravitational potential between the last scattering surface and the present, i.e., not at the time of the emission of the CMB but during the travel of the CMB light from the last scattering surface to the observer on Earth. The integrated Sachs-Wolfe effect (ISW) is important only at low multipoles $\ell < 10$, where the observations are also suffering from cosmic variance.

Whereas superhorizon scales are insensitive to microphysical effects, this is not so for subhorizon scales. Overdensities at these scales have already entered the horizon before decoupling and have started undergoing gravitational collapse. However, before decoupling, the baryonic matter and the photons are still strongly coupled (through Thomson scattering of the photons off the

---

[*]Note the conversion rule: $1\,{\rm rad} = 180°/\pi$.

electrons). This means that, as the baryons move into the potential well of a collapsing overdensity, they drag the photons along with them. As a result the radiation becomes compressed at the collapsing overdensity. However, radiation is not pressureless. As it is being compressed, radiation pressure opposes the collapse and the overdensity bounces back, only to try to recollapse again later and so on. Thus, after horizon entry, the baryon-photon plasma of an overdensity undergoes a series of oscillations, called *acoustic oscillations* because the characteristic scale of the bounce is the *sound horizon*, which is a little less than the cosmological horizon (by a factor of $1/\sqrt{3}$).* As you would expect, the amplitude of these oscillations diminishes with time.

The last scattering of the photons produces a snapshot of the state of affairs at t_{dec}. At this moment, some subhorizon scales are in the compression phase of the acoustic oscillations and some are in the depression phase. The smaller the scale considered, the earlier it has entered the horizon and started the acoustic oscillations. The snapshot at decoupling catches the various scales at a different phase of the oscillations. The compressed are hotter while the depressed are cooler. As we consider larger and larger multipoles, that is smaller and smaller angles (hence smaller and smaller scales), we pass through alternating compressed and depressed multipoles, whose peak amplitude decreases because larger values of ℓ correspond to scales that entered the horizon earlier and the acoustic oscillations are increasingly damped. Thus, for $\ell > \ell_H$ (equivalently $\theta < \theta_H$) the CMB spectrum features a series of peaks and troughs. The location and amplitude of these peaks reveals much of the underlying cosmology and the physics of inflation.

The mere existence of these acoustic peaks is an indication in favour of inflation, because it suggests that there are overdensities beyond the horizon, which start to collapse upon horizon entry. As discussed earlier, in order to form superhorizon overdensities one needs an acausal mechanism, and inflation is the only game in town. The principle contender to inflation for the formation of structures in the Universe used to be cosmic strings. That however, corresponded to a causal mechanism, which would not produce peaks in the CMB spectrum. When the acoustic peaks were observed the cosmic string paradigm collapsed, which demonstrates how closely cosmology is connected to the real world.

The latest CMB observations produced the CMB spectrum shown in Fig. 5.2. The distinction between superhorizon and subhorizon scales is evident. At low multipoles one finds the Sachs-Wolfe plateau. However, when moving towards higher multipoles, the amplitude of the spectrum rises until it peaks at about $\ell_1 \simeq 220$ (angle $\theta_1 = 0.8°$). This corresponds to the scale of the sound horizon at decoupling, which is slightly smaller in size than the cosmological horizon at decoupling at $\ell_H \simeq 130$ ($\ell_1 \simeq \sqrt{3}\,\ell_H$). The first peak is at maximum compression.

The location of the peaks is roughly given by

$$\ell_n = \frac{\pi}{Q}\left(n - \frac{1}{8}\right), \tag{5.20}$$

where $Q \simeq 0.01$. The location of the first peak is shifted by the curvature of space. Indeed, as shown in Fig. 5.3, a positively curved universe ($k > 0$) would result in a larger value of θ_H, therefore a smaller value of $\ell_H = \pi/\theta_H$, which also means that the first peak would follow ℓ_H and be at a smaller ℓ as well. Conversely, a negatively curved universe ($k < 0$) would result in a smaller value of θ_H and so a larger value of ℓ_H, which also means that the first peak would be at a larger ℓ as well. The location of the first peak strongly suggests that the Universe is close to being spatially flat ($k = 0$).

The amplitude of the first peak is a diagnostic of the baryon density ρ_B. An increase in ρ_B increases the height of the first peak and decreases the height of the second peak at $\ell_2 \simeq 520$.

---

*The speed of sound is $c_s^2 = \partial p/\partial \rho$ (cf. Eq. (2.16)). For radiation, $p_r = \frac{1}{3}\rho_r c^2$, which means that $c_s = c/\sqrt{3}$. The sound horizon is given by $D_s = \int_{z_{\text{eq}}}^{\infty} [c_s^2/H(z)]dz$, where $z_{\text{eq}} \simeq 3600$, cf. Sec. 2.8.2.

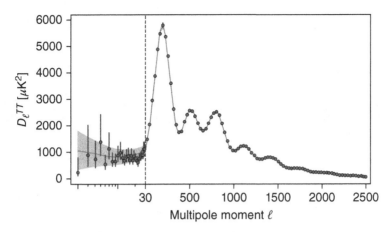

FIGURE 5.2 CMB temperature perturbation spectrum. The graph is a plot of $\mathcal{D}_\ell = \ell(\ell+1)C_\ell/2\pi$ as a function of the multipole moment ℓ of spherical harmonics and also of the pitch angle in the sky. For angles larger than 1.5° or equivalently for $\ell < 30$ the scaling of the horizontal axes changes so as to feature the Sachs-Wolfe plateau ($2 \leq \ell \leq 40$) more clearly. The shaded area depicts the region under the influence of cosmic variance, which suggests that deviations from the theoretical curve (depicted by the solid line) within the shaded area are due to poor statistics and do not have a physical meaning. The effect of cosmic variance diminishes (virtually disappears) for large ℓ. For $\ell \geq 30$ we have the clear depiction of seven acoustic peaks. The line between the data points is not the line that connects the dots, although you would be excused for thinking so. It is the theoretical line produced from inflation. The agreement with the data is spectacular. Reproduced with permission from *Astronomy & Astrophysics*, © ESO; original source ESA and the Planck Collaboration [A+16].

However, a growth of the CDM density ρ_{CDM} decreases the height of both ℓ_1 and ℓ_2. The existence of the second peak alone implies that $\rho_m < \rho_c$ and $\rho_{\mathrm{CDM}} > \rho_B$ today. Since $\rho_m < \rho_c$ but the Universe is flat (i.e., with density ρ_c), this indicates (late) dark energy with density $\rho_{\mathrm{DE}} = \rho_c - \rho_m$, independent from the supernovae observations. The height and location of the first and second peaks are also used to estimate the value of the Hubble constant H_0.

The third peak ($\ell_3 \simeq 820$) is not sensitive to either ρ_B or ρ_{CDM}. It is, however, sensitive to the spectral index n_s of the density perturbations (see later), which is of prime interest to inflation model building. Indeed, defining the ratio of the heights of the third over the first peak $r \equiv \mathcal{H}_3/\mathcal{H}_1$ we have

$$\frac{\Delta r}{r} \simeq 1 - \left(\frac{\ell_3}{\ell_1}\right)^{1-n_s} \simeq (n_s - 1)\ln\left(\frac{\ell_3}{\ell_1}\right), \tag{5.21}$$

where in the last equation we considered $|(\ell_3/\ell_1)^{1-n_s} - 1| \ll 1$.* Thus, the height of the third peak decreases for $n_s < 1$.

For higher multipoles the height of the peaks decreases. At $\ell > 2500$ or so, the signal is corrupted by foreground effects (e.g., the *Sunyaev-Zel'dovich* effect). Also, the last scattering surface has a minute thickness (because there is photon diffusion at any epoch), which is called the *Silk scale*. This corresponds to $\ell \sim 10^3$, which implies that the CMB anisotropy is wiped out at smaller scales (larger multipoles).

The existence of the Sachs-Wolfe plateau and the acoustic peaks confirms that the density perturbations are predominantly scale-invariant and adiabatic. Higher-order correlators between

---

*Recall that $x \simeq \ln(1+x)$ when $|x| \ll 1$.

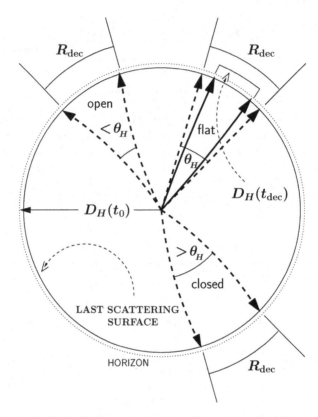

FIGURE 5.3 The scale of the horizon at the time of last scattering (decoupling) has grown to $R_{\text{dec}} = (1 + z_{\text{dec}})D_H(t_{\text{dec}})$ by the Universe expansion until today. In a flat Universe, this scale is viewed at angle $\theta_H \simeq R_{\text{dec}}/D_H(t_0)$ because the radius of the last scattering surface is very close to the radius of the horizon at present $D_H(t_0)$. This is depicted with straight dashed arrows. When space is not flat but curved the viewing angle is changed. When the geometry is spherical (closed Universe) the viewing angle is bigger than θ_H, while when the geometry is hyperbolic the viewing angle is smaller than θ_H. These two cases are depicted by curved dashed arrows. The location of the first peak is at multiple $\ell_1 \simeq \sqrt{3}\ell_H$, with $\ell_H = \pi/\theta_H$ in a flat Universe. In a curved Universe, because the viewing angle changes, so does the location of the first peak, which is moved to a smaller multipole value in a closed Universe and a larger multipole value in an open Universe, because it is inversely proportional to the viewing angle.

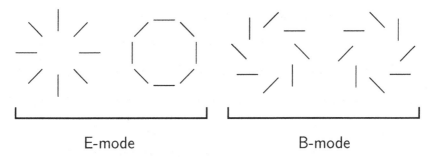

FIGURE 5.4   The configurations of the polarisation vectors for E-mode and B-mode.

the perturbations (see later) also show that they are predominantly Gaussian too, as suggested by inflation. The form of the CMB spectrum is a strong confirmation of the standard model of cosmology (Hot Big Bang + cosmic inflation). As suggested above, the morphology of the peaks is used to fit the cosmological parameters, such as H_0.

5.1.4   Polarisation

Apart from the CMB temperature perturbations, more information is encoded in the polarisation of CMB light. There are two types of such polarisation called E-mode polarisation and B-mode polarisation, which are dubbed "electric" and "magnetic," respectively. The E-mode is due to radiation being polarised when light is dragged by the baryons along the infall into an overdensity. The E-mode produces arrangements of polarisation vectors, which are oriented radially or tangentially with respect to the infalling overdensity, respecting axial symmetry, as shown in Fig. 5.4. In contrast, the B-mode, whose polarisation vectors follow "swirl" patterns as shown in Fig. 5.4, cannot be produced by the overdensities. B-mode polarisation has to do with deformations of space(time) different from the ones caused by density perturbations that are due to gravitational waves. As we discuss in the following chapter, inflation generates an almost scale-invariant spectrum of gravitational waves quantum-mechanically in the same way it gives rise to the density perturbations (through particle production). In fact, many cosmologists believe that the detection of such a primordial gravitational-wave background would be a smoking gun for the inflation theory. This is why there is an effort to detect the B-mode polarisation in the CMB, but in contrast to the E-mode, this has proven too difficult so far. At present, only an upper bound exists for the gravitational-waves contribution. It is nevertheless an important constraint that has lead to the exclusion of many classes of inflation models, which would produce too much gravitational radiation. Note that gravitational waves generate both E-mode and B-mode polarisation.

5.2   Baryon acoustic oscillations

As we have said, the collapsing overdensities are the seeds for the formation of structures in the Universe, such as galaxies and galaxy clusters. Therefore, it is expected that the state of the overdensities at decoupling determines, in principle, the characteristics of the produced structures. In particular, the maximum compression of the baryonic material, which corresponds to the first peak in the CMB spectrum, is expected to result in a corresponding feature in the distribution of matter, because there is a special scale; the scale of the sound horizon at decoupling. Sure enough, the observations suggest that the two-point correlation function between galaxies (which expresses the probability to find another galaxy at a fixed distance from a given galaxy) features a bump at the scale ~ 150 Mpc, exactly as expected. Indeed, it is easy to estimate this scale. For the comoving size of the horizon at decoupling we have

$$R_{\text{dec}} \simeq (1 + z_{\text{dec}})D_H(t_{\text{dec}}) \simeq 150 \,\text{Mpc} \,, \tag{5.22}$$

where $z_{\text{dec}} \simeq 1090$ and $D_H(t_{\text{dec}}) \sim ct_{\text{dec}} \simeq 0.14 \,\text{Mpc}$. The above is called the *acoustic scale*.

The acoustic scale is a standard yardstick, which can be used to determine the geometry of the Universe and its recent evolution. In particular, Baryon acoustic oscillations (BAO) observations provide evidence for the existence and characteristics of dark energy, which is independent of supernovae observations (see Fig. 2.10). The effect has been predicted in 1970 by Rashid A. Sunyaev and Ya. B. Zel'dovich [SZ70] and independently in the same year by P. James E. Peebles and J.T. Yu [PY70].

5.3   The curvature perturbation

We are in a position now to define the primordial curvature perturbation and relate it with the density perturbations and the perturbation in the temperature of the CMB. The primordial density perturbations are responsible for the formation of structure in the Universe. We briefly discuss this process in this section. Since they are a product of inflation and since they are strongly constrained by CMB observations, we talk about the density perturbations and the associated primordial curvature perturbation in some detail.

5.3.1   The two-point correlator

From the Friedmann equation (2.10) it is evident that spatial perturbations in the density would result in spatial perturbations in the rate of expansion. Thus, the local value of the scale factor is in fact dependent on position after all, $a(\boldsymbol{x}, t)$. To distinguish it from the homogeneous average scale factor $a(t)$, we can lump its spatial dependence in a function $\zeta(\boldsymbol{x}, t)$ which is defined as

$$a(\boldsymbol{x}, t) \equiv a(t)e^{\zeta(\boldsymbol{x}, t)}. \tag{5.23}$$

The function ζ is called the *curvature perturbation*. From the above definition, it is evident that $\zeta = \delta a/a = H\delta t = \delta N$, where we considered Eq. (4.25).

The curvature perturbation is a real function. In order to make use of it, we calculate its two-point correlator in momentum space. Using the inverse transformation to Eq. (5.14), we analyse it in Fourier components, which are complex functions given by

$$\zeta_{\mathbf{k}} \equiv \zeta(\boldsymbol{k}) = \int \zeta(\boldsymbol{x}) \, e^{i\boldsymbol{k} \cdot \boldsymbol{x}} d^3x \,, \tag{5.24}$$

where we have suppressed the time dependence. As we show in Appendix B, the two-point correlator in momentum space is

$$\langle \zeta_{\mathbf{k}} \zeta_{\mathbf{k}'} \rangle = (2\pi)^3 \delta^{(3)}(\boldsymbol{k} + \boldsymbol{k}') \frac{2\pi^2}{k^3} \mathcal{P}_\zeta(k) \,, \tag{5.25}$$

where $\delta^{(3)}(\boldsymbol{k})$ is the 3-D Dirac delta function, and we have defined the *spectrum* $\mathcal{P}_\zeta(k)$ of the curvature perturbation. We can also find the two-point spatial correlator. From Eq. (5.14), we have

$$\zeta(\boldsymbol{x}) = \frac{1}{(2\pi)^3} \int \zeta_{\mathbf{k}} e^{-i\boldsymbol{k} \cdot \boldsymbol{x}} d^3k \,. \tag{5.26}$$

Using this, in Appendix B we show that

$$\langle \zeta^2(\boldsymbol{x}) \rangle = \int_0^\infty \mathcal{P}_\zeta(k) \frac{dk}{k} \,. \tag{5.27}$$

From Eq. (5.27) we see that $\langle \zeta^2 \rangle = \int_0^\infty \mathcal{P}_\zeta (d \ln k)$. If the scale dependence of the curvature perturbation is small (i.e., ζ is approximately scale-invariant) then $\langle \zeta^2 \rangle \sim \mathcal{P}_\zeta$ in a logarithmic interval. In Appendix B it is also shown that

$$\langle |\zeta_{\mathbf{k}}|^2 \rangle = \frac{2\pi^2}{k^3} \mathcal{P}_\zeta (k) . \tag{5.28}$$

Therefore, we find $\langle \zeta^2 (\boldsymbol{x}) \rangle \sim \mathcal{P}_\zeta \sim k^3 \langle |\zeta_{\mathbf{k}}|^2 \rangle$.

5.3.2 Curvature perturbation and observables

We can now connect the curvature perturbation with the density perturbations. The formal relation is [LL09]

$$\zeta = -H \frac{\delta\rho}{\dot{\rho}} = \frac{\delta\rho}{3 (\rho + p/c^2)} , \tag{5.29}$$

where we have also used the continuity equation (2.12). The above is also valid individually for independent fluids, so that the total curvature perturbation is $\zeta = \sum_i \zeta_i$, where $\zeta_i = -H(\delta\rho/\dot{\rho})_i$.

If two or more fluids contribute to the curvature perturbation then it is possible to generate *isocurvature* perturbations, for which $\delta\rho = 0$. Recalling that $\delta\rho = \sum \delta\rho_i$, the isocurvature perturbation corresponds to density perturbations of individual fluids which exactly cancel each other. Obviously, in general, the curvature perturbation has two components: the adiabatic one, for which the density contrast is exactly the same for all constituent fluids, and the isocurvature one, for which the density contrasts of all components exactly cancel each other. Because different components are characterised by different barotropic parameters, the isocurvature perturbation corresponds to a local variation of the barotropic parameter of the Universe $w = w(\boldsymbol{x})$. Observations put a stringent bound on the primordial isocurvature perturbation, which has to be less than 1% or so. Therefore, the curvature perturbation is predominantly adiabatic.

To simplify our notation we define the *density contrast* as

$$\delta \equiv \frac{\delta\rho}{\rho}. \tag{5.30}$$

Analysing the above in Fourier components, we can write Eq. (5.15) as $\Phi_{\mathbf{k}} = \frac{3}{2} \left(\frac{aH}{k} \right)^2 \delta_{\mathbf{k}}$. It can be shown that, for superhorizon scales we have $\Phi_{\mathbf{k}}/c^2 = \frac{3(1+w)}{5+3w} \zeta_{\mathbf{k}}$. Combining these, during matter domination ($w = 0$), we obtain

$$\delta_{\mathbf{k}} = \frac{2}{5} c^2 \left(\frac{k}{aH} \right)^2 \zeta_{\mathbf{k}} . \tag{5.31}$$

This means that, at horizon crossing when $k = aH/c$ we have $(\delta_{\mathbf{k}})_H = \frac{2}{5}(\zeta_{\mathbf{k}})_H$. Integrating over space we obtain

$$\langle \delta_H^2 \rangle \simeq \frac{4}{25} \langle \zeta^2 \rangle \simeq \left(\frac{2}{5} \right)^2 \mathcal{P}_\zeta , \tag{5.32}$$

where we used $\langle \zeta^2 \rangle \sim \mathcal{P}_\zeta$. Thus, the Sachs-Wolfe result in Eq. (5.18) becomes

$$\left. \frac{1}{2} \frac{\delta\rho}{\rho} \right|_H \simeq \frac{1}{5} \sqrt{\mathcal{P}_\zeta} \simeq \left. \frac{\Delta T}{T} \right|_{\text{CMB}} \simeq 10^{-5} . \tag{5.33}$$

The spectrum of the curvature perturbation is parametrised as follows:

$$\mathcal{P}_\zeta (k) = A_s \left(\frac{k}{k_*} \right)^{n_s - 1} , \tag{5.34}$$

where A_s is a constant, k_* corresponds to the *pivot scale* and n_s is the *scalar spectral index*. The above is motivated by the original proposal of Edward Harrison (1970) and independently by Yakov Zel'dovich (1972), well before inflation was thought of, who simply parametrised the curvature perturbation as a power-law and considered further that the spectrum can be exactly scale-invariant $n_s = 1$, i.e., independent of scale, which is called the *Harrison-Zel'dovich spectrum*. However, the spectrum does not have to be a power-law, so in principle the spectral index is not constant $n_s = n_s(k)$. Then, following Eq. (5.34), the definition of the spectral index is[*]

$$n_s(k) - 1 \equiv \frac{d \ln \mathcal{P}_\zeta}{d \ln k} \,. \tag{5.35}$$

In fact, the variation of the spectral index (running) is also considered as[†]

$$n_s' \equiv \frac{d n_s}{d \ln k} \,. \tag{5.36}$$

The pivot scale deserves to be briefly discussed here. Obviously, from Eq. (5.34) $A_s = \mathcal{P}_\zeta(k_*)$ regardless of the spectral index. This is the scale at which the CMB observations constrain the parameters, which characterise the curvature perturbation, such as A_s or $n_s(k_*)$. Although it is important to remember that the values of parameters such as A_s or n_s depend on the choice of k_*, this choice is largely arbitrary. This is not a problem because the models which aim to explain the curvature perturbations and the resulting CMB temperature anisotropy, are tested by contrast to the data of their predictions on the same scale k_*, whatever this is. The latest observations have chosen the pivot scale $k_* = 0.05 \, \mathrm{Mpc}^{-1}$ because it roughly lies in the middle of the logarithmic range of the scales probed (taking the normalisation $a_0 = 1$).

5.3.3   Tensors

Perturbations in spacetime are of three kinds. First, we have the scalar perturbations, which are associated with the density perturbations and about which we have been talking all along. We also have the vector perturbations, which describe the vorticity of the fluid flow. These are decaying in time as $1/a$ so they play no role in structure formation and are therefore ignored. Finally, we have tensorial perturbations, which are associated with gravitational waves. This is a rather demanding topic, which heavily employs general relativity (there are no tensor perturbations in Newtonian theory). For our purposes it suffices to say that the analogue of Eq. (5.25) is

$$4 \langle h_+(\boldsymbol{k}) h_+(\boldsymbol{k}') \rangle = 4 \langle h_\times(\boldsymbol{k}) h_\times(\boldsymbol{k}') \rangle = (2\pi)^3 \delta^{(3)}(\boldsymbol{k} + \boldsymbol{k}') \frac{2\pi^2}{k^3} \mathcal{P}_h(k) \,, \tag{5.37}$$

where h_+ and $h_\times$ are the components of the tensor perturbation of the metric. Similarly to Eq. (5.34) we parametrise the spectrum as

$$\mathcal{P}_h(k) = A_t \left(\frac{k}{k_*} \right)^{n_t} , \tag{5.38}$$

where the tensor spectral index is defined as[‡]

$$n_t(k) \equiv \frac{d \ln \mathcal{P}_h}{d \ln k} \,. \tag{5.39}$$

---

[*]Strictly speaking, one cannot take the logarithm of a dimensionful quantity, so Eq. (5.35) should feature $d \ln(k/k_*)$ instead of just $d \ln k$. However, independence of the choice of normalisation (a normalisation value different from k_* would not make any difference) implies we can omit displaying the normalisation of k without problem, as long as we keep in the back of our mind that some normalisation must be considered such that the argument of the logarithm is dimensionless as it should be. The same is true for the normalisation of H inside the logarithm in Eq. (6.12).

[†]Also, the "running of the running" is considered by some authors, but I think it is getting ridiculous.

[‡]Note that there is no "-1" term in the exponent of Eq. (5.38).

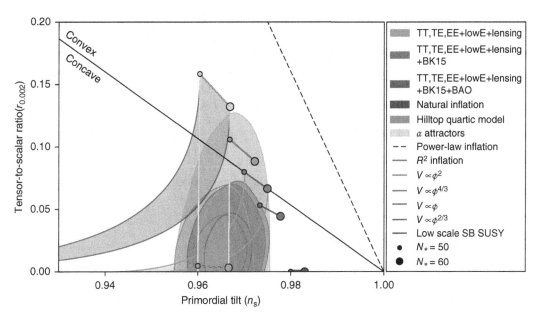

FIGURE 5.5 Likelihood contours over the values of the tensor-to-scalar ratio r and the scalar spectral index n_s, at pivot scale $k_* = 0.002 \, \mathrm{Mpc}^{-1}$. The concave contours depict the parameter space with all observational constraints taken into account. The allowed parameter space is shown at 1-σ − 68% confidence level (dark shaded region) and at 2-σ − 95% confidence level (light shaded region). The correlation of the observables does not allow large values of n_s, while negligible tensors ($r \simeq 0$) are not excluded. The lines and blobs on the diagram have to do with inflation physics. We discuss them in the following chapters. In particular, the line segments ending in blobs outside the Planck contours correspond to models of chaotic inflation discussed in Sec. 7.2.1, except the one which lies on the bottom line, with very small value of r. This corresponds to hybrid inflation in supergravity discussed in Sec. 7.2.4. One other line segment ending in blobs, which lies near the best fit point of the Planck findings (within the Planck contours) is the Starobinsky R^2 inflation, discussed in Sec. 7.3.4. The band delineated by the vertical lines linking the blobs of Starobinsky inflation with the ones of quadratic chaotic inflation corresponds to the T-model α-attractor inflation discussed in Sec. 7.3.5. The curved band which lies mostly outside the Planck contours corresponds to natural inflation, discussed in Sec. 7.2.3. A similar curved band which lies mostly inside the Planck contours corresponds to quartic hilltop inflation discussed in Sec. 7.2.2. and in exercise 2 of Chapter 7. Finally, the dashed line, which lies way off the Planck observations, corresponds to power-law inflation discussed in Sec. 8.2. Reproduced with permission from *Astronomy & Astrophysics*, © ESO; original source ESA and the Planck Collaboration [A$^+$18b].

Inflation generates an almost scale-invariant spectrum of tensor perturbations (gravitational waves), which has not been observed yet. However, CMB observations put stringent bounds on the production of gravitational waves, which severely constrain inflation model building. The associated parameter constrained by CMB observations is the tensor-to-scalar ratio, defined as

$$r \equiv \frac{\mathcal{P}_h}{\mathcal{P}_\zeta} \, . \tag{5.40}$$

It turns out that the observed values of n_s and r are related (see Fig. 5.5). The latest observations suggest [A$^+$18a]

$$n_s = 0.9649 \pm 0.0042 \quad (\text{at 1-}\sigma) \quad \text{and} \quad r_{0.002} < 0.056 \quad (\text{at 2-}\sigma) \, , \tag{5.41}$$

where the value of n_s is for negligible tensors (so $r \to 0$) and the bound on r is calculated at pivot scale $k_* = 0.002\,\mathrm{Mpc}^{-1}$, because the sensitivity to tensors is maximised there. We see that the spectral index is close to unity so the spectrum is almost scale-invariant (Harrison-Zel'dovich spectrum) but not quite. In fact, a Harrison-Zel'dovich spectrum is excluded by more than 7-σ. As we discuss later, this in exactly what one expects from inflation.

For the amplitude of the scalar spectrum we have [A$^+$18a]

$$\ln(10^{10}A_s) = 3.043 \pm 0.014 \quad (\text{at } 1\text{-}\sigma) \quad \Rightarrow \quad \sqrt{\mathcal{P}_\zeta} = (4.579 \pm 0.032) \times 10^{-5} \tag{5.42}$$

$$\Rightarrow \quad \delta_H = (1.832 \pm 0.013) \times 10^{-5}. \tag{5.43}$$

We show later that the observed amplitude of the primordial curvature perturbation reveals information about the energy scale of inflation (COBE constraint).

5.4 Structure formation

Let us follow now the fate of overdensities after horizon reentry. As we have discussed already, the overdensities, when they enter the horizon (that is the horizon grows to a larger size than the overdensities) become causally connected and collapse gravitationally through the process of gravitational instability, which amounts to accelerated growth of the overdensity. Naïvely, this accelerated growth would be exponential because it is a runaway effect; the more prominent the overdensity becomes the more intense its attraction to the surrounding matter and the faster its growth. However, the growth of overdensities is counteracted by the effects of pressure and of the expansion of the Universe, which tends to spread them out. Let us consider pressure first.

5.4.1 The Jeans length

Gravitationally collapsing baryonic matter is counteracted by pressure. If the two forces pressure and gravity are balanced then we have hydrostatic equilibrium. Stars are a good example of this. Let us estimate the critical radius r of an object with density ρ, where gravity is balanced by pressure. The pressure per unit mass is roughly $(1/\rho)\partial(\delta p)/\partial r \sim \delta p/\rho r$. The gravitational pull per unit mass is $G\delta M/r^2 \sim Gr\delta\rho$, where $\delta M \sim r^3\delta\rho$. Equating the two we find $r^2 \sim (G\rho)^{-1}\delta p/\delta\rho \sim c_s^2/G\rho$, where we considered Eq. (2.16). The formal result is the *Jeans length* given by

$$\lambda_J = \frac{2\pi c_s}{\sqrt{4\pi G\rho}}. \tag{5.44}$$

When the size of the object is bigger than λ_J, then gravity wins and the object collapses gravitationally. Conversely, when the size of the object is smaller than λ_J pressure wins and the object expands.

Consider now a collapsing overdensity, which has entered the horizon and started to collapse. Its size becomes smaller until it reaches the Jeans length, when the pressure halts the collapse and the overdensity bounces back and engages in a series of acoustic oscillations, which are gradually damped. If we consider the flat expanding Universe, the Jeans length becomes

$$\lambda_J = 2\pi\sqrt{\frac{2}{3}}\frac{c_s}{H} = \pi\sqrt{\frac{2}{3}}\sqrt{w}|1 + 3w|D_H, \tag{5.45}$$

where we employed the flat Friedmann equation $3H^2 = 8\pi G\rho$ (cf. Eq. (2.38)) and used that $c_s = \sqrt{w}c$ (cf. Eq. (2.16)) and Eq. (2.57). For matter domination $w = 0$ and λ_J is negligible so collapse occurs largely unimpeded by pressure. In the radiation era however $w = \frac{1}{3}$ and $\lambda_J = \frac{2}{3}\pi\sqrt{2}\,D_H \gtrsim D_H$, i.e., the Jeans length is larger than the horizon, which means the gravitational collapse of baryonic matter is impossible. However, even just after the onset of the matter era the baryonic density perturbations do not collapse because the density of photons is still larger than the density of baryons $\rho_B \lesssim \frac{1}{6}\rho_m$, where the density of matter includes both baryonic matter and CDM $\rho_m = \rho_B + \rho_{\text{CDM}}$. Hence at decoupling, the baryonic overdensities are still undergoing acoustic oscillations, which as we have discussed, result in the acoustic peaks in the CMB spectrum.

However, the story is different for CDM, which is always pressureless. Hence, after an overdensity enters the horizon its CDM part is collapsing unimpeded by pressure. Still, even the collapse of CDM overdensities is not free; it is counteracted by the expansion of the Universe.

5.4.2 Linear growth of overdensities

Assuming pressure is negligible, in an expanding Universe the fate of the collapsing overdensites, when they are small $\delta \ll 1$, is determined by the expression

$$\ddot{\delta} + 2H\dot{\delta} = 4\pi G\rho_m\delta = \frac{3}{2}\Omega_m H^2\delta\,, \tag{5.46}$$

where $\Omega_m = \rho_m/\rho$ and in the last equation we used the flat Friedmann equation (2.38).*

If we are in the radiation era or a dark energy dominated era, then $\Omega_m \ll 1$ because matter is not the dominant component of the Universe density. Then the equation simplifies to $\ddot{\delta} + 2H\dot{\delta} = 0$. Let us consider radiation domination first. In the radiation era $H = 1/2t$ and the solution is $\delta = \delta_i + \dot{\delta}_i t_i \ln(t/t_i)$, where the subscript "$i$" refers to an initial time, corresponding to horizon entry. We see that the growth of the overdensity is at best logarithmic and requires some initial variation $\dot{\delta}_i \neq 0$. Now consider dark-energy domination. For simplicity, consider $w = -1$ so that $H \equiv H_\Lambda =$ constant, and dark energy is taken to be vacuum density. The solution becomes $\delta = C_1 + C_2 e^{-2H_\Lambda t}$, with $C_1 = \delta_i + \dot{\delta}_i/2H_\Lambda$ and $C_2 = -(\dot{\delta}_i/H_\Lambda)e^{2H_\Lambda t_i}$. Ignoring the decay mode, we see that the overdensity remains constant (initially, it grows a little but only if $\dot{\delta}_i \neq 0$). In both cases the growth of the overdensity is at best minimal. This is why structures cannot form in the radiation era and also why dark energy impedes structure formation, so its reign cannot begin too early.

Consider now matter domination, during which $\Omega_m = 1$ and Eq. (5.46) becomes $\ddot{\delta} + 2H\dot{\delta} = \frac{3}{2}H^2\delta$. In the matter era $H = 2/3t$, which means that the solution is $\delta = D_1 t^{2/3} + D_2^{-2}$, where $D_1 = \frac{3}{5}t_i^{-2/3}(\delta_i + \dot{\delta}_i t_i)$ and $D_2 = \frac{1}{5}t_i(2\delta_i - 2\dot{\delta}_i t_i)$. We see that the growing mode is $\delta \propto t^{2/3} \propto a(t)$. In fact, even if $\dot{\delta}_i = 0$ we find $\delta \simeq \delta_i(t/t_i)^{2/3}$. The behaviour $\delta \propto a$ can be understood as follows.

A collapsing overdensity corresponds to a locally closed Universe, which is collapsing to a Big crunch. Then the Friedmann equation (2.10) suggests

$$\left.\begin{array}{l} 3H^2 = 8\pi G\rho \\ 3H^2 = 8\pi G(\rho + \delta\rho) - 3kc^2/a^2 \end{array}\right\} \Rightarrow \delta = \frac{\delta\rho}{\rho} = \frac{3kc^2}{8\pi G\rho a^2} \propto \frac{1}{\rho a^2} \propto a\,, \tag{5.47}$$

---

*Equation (5.46) remains invariant if we switch to momentum space. Considering also the effects of pressure renders the equation

$$\ddot{\delta}_{\mathbf{k}} + 2H\dot{\delta}_{\mathbf{k}} + \left(\frac{c_s^2 k^2}{a^2} - 4\pi G\rho_m\right)\delta_{\mathbf{k}} = 0\,.$$

This suggests that the mode $\delta_{\mathbf{k}}$ undergoes oscillations when its wavenumber is larger than k_J, which is determined by $(c_s k_J/a)^2 = 4\pi G\rho_m$. The oscillations of the baryon-photon plasma have frequency $c_s k/a$. The Jeans length in Eq. (5.44) is readily obtained as $\lambda_J = 2\pi/k_J$.

where we considered that on average the Universe is flat ($k = 0$) with density ρ, but the local universe is closed ($k > 0$) with density $\rho + \delta\rho$. We also used that in matter domination $\rho = \rho_m \propto a^{-3}$.

We see that the growth of overdensities is not exponential but linear $\delta \propto a$. This means that we need a lot of time to go from $\delta \sim 10^{-5}$ (cf. Eq. (5.43)) to $\delta \sim 1$, when the growth becomes nonlinear and structure is formed. Thus, at decoupling, only after which baryonic matter can collapse, the density contrast needs to be already spiked $\delta \gtrsim 10^{-3}$. This is another supportive argument for the existence of non-baryonic, pressureless dark matter. CDM overdensities begin to collapse immediately after the onset of the matter era, while baryonic matter is still locked in acoustic oscillations and cannot collapse. Only after decoupling can baryonic matter collapse, falling in the preexisting potential wells of the CDM. If it were not for CDM then structures would not have enough time to form to the extent observations suggest.

5.4.3   Turnaround and virialization

As we have seen, a collapsing overdensity behaves as a locally closed universe. As such, it reaches a maximum expansion and then it begins to collapse, breaking off from the Hubble flow of the rest of the Universe. The linear approximation breaks down when $\delta \sim 1$. The fate of the overdensity is followed through numerical simulations which suggest that the turnaround moment occurs when $\delta \simeq 4.5$.

After turnaround, the overdensity collapses. A closed Universe ends up in a Big Crunch singularity, so naïvely we would expect the collapsing overdensity to end up as a black hole. However, this is only under the crucial assumption that the overdensity is perfectly isotropic, meaning spherical. In reality, the overdensity is a nonspherical self-gravitating system, whose asphericity grows as it collapses. As a result, the collapsing baryons and dark matter particles redistribute their kinetic energy into random motions, a process which is called virialization, or violent relaxation. When an overdensity is virialized its radius has shrunk by a factor of 2 from turnaround and $\delta \simeq 180$ or so. Further collapse of CDM is not happening because its gravitational pull is kinetically balanced by the random motions of the CDM particles. This is why CDM structures correspond to large halos. However, the baryonic component of the overdensity is able to radiate away some of its kinetic energy. Dissipation of energy means that baryonic matter condenses further inside the overdensity, forming say a galaxy surrounded by a much larger CDM halo.

Apart from baryonic matter and CDM, an adiabatic overdensity is also comprised of photons and neutrinos. The photons are relativistic, so once the overdensity becomes causally connected (subhorizon) photons stream out at the speed of light. The same is true for neutrinos for most of the matter era, because they too are relativistic. The effect is called *free-streaming*. Recently however, the neutrinos may have ceased to be relativistic. Indeed, a particle species becomes non-relativistic when $mc^2 < 3k_B T$. Today $3k_B T = 2.4 \times 10^{-4}$ eV, and we have compelling evidence (neutrino oscillations) that (at least some of) the neutrino masses are not zero. The cosmological bound on neutrino masses is $\sum m_\nu c^2 < 0.12\,\mathrm{eV}$. This allows the possibility that most of the neutrinos have recently become heavy. The contribution to the density budget from massive neutrinos is given by Eq. (3.16). Thus, neutrinos today contribute to dark matter by a small amount with $\Omega_\nu(t_0) \lesssim 10^{-3}$.

5.4.4   The cosmic web

As shown in Fig 4.12, the earlier an overdensity enters the horizon the smaller its size. Thus, as time passes, larger and larger overdensities become causally connected, collapse and virialize giving rise to progressively larger and larger structures. Therefore, structure in the Universe builds up from small to large, i.e., first the galaxies then the galaxy clusters. This scenario is called

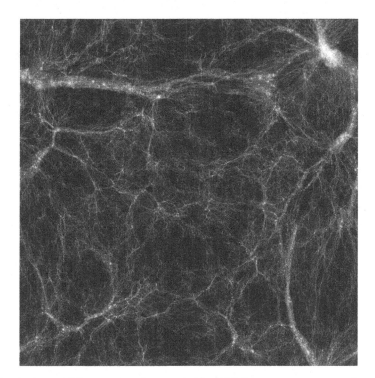

FIGURE 5.6 An image of the cosmic web generated by ESO Supernova supercomputer numerical simulation of structure formation. Reproduced with permision. © : Maureen Teyssier (Rutgers University) and Andrew Pontzen (University College London).

bottom-up. Bigger objects also form through tidal interactions and mergers of smaller objects. The largest structures are the partially-bound galaxy superclusters as even larger structures have not had the time to form. Since matter domination is ending with the advance of dark energy, structure formation will not continue further, because as shown, the overdensities remain constant during dark energy domination, if the latter is not too different from vacuum density.*

The spectrum of density perturbations is nearly scale-invariant. Thus, the density field of the Universe forms a landscape filled with mountains of comparable height and all base sizes, superimposed on each-other. The neighbouring peaks in this landscape are connected by saddlebacks, with height lower than the peaks. They source the formation of filament-like structures. Walls (pancakes) also connect the filaments giving rise to a bubbly structure of characteristic size $\sim 100\,\text{Mpc}$. This is called the *cosmic web* (see Fig. 5.6), and it corresponds to an intermediate stage before the final virialization.

---

*Of course, if dark energy were phantom, even the existing structures will be dissolved by the super-acceleration.

EXERCISE

1. Suppose that the Universe is spatially flat and it is dominated by a dark energy substance with constant barotropic parameter $-1 < w < -\frac{1}{3}$.

   (a) Investigate the growth of overdensities in the linear regime, where $\delta \ll 1$ and show that

   $$\delta(t) = \hat{C}_1 t^{\frac{3w-1}{3(1+w)}} + \hat{C}_2 \,,$$

   where $\hat{C}_1$ and $\hat{C}_2$ are integration constants.

   (b) Find the value of $\hat{C}_1$ and $\hat{C}_2$ in terms of initial conditions, considering that at some initial time t_i, the overdensity and its growth rate was δ_i and $\dot{\delta}_i$, respectively.

   (c) Express the solution in terms of initial conditions only. Comment whether overdensities grow in time.

6

The Inflationary Paradigm

6.1   Natural units.. 109

6.2   Dark energy as a scalar field........................ 111
Density and pressure • Klein-Gordon equation • The inflationary parameter ϵ • Slow-roll approximation

6.3   Reheating 116
Perturbative reheating • Solitons • Preheating • Other reheating mechanisms

6.4   The inflationary attractor 125

6.5   Particle production revisited 125
Light scalar field superhorizon perturbations • From the field perturbations to the curvature perturbation • Scalar spectral index in the inflationary paradigm • Tensors in the inflationary paradigm • Beyond the inflaton paradigm

6.6   Inflationary e-folds revisited...................... 133

We are now in a position to consider how the idea of cosmic inflation can be realised in modern theoretical physics. First we discuss a change in the unit system we have been considering so far and the reasons behind this change.

6.1   Natural units

As inflation takes place in the very early Universe, before the onset of the Hot Big Bang, the physics of inflation deals with high-energy scales, because at the onset of the Universe history the density is huge. High energy physics uses a very convenient unit system, which simplifies the equations considerably and which we adopt from now on. It is called the *natural* unit system.

First, the speed of light is taken to be unity $c \equiv 1$, meaning that all speeds appear as fractions of lightspeed. For example, speed $v = 0.3$ really means $v = 0.3\,c$. It also means that length and time have the same units, so that their ratio is dimensionless. Furthermore, in view of Einstein's famous equation $E = mc^2$, energy and mass also have the same units mirroring the fact that they are really the same thing. Similarly for energy density, which equals the density, i.e., $\epsilon = \rho$. The above convention is adopted in relativity, but natural units go beyond this.

In natural units we also take the reduced Planck constant to be unity $\hbar \equiv 1$. This removes the distinction between the classical and the quantum phenomena, because the latter are always characterised by factors of $\hbar$. For example, the condition for a (nonconformal) field undergoing particle production during inflation (an inherently quantum process) is simply $m < H$ (cf. Eq. (4.18)). Note, that the Hubble parameter H is comparable with the mass m. The latter has units of energy, which means that time has units of inverse energy because H has dimensions of inverse time (i.e., [energy]$= [H] = 1/$[time]). To make this even clearer consider the energy of a photon of frequency f: $E = hf$, which in natural units becomes $E = 2\pi f$, where we considered

$h = 2\pi\hbar$. As we know, the frequency has units of inverse time, which are now equivalent to units of energy.

Additionally, the Boltzmann constant is taken to be unity $k_B \equiv 1$, so that temperature has units of energy. Thus, the Hawking temperature of de Sitter space is now $T_H = H/2\pi$ (cf. Eq (4.27)) while the density of radiation in Eq. (3.8) is written as

$$\rho_r = \frac{\pi^2}{30} g_* T^4 \,, \tag{6.1}$$

where g_* is the number of effective relativistic degrees of freedom. Schematically, the dimensions in natural units are*

$$[\text{length}] = [\text{time}] = [\text{energy}]^{-1} \quad \text{and} \quad [\text{mass}] = [\text{temperature}] = [\text{energy}]$$

$$[\text{energy density}] = [\text{density}] = [\text{pressure}] = [\text{energy}]^4.$$

Usually, when inputting numbers, energy is expressed in GeV (giga-electron-volt), which is a billion times the energy that an electron gains when propelled by an electric potential of 1 Volt.

Using natural units also enables us to dispense with Newton's gravitational constant, using the Planck mass instead, because $M_P = 1/\sqrt{G}$ (cf. Eq. (2.35)). In this book we will consider the reduced Planck mass $m_P \equiv M_P/\sqrt{8\pi}$, such that $8\pi G = m_P^{-2}$, where

$$m_P = 2.4357 \times 10^{18} \, \text{GeV} \,.$$

In terms of the reduced Planck mass the flat Friedmann equation (2.38) becomes

$$\rho = 3m_P^2 H^2 \,. \tag{6.2}$$

Similarly, the acceleration equation (2.21) is now

$$\frac{\ddot{a}}{a} = -\frac{\rho + 3p}{6m_P^2} \,. \tag{6.3}$$

The continuity equation (2.12) is largely unmodified,

$$\dot{\rho} + 3H(\rho + p) = 0 \,, \tag{6.4}$$

while the equation of state (2.15) becomes simply

$$p = w\rho \,. \tag{6.5}$$

The vacuum density in Eq. (2.31) is now $\rho_\Lambda = m_P^2 \Lambda$, which demonstrates that the cosmological constant has dimensions of energy-squared $[\Lambda] = [\text{energy}]^2$.

Thus, using natural units means $c = \hbar = k_B = 1$ and $8\pi G = m_P^{-2}$. We could go a step further and assume that $m_P = 1$ as well, in which case all quantities become dimensionless. Such units are called *geometric* and many cosmologists use them. However, using geometric units results in the loss of the ability to check at a glance whether an expression is dimensionally correct. This is very useful in spotting mistakes, so we will stick with natural units and put up with factors of m_P appearing all over the place.

---

*Recall that [density]=[mass]/[length]^3.

6.2 Dark energy as a scalar field

To have accelerated expansion, general relativity requires that spacetime is dominated by an exotic substance with $p < -\frac{1}{3}\rho$. Such a substance is defined as dark energy and may well be a homogeneous scalar field.

A scalar field is defined as a spin-zero field, assuming a unique value (or set of values) at each point in space regardless of spatial orientation. This means that even if a scalar field is characterised by a set of many values at a given point in space (e.g., a complex scalar field corresponds to two different values, its real and its imaginary part), these values are unaffected by a rotation of the spatial coordinate system. That is not so for a vector (or a tensor) field, which is also characterised by many values at a given point in space (the field components), but these values *are* indeed modified by a rotation of the spatial coordinate system (because the values of the components of the field change accordingly).

Until recently the existence of fundamental scalar fields was only theorised, in contrast to fundamental vector fields (like the photon), which have been known to exist for some time. This changed with the discovery of the Higgs particle at CERN in 2012. As we have already discussed, a particle is a quantum of the corresponding field. For example, the photon particle is the quantum of the electromagnetic field. The fact that the Higgs particle exists means that the corresponding Higgs field, which is a fundamental scalar field in the standard model of particle physics, is also there.

The experimental confirmation that fundamental scalar fields exist makes it more credible that such a scalar field is also behind dark energy, especially in the early Universe, where the existing fundamental theories consider a multitude of scalar fields. As we discuss in Chapter 9 a scalar field can also be responsible for the observed late acceleration of the Universe expansion, while dark matter may well be a scalar particle as well (e.g., the axion).

Let us see how a scalar field can be the dark energy. For simplicity, we will consider a single component scalar field, unless explicitly stated otherwise.

6.2.1 Density and pressure

Once inflation begins, spatial inhomogeneities are diminished exponentially and soon become negligible. We say that they are "inflated away," similarly to the curvature of space. Thus, all classical quantities become homogenised.

A homogeneous scalar field can be treated as a perfect fluid with a barotropic equation of state and density and pressure given by (see Appendix C)

$$\begin{aligned} \rho_\phi &= \rho_{\text{kin}} + V \\ p_\phi &= \rho_{\text{kin}} - V \end{aligned} \quad \text{where} \quad \rho_{\text{kin}} \equiv \frac{1}{2}\dot{\phi}^2 \text{ and } V = V(\phi)\,, \tag{6.6}$$

where V is the potential density of the scalar field and ρ_{kin} is called the kinetic density, because it is due to the variation of the field, which can be understood as "motion" in field space (see Sec. 6.2.2).

The above suggests that the dimensions of the scalar field are $[\phi] = [\text{energy}]$. Equation (6.6) suggests that the barotropic parameter of a homogeneous scalar field is in principle variable and depends on the relative importance of its kinetic and potential density because

$$w_\phi = \frac{p_\phi}{\rho_\phi} = \frac{\rho_{\text{kin}} - V}{\rho_{\text{kin}} + V}\,. \tag{6.7}$$

Thus, if the potential density of the homogeneous scalar field dominates its kinetic density, then its pressure becomes negative. It is straightforward to see that the homogeneous scalar field becomes a dark energy substance with $w_\phi < -\frac{1}{3}$ when $\rho_{\text{kin}} < \frac{1}{2}V$. If $\rho_{\text{kin}} \ll V$ then we have $w_\phi \approx -1$, and the field behaves like vacuum density.

In order to have accelerated expansion, one needs dark energy to dominate the Universe content. Thus, the above suggest that

---

Inflationary paradigm: the Universe undergoes accelerated expansion when it is dominated by the potential density of a homogeneous scalar field

---

If this occurs in the early Universe, we have a period of inflation and ϕ is called the *inflaton* field.

In fact, if $\rho_{\rm kin} \ll V$ then we have $w_\phi \approx -1$ and the resulting inflation is quasi-de Sitter. One way to understand this is that, the condition $\frac{1}{2}\dot\phi^2 = \rho_{\rm kin} \ll V$ implies that in quasi-de Sitter inflation $\dot\phi$ is very small, which means that ϕ hardly varies. Consequently, $\rho_\phi \simeq V \simeq {\rm constant} \equiv \Lambda_{\rm eff} m_P^2$, meaning that the potential density of the scalar field gives rise to an effective cosmological constant $\Lambda_{\rm eff}$. This situation, however, cannot continue indefinitely because inflation has to end for the Hot Big Bang to begin, so $\Lambda_{\rm eff}$ is not a true constant of Nature. After the end of inflation, the density of the scalar field ρ_ϕ decays into the density of the Hot Big Bang. We discuss this in Sec. 6.3.

6.2.2 Klein-Gordon equation

The dynamics of the Universe during inflation are completely determined by the evolution of the inflaton field. The equation of motion of the inflaton is straightforward to obtain by inserting Eq. (6.6) into Eq. (2.12). We find

$$\ddot\phi + 3H\dot\phi + V' = 0 \,, \tag{6.8}$$

where the prime denotes derivative with respect to ϕ: ($' \equiv \frac{\partial}{\partial\phi}$). The above is called the Klein-Gordon equation, which is the field equation of a bosonic field in quantum field theory (QFT) (see Appendix C for a formal derivation). It is evident that Eq. (6.8) resembles the equation of motion of a particle with potential density V in field space (where the coordinate is ϕ) with friction given by $3H$, i.e., determined by the rate of the Universe expansion and called Hubble friction. In view of Eq. (6.2) we see that, during inflation, friction experienced by the inflaton field is greater the larger the potential density is because $\rho_\phi \simeq V$.

It helps our intuition to imagine the varying inflaton expectation value as a ball rolling down a potential landscape, e.g., a hill, in field space, subject to friction which is intensified on "high ground." For the potential density to be dominant over the kinetic density the form of the potential is either plateau like (called the *inflationary plateau*) so the rolling field has only a slight incentive to move, or the field is climbing down from a high potential value so that Hubble friction is very large and the field moves slowly (i.e., with small $\rho_{\rm kin}$) despite the steepness of the potential slope. In both cases, the slow-movement (variation) of the inflaton field is eventually terminated, either because there is an edge to the inflationary plateau (a cliff), or simply because the field rolls down enough that friction cannot restrain it efficiently. The inflaton then moves (varies) faster, its kinetic density $\rho_{\rm kin}$ increases, and its pressure ceases to be negative enough, so inflation ends. A schematic view of the form of the inflationary potential is shown in Fig. 6.1.

6.2.3 The inflationary parameter ϵ

It is very useful to determine when inflation is possible. By definition, inflationary expansion requires $\ddot a > 0$. Using that $H \equiv \dot a/a$ we readily find

$$\frac{\ddot a}{a} = \dot H + H^2 \,. \tag{6.9}$$

This motivates us to define the parameter

$$\epsilon \equiv -\frac{\dot H}{H^2} \,, \tag{6.10}$$

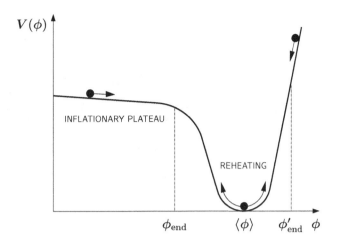

FIGURE 6.1 Schematic view of the form of a scalar potential $V(\phi)$, which can support inflation. The varying expectation value of the inflaton field ϕ is represented by a rolling ball subject to the Hubble friction. There are two possibilities. Inflation can happen along a relatively flat part of the potential called the inflationary plateau. In this case inflation is terminated when the inflaton field reaches a value ϕ_{end} where the plateau ends and the potential becomes steep and curved. The other possibility is a steep slope which corresponds to high potential density $V(\phi)$ resulting in substantial Hubble friction. The field rolls towards values with smaller $V(\phi)$ until a value ϕ'_{end} when the Hubble friction becomes too small to halt the variation of the field so the inflaton kinetic density grows large and inflation ends. In both cases, after the end of inflation the field oscillates around its vacuum expectation value $\langle\phi\rangle$, corresponding to massive inflaton particles, which decay into the thermal bath of the Hot Big Bang through the reheating process.

which implies that the condition for accelerated expansion is simply $\epsilon < 1$. Using Eqs. (6.2), (6.4), (6.5), and (6.10) it is easy to show

$$\epsilon = \frac{3}{2}(1+w),\tag{6.11}$$

which implies that $0 < \epsilon \ll 1$ guarantees quasi-de Sitter inflation with $w \approx -1$.

Equation (4.25) defines the remaining e-folds of inflation as $\exp(N) \equiv a_{\text{end}}/a$, where "end" denotes the end of inflation. From Eq. (4.24), we have $dN = -Hdt$ so the remaining e-folds of inflation are

$$N = \int_{\text{end}} Hdt = -\int_{\text{end}} \frac{d\ln H}{\epsilon},\tag{6.12}$$

where we also considered Eq. (6.10).* Eqs. (4.24) and (6.10) readily suggest that

$$\frac{dH}{H} = \epsilon dN,\tag{6.13}$$

which means that, if $0 < \epsilon \ll 1$, one needs numerous e-folds $\Delta N \sim 1/\epsilon \gg 1$ to achieve significant variation in H. This means that ϵ is a measure of the robustness of H, because H remains almost invariant when $0 < \epsilon \ll 1$.

---

*As explained before (cf. Eq. (5.35)), some normalisation of H is necessary for the argument in the logarithm of Eq. (6.12) to be dimensionless. However, because the choice of normalisation is of no importance, we choose to omit it for simplicity.

Taking the time derivative of Eq. (6.2) and inserting it in Eq. (6.4) we readily obtain

$$-2\dot{H}m_P^2 = \rho + p\,. \tag{6.14}$$

During inflation, when the Universe is dominated by a homogeneous scalar field the above equation becomes

$$-\dot{H}m_P^2 = \rho_{\rm kin}\,, \tag{6.15}$$

where $\rho_{\rm kin} = \frac{1}{2}\dot{\phi}^2$ and we considered Eq. (6.6). This finding shows that the variation of H is solely due to the kinetic density of the inflaton field. Combining Eq. (6.15) with Eqs. (6.2) and (6.10) we find

$$\epsilon = \frac{3\rho_{\rm kin}}{\rho_{\rm kin} + V}\,. \tag{6.16}$$

This suggests

$$0 < \epsilon \ll 1 \Leftrightarrow \rho_{\rm kin} \ll V\,. \tag{6.17}$$

Hence, quasi-de Sitter inflation is possible only if $\rho_{\rm kin} \ll V$. In the opposite limit $\rho_{\rm kin} \gg V$ we find $\epsilon \simeq 3$ and $w_\phi \simeq 1$ (cf. Eq. (6.7)).

So far, the expressions we derived in this section are always valid; not necessarily only in inflation when $\epsilon < 1$ but also when $\epsilon \geq 1$ or even when we consider phantom dark energy, in which case $\epsilon < 0$ and $\dot{H} > 0$ (super-inflation). Clearly though, in the latter case, dark energy cannot be a homogeneous scalar field with $\rho_{\rm kin} > 0$, meaning Eqs. (6.15) and (6.16) are not applicable in the case of phantom dark energy. Let us now focus on slow-roll quasi-de Sitter inflation and define the *slow-roll parameters* ε and η.

6.2.4 Slow-roll approximation

We are ready to discuss how we can obtain quasi-de Sitter inflation in the context of the inflationary paradigm. Equation (6.8) can be recast as

$$\ddot{\phi} + V' = \frac{1}{\dot{\phi}}\frac{d}{dt}(\rho_{\rm kin} + V)\,, \tag{6.18}$$

where $\rho_{\rm kin} \equiv \frac{1}{2}\dot{\phi}^2$. As we have said, in quasi-de Sitter inflation $\rho_{\rm kin} \ll V$ so that the term $\rho_{\rm kin}$ can be ignored inside the parenthesis of the above equation. Then, the equation suggests that the term $\ddot{\phi}$ is negligible compared with the term V'. Going back to Eq. (6.8), we see that in quasi-de Sitter inflation the inflaton equation of motion reduces to

$$3H\dot{\phi} \simeq -V'(\phi)\,. \tag{6.19}$$

This is called the *slow-roll* equation of motion because it corresponds to the condition $\rho_{\rm kin} \ll V$, which describes the case of a slowly-rolling inflaton field down its potential in field space. Under the slow-roll approximation the kinetic density of the inflaton is negligible and we can write the flat Friedmann equation (6.2) as

$$V \simeq 3m_P^2 H^2\,. \tag{6.20}$$

Because the CMB observations suggest that inflation is quasi-de Sitter, inflation model-building focuses on slow-roll inflation models. Having said that, this is not the only kind of inflation one can obtain through a scalar field and we also mention some different possibilities later on.

Using Eqs. (6.19) and (6.20), it is easy to show that in slow-roll inflation we have $\epsilon \simeq \varepsilon$, where

$$\varepsilon \equiv \frac{1}{2}m_P^2\left(\frac{V'}{V}\right)^2\,. \tag{6.21}$$

Then, in view of the above and Eq. (6.12), it is straightforward to show that

$$N = \frac{1}{m_P^2} \int_{\phi_{\text{end}}}^{\phi} \frac{V}{V'} d\phi = \frac{1}{m_P} \int_{\phi}^{\phi_{\text{end}}} \frac{|d\phi|}{\sqrt{2\varepsilon}}. \tag{6.22}$$

This implies that $dN\sqrt{2\varepsilon}\, m_P = -|d\phi|$,* which means that a small ε has the consequence that it takes many e-folds for an appreciable variation of the inflaton field, i.e., during slow-roll inflation the inflaton ϕ hardly moves (varies) so that its potential density $V(\phi)$ remains roughly constant $\rho_\phi \simeq V \simeq m_P^2 \Lambda_{\text{eff}}$ giving rise to quasi-de Sitter inflation.

Thus, we find that, in order to achieve slow-roll inflation, the inflaton's scalar potential must be such that $\varepsilon \ll 1$. This, however, is not enough. We need inflation to last long enough so that our observable Universe exits and reenters the horizon. Only then are the horizon and flatness problems of the Hot Big Bang overcome. Therefore, ε needs to stay small for an appreciable period of time, at least 60 e-folds or so.

For this reason, we also define

$$\eta \equiv m_P^2 \frac{V''}{V}. \tag{6.23}$$

Using Eqs. (6.19) and (6.20) and after a little algebra it can be shown that

$$\frac{\dot{\varepsilon}}{\varepsilon H} = -\frac{d\ln\varepsilon}{dN} = 2(2\varepsilon - \eta), \tag{6.24}$$

which demonstrates that in order to keep ε small for a large number of inflationary e-folds, we need $|\eta| \ll 1$, for otherwise the variation of ε is substantial and the condition $\varepsilon \ll 1$ cannot be maintained for long.

Using Eq. (6.20) we can show that

$$\eta \simeq \frac{1}{3} \frac{V''}{H^2}. \tag{6.25}$$

Hence,

$$|\eta| \ll 1 \Leftrightarrow |V''| \ll H^2. \tag{6.26}$$

Using the analogy of the particle in field space rolling down its potential, we can understand why V'' is, in fact, the mass-squared of the field. Indeed, consider at first that friction is negligible. Then compare the field equation (6.8) with the one of a harmonic oscillator. For example, consider the equation of motion of a pendulum, when displaced slightly off equilibrium. The potential energy V of the pendulum is $V \simeq \frac{1}{2} m^2 \phi^2$, where ϕ now measures the displacement and m is the pendulum mass. Thus, $V'' = m^2$. Then reintroduce friction by considering a counteracting force proportional to the velocity of the pendulum (as if, for example, the pendulum is immersed in water). This reproduces the friction term in the equation of motion Eq. (6.8), with $V' = m^2 \phi$.

We can also understand that $V'' = m^2$ as follows. We consider the fact that when quantising a field the energy and momentum operators (in natural units: $c = \hbar = 1$) are $E \to i\partial/\partial t$ and $\mathbf{P} \to -i\nabla$. Then we take the flat spacetime limit, when Hubble friction in Eq. (6.8) is negligible, so that the equation becomes $\ddot{\phi} = -V'$. In flat spacetime, $E^2 = P^2 + m^2$ (where $P^2 \equiv \mathbf{P} \cdot \mathbf{P}$), so that $m^2 = E^2 - P^2$. Multiplying this with ϕ and quantising the field we obtain $m^2 \phi = -\ddot{\phi} + \nabla^2\phi$. Assuming that ϕ is homogeneous implies $\nabla^2\phi \to 0$. Thus, we obtain $m^2\phi = -\ddot{\phi} = V'$. Taking the derivative of this we find $V'' = m^2$. This result corresponds to flat spacetime but remains valid in curved spacetime, when the friction term is introduced in the Klein-Gordon.

---

*Note that dN is always negative while inflation progresses because the number of remaining e-folds of inflation decreases. However, $d\phi$ can be positive or negative, depending on $V(\phi)$, while $d\phi/V' < 0$ always.

Thus, in view of Eq. (6.26), the requirement $|\eta| \ll 1$ means that the inflaton field must be light $m < H$ for inflation to be of the slow-roll type. This is not always easy to attain. For example, in supergravity and string theories scalar fields are expected to have $m \sim H$, so the flatness of the inflaton potential is spoilt unless it is protected by some symmetry. This rather serious issue (if you subscribe to supergravity and string theory) is called the η-problem [DRT95, DRT96] because $\eta \simeq \frac{1}{3}(m/H)^2 \sim 1$, cf. Eq. (6.25).

From the above discussion, we conclude that slow-roll inflation can occur only if

$$\varepsilon, |\eta| < 1 . \tag{6.27}$$

Slow-roll inflation is terminated when at least one of the slow roll parameters ε and η becomes comparable to unity (hence the η-problem). However, the condition for the end of inflation is formally $\epsilon = 1$. Since $\epsilon \simeq \varepsilon$ in slow-roll inflation we usually consider $\varepsilon \simeq 1$ as the condition, which determines the value of the inflaton when inflation ends ϕ_{end}.

6.3   Reheating

Reheating is the process which transforms the inflationary energy density into the matter and radiation of the Hot Big Bang after the end of inflation. For example, in quasi-de Sitter inflation, reheating transforms the effective vacuum density corresponding to the effective cosmological constant $\rho = \Lambda_{\mathrm{eff}} m_P^2$ into radiation. In the inflationary paradigm where inflation is due to a homogeneous scalar field, reheating occurs because the inflaton field decays into the particles of the thermal bath of the Hot Big Bang. In this section we take a closer look at this process. As we discuss in the next chapter, the process of reheating and its efficiency has an important influence on inflationary observables. It is also a rather complicated process, which can be a combination of different mechanisms, and it can be highly model dependent [STB95, KLS97].

6.3.1   Perturbative reheating

After the end of inflation the inflaton field, typically, oscillates around the bottom of its potential. This is because the end of slow-roll results in $\eta > 1 \Rightarrow m \simeq \sqrt{V''} > H$, i.e., the friction term in Eq. (6.8) is negligible. By Taylor expanding the potential around the inflaton's vacuum expectation value (VEV) $\phi_0 \equiv \langle \phi \rangle$ we find

$$V(\phi) = V(\phi_0) + V'(\phi_0)\,(\phi - \phi_0) + \frac{1}{2}V''(\phi_0)\,(\phi - \phi_0)^2 + \cdots \Rightarrow V \simeq \frac{1}{2}m^2\hat{\phi}^2 , \tag{6.28}$$

where at the minimum $V'(\phi_0) = 0$, the residual potential density is negligible* $V(\phi_0) \approx 0$, $m^2 \equiv V''(\phi_0)$ and we have defined $\hat{\phi} \equiv \phi - \phi_0$. Considering Eq. (6.8) with negligible friction we have

$$\ddot{\hat{\phi}} + m^2\hat{\phi} = 0 , \tag{6.29}$$

where we Taylor expanded V' as

$$V'(\phi) \simeq V'(\phi_0) + V''(\phi_0)\,(\phi - \phi_0) = m^2\hat{\phi} . \tag{6.30}$$

Thus, the inflaton $\hat{\phi}$ is oscillating in a quadratic potential so the solution of Eq. (6.29) suggests that its expectation value is harmonically oscillating with frequency m:

$$\hat{\phi} = \Phi \sin(mt + b) , \tag{6.31}$$

---

*Otherwise there would be a second bout of inflation, with vacuum density $V(\phi_0)$.

where Φ is the amplitude of the oscillation and b is a constant phase determined by the initial conditions. The kinetic and potential densities of the oscillating field are

$$\left. \begin{array}{l} \rho_{\text{kin}} = \frac{1}{2}m^2\Phi^2\cos^2(mt+b) \\[2mm] V = \frac{1}{2}m^2\Phi^2\sin^2(mt+b) \end{array} \right\} \Rightarrow \left\{ \begin{array}{l} \rho_\phi = \rho_{\text{kin}} + V = \frac{1}{2}m^2\Phi^2 \\[2mm] \overline{\rho_{\text{kin}}} = \overline{V} = \frac{1}{4}m^2\Phi^2 = \frac{1}{2}\rho_\phi\,, \end{array} \right. \tag{6.32}$$

where the overline denotes average per oscillation.

In the above we have neglected the effect of Hubble friction, which is negligible during a single oscillation. Over many oscillations, however, the full equation of motion is

$$\ddot{\hat{\phi}} + 3H\dot{\hat{\phi}} + m^2\hat{\phi} = 0\,. \tag{6.33}$$

Now, the oscillation is quasi-harmonic with varying amplitude, i.e.,

$$\hat{\phi} = \Phi(t)\sin(mt+b)\,. \tag{6.34}$$

This still means that, on average, the energy of the inflaton is divided equally between potential and kinetic, i.e., $\overline{\rho_{\text{kin}}} \approx \overline{V} \approx \frac{1}{2}\rho_\phi$. In view of Eq. (6.18), we have $\dot{\phi}(\ddot{\phi}+V') = \dot{\rho}_\phi$ while $3H\dot{\phi}^2 = 6H\rho_{\text{kin}}$. Thus, multiplying Eq. (6.8) with $\dot{\phi}$ we can write

$$\dot{\rho}_\phi + 3H\rho_\phi = 0 \;\Rightarrow\; \rho_\phi \propto a^{-3}\,, \tag{6.35}$$

where we considered that on average $\overline{\rho_{\text{kin}}} = \frac{1}{2}\rho_\phi$. Therefore, we see that the inflaton density dilutes as pressureless matter when oscillating around the potential minimum. Because $\rho_\phi = \frac{1}{2}m^2\Phi^2$ we find $\Phi \propto a^{-3/2}$.

The above can be understood by considering that an oscillating homogeneous scalar field behaves as a collection of massive particles (in this case, inflatons). It is in fact a condensate of scalar boson particles. Because the condensate is homogeneous (no spatial gradients) a Fourier transform would concentrate power at the zero mode with $\boldsymbol{k} = 0$. Thus, the particles have zero momentum, i.e., they are motionless (cold) so the oscillating inflaton condensate corresponds to pressureless nonrelativistic matter. Thus, Eq. (2.40) suggests $H = 2/3t$.

However, the inflaton particles are not stable but decay with decay rate Γ. This decay eventually transfers the inflation density to the radiation density of the Hot Big Bang. To model this decay, a phenomenological term is introduced in the Klein-Gordon equation (6.8), which becomes

$$\ddot{\phi} + 3H\dot{\phi} + \Gamma\dot{\phi} + V' = 0\,. \tag{6.36}$$

For $\Gamma \ll H$ the decay of the field is negligible and the Hot Big Bang cannot commence (the Universe remains dominated by the oscillating inflaton condensate). For $\Gamma > H$, the decay term dominates the friction term. Thus we expect the decay to occur when $\Gamma \sim H$. Following the above procedure, we can recast the modified Klein-Gordon equation (6.36) as

$$\dot{\rho}_\phi + (3H+\Gamma)\rho_\phi = 0\,. \tag{6.37}$$

This equation is similar to Eq. (3.12). Therefore, its solution is similar to Eq. (3.13) and reads

$$\rho_\phi \simeq \rho_\phi^{\text{end}}\left(\frac{a_{\text{end}}}{a}\right)^3 e^{-\Gamma(t-t_{\text{end}})} \quad \text{for } \Gamma = \text{constant}\,, \tag{6.38}$$

where "end" denotes the end of inflation. Since $\Gamma t \sim \Gamma/H$, we see that when $\Gamma \ll H$ the exponential factor is almost unity and the inflaton density is decreasing as matter $\rho_\phi \propto a^{-3}$. There is a small fraction of inflaton particles which do decay but the bulk of inflaton particles are effectively stable. However, when $\Gamma \gg H$ then the Hubble friction is negligible in Eq. (6.37), which suggests that the inflaton's energy decays exponentially as $\rho_\phi \propto e^{-\Gamma t}$, i.e., extremely efficiently. Hence, once $\Gamma > H$ reheating occurs rapidly by the decay of most of the inflaton particles. Thus, we can say that reheating occurs when $H_{\text{reh}} \sim \Gamma$, where 'reh' denotes reheating. Then, using the Friedmann equation (6.2) and Eq. (6.1) (and taking $g_* \lesssim 100$), the *reheating temperature* is

$$T_{\text{reh}} = \left(\frac{90}{\pi^2 g_*}\right)^{1/4} \sqrt{m_P \Gamma} \sim \sqrt{m_P \Gamma}. \tag{6.39}$$

Reheating is called *prompt* when $\Gamma \gtrsim H_{\text{end}}$ and it occurs immediately after inflation.* With prompt reheating the Hot Big Bang begins right after the end of inflation. In this case, because $\rho_\phi^{\text{end}} = \rho_r^{\text{reh}}$, we have $V_{\text{end}}^{1/4} \sim T_{\text{reh}}$. However, in general, when $\Gamma < H_{\text{end}}$, after the end of inflation but before the onset of the radiation era, the Universe is dominated by the coherently oscillating inflaton condensate, for which $\rho_\phi \propto a^{-3}$.

The reheating temperature is the temperature of radiation at the onset of the radiation era of the Hot Big Bang. However, it is not the largest temperature after the end of inflation. In fact, when the oscillating inflaton condensate is still dominant, there is a subdominant thermal bath because a small fraction of inflaton particles do manage to decay. The temperature of this thermal bath is [KT90]

$$T \sim (m_P^2 \Gamma H)^{1/4}, \tag{6.40}$$

where $H(t) > \Gamma$. During the matter-dominated period of the inflaton oscillations we have $a \propto t^{2/3}$ and $H = 2/3t \propto a^{-3/2}$. In view of Eq. (6.40), this means that $T \propto a^{-3/8}$ and the density of the subdominant radiation dilutes as $\rho_r \propto T^4 \propto a^{-3/2}$, which is much weaker than the behaviour of ρ_r during the radiation era, when $\rho_r \propto a^{-4}$. This is because before reheating is completed, radiation is not an independent fluid, the reason being that more and more radiation is created by the decay of the oscillating inflaton condensate. The maximum temperature may be as large as $T_{\text{max}} \sim (m_P^2 \Gamma H_{\text{end}})^{1/4}$ provided the decay products of the inflaton thermalise immediately after inflation.

What is the value of Γ? This depends on the decay process. If the inflaton field is directly coupled to fermions then

$$\Gamma = \frac{h^2 m}{8\pi}, \tag{6.41}$$

where $h \leq 1$ is the decay coupling. However, h may not be arbitrarily small. Indeed, for tiny h the inflaton decay occurs through gravitational couplings[†] and is given by

$$\Gamma \sim \frac{m^3}{m_P^2}, \tag{6.42}$$

so the effective range of h is

$$\frac{m}{m_P} \lesssim h \lesssim 1. \tag{6.43}$$

For inflation at the energy scale of grand unification (GUT-scale inflation), we have $V_{\text{end}}^{1/4} \sim 10^{16} \, \text{GeV}$ and $T_{\text{reh}} \sim \sqrt{m_P \Gamma} \gtrsim m\sqrt{m/m_P} \sim 10^{12} \, \text{GeV}$, where we have considered $m \sim H_{\text{end}} \sim \sqrt{V_{\text{end}}}/m_P \sim 10^{14} \, \text{GeV}$. This means that reheating occurs deep in the "particle dessert." The maximum temperature of radiation (during the oscillating condensate era) is $T_{\text{max}} \sim (m_P^2 \Gamma H_{\text{end}})^{1/4} \sim m \sim 10^{14} \, \text{GeV}$. For many successful models of inflation the numbers are slightly smaller with $H_{\text{end}} \sim 10^{-5} m_P$ so $T_{\text{reh}} \gtrsim 10^{10} \, \text{GeV}$.

---

*Note that the inflaton field cannot decay before the end of inflation, because when not oscillating it does not have a particle interpretation. Only after the inflaton oscillations begin can it decay perturbatively.

[†]Corresponding to higher-order terms in $V(\phi)$, which are suppressed by m_P (gravitationally suppressed).

This already causes tension with many supergravity theories that predict a stable gravitino particle. The gravitino is the superpartner of the graviton, which is the hypothetical gauge boson that mediates the gravitational force, in the same way that the photon mediates the electromagnetic force, in a theory where gravity is quantised like in supergravity. Stable gravitinos can be the dark matter, but if they exist, their (thermally produced) abundance would be excessive (it overcloses the Universe—like in the monopole problem) if the reheating temperature is larger than $T_{\text{reh}} \sim 10^9\,\text{GeV}$ (which can go down even as low as $T_{\text{reh}} \sim 10^6\,\text{GeV}$ in some supergravity theories). This means that either GUT-scale inflation is excluded or these supergravity theories are excluded and the gravitino (if it exists at all) is not stable. This is an example of how cosmological considerations can have a direct impact on the formulation of fundamental theory.

Before concluding the perturbative reheating section we should mention the possibility that the potential near the minimum is not approximated by a quadratic. This is possible when $V''(\phi_0) = 0$. For example, if near the minimum $V(\phi) \propto \phi^4$ (we have taken $\phi_0 = 0$, i.e., $\hat{\phi} = \phi$) then it can be shown that the density of the oscillating inflaton field scales as $\rho_\phi \propto a^{-4}$, that is like radiation (cf. Sec. 8.5). The oscillating condensate can be still thought of as a collection of inflaton particles but they have a variable mass $m(\phi) = \sqrt{V''} \propto \phi$. Because $\rho_\phi \propto \Phi^4$, where Φ is the oscillation amplitude, we have $\Phi \propto a^{-1}$ and $\bar{m} \propto \Phi \propto a^{-1}$. Thus, before reheating, the number density of the inflaton particles decreases as $n \propto a^{-3}$, with the average mass of each particle scaling as $\bar{m} \propto a^{-1}$, similarly to relativistic particles in a radiation gas.

Finally, we should mention the possibility that the perturbative decay of the inflaton is not complete, which allows a residual oscillating inflaton condensate to survive and much later play the role of dark matter. This is not too different from axion dark matter, which is one of the prime suspects for CDM (cf. Sec. 2.12.3), because axion dark matter is also a coherently oscillating scalar field condensate.

6.3.2 Solitons

Solitons are solitary field configurations comprised of self-reinforcing wave packets, which retain their shape and are stable (they do not spread or dissipate) or quasi-stable. With respect to reheating, solitons can be formed by fragmentation of the coherently oscillating condensate, if the theory allows a localised configuration which has less energy than the free field. In this case, the condensate can fragment into finite-sized blobs with fixed numbers of particles. A particular example is *Q-balls*, whose stability is due to a conserved charge (hence the Q in the name). Another example is *oscillons*, which are metastable self-sustained states, that are not protected by any conservation law (like Q-balls), but their integrity can be maintained for thousands of oscillations. Because the existence of Q-balls or oscillons is highly model dependent, we will not consider them further.

6.3.3 Preheating

Perturbative reheating corresponds to the decay of individual inflaton particles making up the oscillating inflaton condensate. However, the condensate is also acting as a coherent whole. As such, it may undergo non-perturbative, explosive decay, through parametric resonance effects. Such decay is called *preheating* [KLS97].

To illustrate this, consider the following toy-model for the inflaton decay. Suppose that one of the decay products of the inflaton is another scalar particle χ. The coupling between the two scalar fields introduces a cross-term in the scalar potential (recall that V determines the dynamics through the Klein-Gordon equation) of the form

$$\Delta V = \frac{1}{2}g^2\chi^2\phi^2 \,, \tag{6.44}$$

where $g \leq 1$ is a (perturbative) coupling. This means that, during the inflaton oscillations, the equation of motion of the inflaton Eq. (6.33) becomes

$$\ddot{\phi} + 3H\dot{\phi} + m_\phi^2 \phi = -g^2 \chi^2 \phi \,, \tag{6.45}$$

where we considered that $\phi_0 = 0$ and set $m = m_\phi$. Rearranging the above we can write

$$\ddot{\phi} + (m_\phi^2 + g^2 \chi^2)\phi \simeq 0 \,, \tag{6.46}$$

where we ignored the friction term because it is negligible over the timescale of inflaton oscillations. The decay products also satisfy a similar equation

$$\ddot{\chi} + (m_\chi^2 + g^2 \phi^2)\chi \simeq 0 \,, \tag{6.47}$$

since χ is also a scalar field, with mass m_χ. The above corresponds to a system of coupled oscillators with oscillating frequencies. Hence, they are expected to give rise to resonant effects, which result in explosive production of χ-particles. In turn, these χ-particles decay further into the Hot Big Bang thermal bath.

To follow this resonant production of χ-particles, we augment Eq. (6.47) by noting that χ is not necessarily homogeneous (cf. Eqs. (7.7) and (D.7), with $a = 1$)[*]

$$\ddot{\chi} - \nabla^2 \chi + (m_\chi^2 + g^2 \phi^2)\chi \simeq 0 \,. \tag{6.48}$$

Then, using Eq. (5.14), we obtain the corresponding equation of motion of the Fourier modes of the χ-particles[†]

$$\ddot{\chi}_k + \omega^2(t)\chi_k \simeq 0 \,, \tag{6.49}$$

where

$$\omega^2(t) = k^2 + m_\chi^2 + g^2 \Phi^2 \sin^2(m_\phi t + b) \tag{6.50}$$

and we used Eq. (6.31). With a change of variable $(m_\phi t + b = z)$, the above can be written as

$$\chi_k{''} + [A_k - 2q\cos(2z)]\chi_k = 0 \,, \tag{6.51}$$

where the apostrophe denotes derivative with respect to z and we defined

$$A_k = \frac{k^2 + m_\chi^2}{m_\phi^2} + 2q \quad \text{and} \quad q = \frac{g^2 \Phi^2}{4m_\phi^2} \,. \tag{6.52}$$

Equation (6.51) is called the Mathieu equation, whose solutions are stable or unstable depending on the position in the parameter space $\{A_k, q\}$ (see Fig. 6.2).

For a given χ-field mode, an instability leads to exponential growth of the occupation number (defined as the number density of particles of momentum k) of χ-particles

$$n_k \propto \exp(2\mu_k^{(n)}z) \,, \tag{6.53}$$

where $\mu_k^{(n)} \lesssim 0.28$ is called the Floquet index and n denotes the particular resonance band. This growth corresponds to particle production of χ-particles with momentum k.

---

[*]We have ignored the Universe expansion (which is equivalent to ignoring Hubble friction). This is because the Universe hardly expands over the timescale of oscillations.
[†]Recall that $\nabla^2 e^{i\boldsymbol{k} \cdot \boldsymbol{x}} = -k^2 e^{i\boldsymbol{k} \cdot \boldsymbol{x}}$.

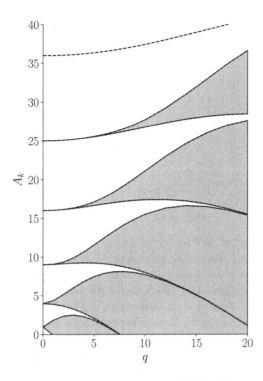

FIGURE 6.2 Instability diagram of the Mathieu equation (6.51). The instability bands are depicted in shaded colour. the diagram shows that the resonance narrows and disappears for small values $q \sim 1$ of the quality factor.

The efficiency of the resonant χ-particle production depends on the value of the quality factor q. When $q > 1$ we have broad resonance, where particle production occurs in bursts corresponding to parts of the oscillations when the *adiabaticity* condition is violated and we have

$$\frac{|\dot{\omega}|}{\omega^2} \gg 1. \tag{6.54}$$

However, as time progresses, q diminishes because the amplitude Φ decreases (cf. Eq. (6.52)). When $q < 1$ the resonance becomes narrow and disappears. Backreaction due to the produced particles undermines resonant production and contributes to shutting down the resonance.

So far we have ignored the effect of the Universe expansion. However, as with perturbative reheating, the expansion does make itself felt in the long run. Consequently, resonant particle production is affected, with particle production occasionally reversed, which results in periods of diminishing n_k. Still, there are more periods of exponential growth, so the net effect is that of exponential growth in the *stochastic resonance*.

It is important to stress here that the above parametric resonance soon becomes inefficient and stops. A fraction of the inflaton energy always survives preheating, and the inflaton field continues oscillating until it decays perturbatively. Assuming the produced χ-particles decay promptly to the relativistic particles of the thermal bath, the radiation density produced by preheating decays as $\rho_r \propto a^{-4}$ after its production. In contrast, after the end of preheating the density of the oscillating inflaton scales as $\rho_\phi \propto a^{-3}$. So, even if only a small fraction of the inflaton's energy escapes preheating, it may eventually come to dominate the Universe until its perturbative decay. Thus, preheating can only produce a significant fraction of the Hot Big Bang thermal bath if perturbative inflaton decay is efficient enough (see Fig. 6.3).

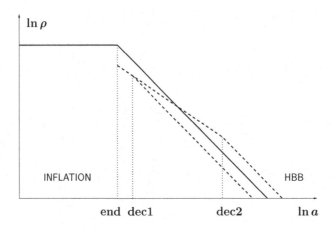

FIGURE 6.3 Logarithmic plot of the density $\rho(a)$. During inflation, ρ is kept roughly constant. Inflation ends at a moment denoted "end." Afterwards, preheating results in the decay of the bulk of the inflation density into relativistic particles whose density scales as a^{-4} (solid line). However, a small fraction of the inflation density is retained by the oscillating inflaton condensate, whose density scales as a^{-3} (dashed line). If the perturbative decay rate is large (dec1) then the oscillating condensate decays before overtaking the density of the radiation due to preheating. In this case (depicted by the lower dashed line), the decay products of perturbative reheating are negligible, and the Universe is reheated by preheating, which is prompt, with $T_{\text{reh}} \sim \rho_{\text{end}}^{1/4}$. If, on the other hand, the perturbative decay rate is small (dec2) then the oscillating condensate decays after it dominates the density of radiation due to preheating. In this case (depicted by the upper dashed line), the radiation due to preheating is negligible, and the Universe is reheated due to the perturbative decay of the inflaton condensate, with reheating temperature $T_{\text{reh}} \sim \sqrt{m_P \Gamma_{\text{dec2}}}$.

6.3.4   Other reheating mechanisms

There are other mechanisms through which the density of the inflaton field can be transferred to the newly created thermal bath of the Hot Big Bang. Here we discuss two other major proposals, called instant preheating and gravitational reheating. The crucial difference with the mechanisms already discussed (perturbative reheating and preheating) is that it is not necessary for the inflaton field to engage into oscillations around the minimum of its scalar potential. As a result the mechanisms discussed in this section are applicable to a distinct class of models of nonoscillatory inflation, with runaway potentials, which we revisit in Sec. 9.7, where we discuss quintessential inflation.

6.3.4.1 Instant preheating

We assume that along the inflaton direction in field space there is a *enhanced symmetry point* (ESP), such that another scalar field corresponding to a direction in field space other than the inflaton becomes light. The two directions are parametrised by the scalar fields ϕ (inflaton) and χ, respectively. The interaction potential at the ESP, taken at $\phi = 0$, is given by Eq. (6.44). In fact, when discussing preheating in the above, we implicitly assumed that the minimum of the inflaton potential is an ESP. But the presence of an ESP does not necessarily imply that this is the minimum of the inflaton potential.

Suppose that, while rolling in field space, the inflaton crosses through the ESP. Then there is a window, in which the adiabaticity condition is broken, which corresponds to the condition in Eq. (6.54), where $\omega^2 = g^2 \phi^2 \equiv m_\chi^2$ is the effective mass-squared of the χ-field, corresponding to the second-derivative of the interaction potential along the χ-direction $\partial^2 \Delta V / \partial \chi^2$.

Equation (6.54) suggests that adiabaticity is violated in the range

$$-\sqrt{\frac{|\dot{\phi}|}{g}} < \phi < \sqrt{\frac{|\dot{\phi}|}{g}}\,. \tag{6.55}$$

This range is called the production window because while the inflaton traverses the ESP at $\phi = 0$ we have the production of χ-particles with occupation number [FKL99]

$$n_k = \exp\left(-\frac{\pi\omega^2}{\dot{\omega}}\right) \simeq \exp\left(-\frac{\pi k^2}{g|\dot{\phi}|}\right)\,, \tag{6.56}$$

where we considered that the produced χ-particles are highly relativistic with[*] $\omega^2 \sim k^2$. Integrating over all momentum space we have

$$n_\chi = \frac{1}{(2\pi)^3}\int n_k\, d^3k = \frac{1}{2\pi^2}\int_0^\infty k^2 n_k dk \simeq \frac{(g|\dot{\phi}|)^{3/2}}{8\pi^3}\,. \tag{6.57}$$

According to Eq. (6.56), the typical momentum of the produced particles is $\sim \sqrt{g|\dot{\phi}|}$. Thus, since they are relativistic, the density of the produced particles is

$$\rho_\chi \sim \sqrt{g|\dot{\phi}|}\, n_\chi \sim (g|\dot{\phi}|)^2\,. \tag{6.58}$$

However, because the effective mass of the χ-particles is $m_\chi = g|\phi|$, we see that, as the inflaton continues its travel away from the ESP and $|\phi|$ grows, the χ-particles become heavy and nonrelativistic, with $\rho_\chi = m_\chi n_\chi$. Indeed, at the edge of the production window in Eq. (6.55) we have $|\phi| = \sqrt{|\dot{\phi}|/g}$, and we see that $\rho_\chi = g|\phi|n_\chi \sim (g|\dot{\phi}|)^2$, which is the result in Eq. (6.58) and we used Eq. (6.57). Thus, after the inflaton exits the production window the χ-particles become nonrelativistic.

The thermal bath of the Hot Big Bang is due to the decay of the χ-particles. In analogy to Eq. (6.41), the decay rate is $\Gamma_\chi = h^2 g|\phi|/8\pi$, so it increases the more the inflaton moves away from the ESP and the heavier the χ-particles become. Thus, the efficiency of the reheating depends on the kinetic density of the inflaton as it crosses the ESP $\frac{1}{2}\dot{\phi}^2$, the strength of the ESP parametrised by the coupling g and the decay rate of the produced χ-particles parametrised by the coupling h. Assuming that inflation has ended (otherwise the products of instant preheating would be diluted by the inflationary expansion) we estimate the reheating temperature as follows.

Assuming for simplicity immediate thermalisation at the moment of the decay of the χ-particles we have

$$T_{\rm reh}^4 \sim \rho_r^{\rm reh} = \rho_\chi^{\rm dec} = m_\chi n_\chi \sim \frac{\Gamma_\chi}{h^2}n_\chi \sim \frac{T_{\rm reh}^2}{h^2 m_P}n_\chi \Rightarrow T_{\rm reh} \sim \sqrt{\frac{n_\chi}{h^2 m_P}} \sim \frac{(g|\dot{\phi}|)^{3/4}}{h\sqrt{m_P}}\,, \tag{6.59}$$

where we used Eqs. (6.1) and (6.57), while considering that the decay of the χ-particles (denoted by "dec") occurred when $H_{\rm reh} \sim \Gamma_\chi \sim h^2 m_\chi$, with $H_{\rm reh} \sim \sqrt{\rho_\chi^{\rm dec}}/m_P \sim T_{\rm reh}^2/m_P$ using the Friedmann equation (6.2). The density of the thermal bath produced by instant preheating is due to a part of the kinetic density of the inflaton.

If the χ-particles also have a bare mass m_0 then Eqs. (6.56) and (6.57) become

$$n_k = \exp\left(-\frac{\pi(k^2 + m_0^2)}{g|\dot{\phi}|}\right) \Rightarrow n_\chi \simeq \frac{(g|\dot{\phi}|)^{3/2}}{8\pi^3}\exp\left(-\frac{\pi m_0}{g|\dot{\phi}|}\right)\,, \tag{6.60}$$

---

[*]Since mass is energy in natural units we have $\omega = E$. For relativistic particles $E^2 = P^2 + m^2 \simeq P^2$, where $\boldsymbol{P} = -i\nabla$ so $P^2 = -\nabla^2$. In Fourier space $\nabla^2 \to -k^2$ ($\nabla^2 e^{i\boldsymbol{k}\cdot\boldsymbol{x}} = -k^2 e^{i\boldsymbol{k}\cdot\boldsymbol{x}}$).

which shows that the production of χ-particles is exponentially suppressed if $m_0 > g|\dot{\phi}|$, where $\frac{1}{2}\dot{\phi}^2$ is the inflaton kinetic density when crossing the ESP.

If the coupling constant h is really small, the decay is delayed, and the χ-particles can become really massive. In this way, they may decay into super-heavy fermions dubbed "fat WIMPzillas." They may also give rise to primordial black holes (PBHs) if m_χ grows up to $m_\chi \sim m_P$ because then the Compton wavelength (cf. Eq. (4.17)) of the χ-particles becomes comparable to their Schwarzschild radius $m_\chi^{-1} \sim 2Gm_\chi \sim m_\chi/m_P^2$. Once formed, PBHs would dominate the Universe resulting in a matter dominated period, until they evaporate quantum-mechanically, thereby reheating the Universe.

However, the varying (rolling) inflaton field may be halted by the production of the χ-particles, the energy of which is taken from the kinetic energy of the inflaton. Indeed, once the field exits from the production window it climbs a linear effective potential $\rho_\chi = g|\phi|n_\chi$, where n_χ is constant outside the range in Eq. (6.55). Thus, the maximum displacement of the field away from the ESP is

$$\rho_\chi^{\max} = \rho_{\text{kin}} \Rightarrow g|\phi_{\max}|n_\chi = \frac{1}{2}\dot{\phi}^2 \Rightarrow |\phi_{\max}| = 4\pi^3 \frac{|\dot{\phi}|^{1/2}}{g^{5/2}}, \tag{6.61}$$

where we used Eq. (6.57). If the χ-particles do not decay until the value $|\phi_{\max}|$ is reached then the inflaton field momentarily stops rolling (all its kinetic density has been exhausted by the production of the χ-particles) and turns back towards the ESP. When ϕ crosses the production window in Eq. (6.55) again there is a second burst of particle production, which increases n_χ and makes the potential $g|\phi|n_\chi$ outside the production window steeper. This means that the value $|\phi_{\max}|$ is now smaller so the χ-particles do not decay again, the inflaton reverses direction, comes back towards the ESP and crosses the ESP producing yet again more χ-particles, which increase n_χ even further and steepen more the linear potential. Following this process, the inflaton field gets trapped at the ESP and the χ-particles become massless.

6.3.4.2 Gravitational reheating

In Sec. 4.3.4 we have described how the existence of an event horizon during inflation results in particle production of light, nonconformal fields. The density of these particles is determined by the Hawking temperature of de Sitter space $T_H = H/2\pi$ and is given by [For87b]

$$\rho_r^{\text{gr}} = q\frac{\pi^2}{30}g_*^{\text{gr}}\left(\frac{H}{2\pi}\right)^4 = \frac{qg_*^{\text{gr}}}{480\pi^2}H^4 \sim 10^{-2}H^4, \tag{6.62}$$

where we used Eq. (6.1) and introduced $q \sim 1$ because the spectrum is not exactly thermal. g_*^{gr} is the effective relativistic degrees of freedom of all of the produced particles, with "gr" denoting this gravitational production mechanism. In current fundamental theories, we expect $g_*^{\text{gr}} = \mathcal{O}(100)$, but it might be (much) larger.

The above density is an unavoidable by-product of accelerated expansion, because it gives rise to the event horizon. ρ_{gr} is negligible compared to the radiation produced by the other reheating mechanisms discussed in this chapter, so it is usually just ignored. However, in the absence of any other reheating mechanism, gravitational reheating is the last resort.

Both gravitational reheating and instant preheating do not require the decay of the inflaton field (only part of the inflaton's kinetic density is used by instant preheating), which means that the inflaton can survive after inflation has ended and possibly play the role of dark energy near the present time. Thus, the story of the inflaton may not be over once inflation ends. It may lie dormant during the Hot Big Bang in order to strike again when the time is right. This possibility cannot be realised under the usual perturbative reheating and preheating mechanisms, in which the unfortunate inflaton does decay at the end of inflation, never to be seen or heard of again.

6.4   The inflationary attractor

Solutions to the Klein-Gordon equation (6.8) can be shown to converge in time in an attractor solution, when the scalar field dominates the Universe. If we are considering inflation and the inflaton field, this solution is the inflationary attractor. We can understand this as follows.

We assume that the density is dominated by a scalar field. Then, from Eqs. (6.2) and (6.6) we find

$$3H^2 m_P^2 = \rho_\phi = \rho_{\text{kin}} + V \,. \tag{6.63}$$

Combining this with Eq. (6.15) (which is valid without assuming slow-roll), we find

$$3H^2 + \dot{H} = V/m_P^2 \,. \tag{6.64}$$

Suppose that we have a solution $\phi = \phi(t)$ of the Klein-Gordon equation (6.8). This solution would correspond to $\rho_{\text{kin}} = \frac{1}{2}\dot{\phi}^2 = \rho_{\text{kin}}(t)$ and, of course, $V = V(\phi) = V(\phi(t)) = V(t)$. In view of Eq. (6.63), we see that our solution would result in a particular function $H = H(t)$, which would be a solution of Eq. (6.64). Consider a small perturbation to this solution of the form δH such that $|\delta H| \ll H$. Putting $H \to H + \delta H$ in Eq. (6.64), we find

$$
\begin{aligned}
V/m_P^2 &= 3(H + \delta H)^2 + (\dot{H} + \delta \dot{H}) \\
&\simeq 3(H^2 + 2H\delta H) + \dot{H} + \delta \dot{H} \Rightarrow \\
\delta \dot{H} + 6H\delta H \simeq 0 \quad &\Rightarrow \quad \delta H \propto a^{-6} \,,
\end{aligned}
\tag{6.65}
$$

where we ignore higher-order terms in the perturbation and took into account that $H(t)$ is a solution of Eq. (6.64). Combining Eq. (6.63) with Eq. (2.17) and also taking Eq. (6.7) into account we find

$$3H^2 m_P^2 = \rho_\phi \propto a^{-3(1+w_\phi)} \Rightarrow H \propto a^{-\frac{3\rho_{\text{kin}}}{\rho_{\text{kin}}+V}} = a^{-\epsilon} \,, \tag{6.66}$$

where we also used Eq. (6.16). In slow-roll we have $\rho_{\text{kin}} \ll V$ and H remains almost constant. In the opposite limit $\rho_{\text{kin}} \gg V$ we see that $H \propto a^{-3}$. Thus, $H(t)$ never decreases as quickly as $\delta H \propto a^{-6}$ found above. In fact, we obtained $\delta H/H \propto a^{-3}$ at least. This means that the perturbation of our solution $H(t)$ diminishes with time, so perturbed solutions tend to merge into an attractor solution. If we consider slow-roll inflation where $H \simeq$ constant, then we find $\delta H/H \propto a^{-6} \propto e^{6H\Delta t} \propto e^{6\Delta N}$, where $a \propto e^{Ht}$ and $\Delta N = H\Delta t$ is the elapsing e-folds. Thus, we see that, in slow-roll the inflationary attractor solution is approached exponentially. Note however, that in our proof we have not assumed slow-roll inflation or even that expansion is inflationary. We only assumed that the Universe is dominated by a scalar field.

The inflationary attractor complements the no-hair theorem discussed in Sec. 4.3.5 in erasing the memory of initial conditions. Indeed, even when the evolution of the system in the beginning of inflation starts with different (perturbed) values for the degrees of freedom (the inflaton field and its variation), the inflationary attractor suggests that these differences are ironed out and the system approaches (exponentially in the case of slow-roll) a common evolution. The perturbations of the field, which are responsible for the curvature perturbation, are not memories of perturbations in the initial conditions, but are generated quantum-mechanically, along the duration of inflation. We take a look in more detail at how this is done in what follows.

6.5   Particle production revisited

6.5.1   Light scalar field superhorizon perturbations

As we have discussed in Chapter 4, inflation amplifies the quantum fluctuations of light fields and turns them into classical perturbations, of superhorizon size. They, in turn, may source perturbations in the density of the Universe, which are responsible for the formation of structures, such as galaxies and galactic clusters. We can now investigate a bit more quantitatively how this is achieved in the context of the inflationary paradigm.

A "light" field is a field whose mass m is smaller than H (i.e., its Compton wavelength $1/m$ is larger than the horizon size $\sim H^{-1}$), so that its quantum fluctuations are able to reach and exit the horizon, being caught by the superluminal expansion of space during inflation. As already discussed, this is true for any light (and non-conformal) field, but in particular it is true for the inflaton field, in slow-roll inflation. Indeed, the inflaton field in slow-roll inflation is appropriately light (because $m \sim \sqrt{V''} \ll H$ cf. Eq. (6.26)) to obtain a superhorizon spectrum of perturbations.

When each scale exits the horizon during inflation, the inflaton and any other light scalar field obtain a perturbation whose typical amplitude is given by the Hawking temperature in Eq. (4.27)

$$\delta\phi = \frac{H}{2\pi}. \tag{6.67}$$

Indeed, a heuristic way to understand this is as follows. Being light, the scalar field is effectively massless, so its energy density is really its kinetic density, which is

$$\rho_\phi = \frac{1}{2}\left(\frac{\delta\phi}{\delta t}\right)^2 \Rightarrow \Delta E = \rho_\phi \times \mathcal{V} \sim \left(\frac{\delta\phi}{\delta t}\right)^2 H^{-3} \Rightarrow \Delta E \cdot \Delta t \sim \delta\phi^2 H^{-2}, \tag{6.68}$$

where $\mathcal{V} \sim H^{-3}$ is the horizon volume, and we used $\Delta t = \delta t \sim H^{-1}$ (Hubble time). In view of the uncertainty relation in Eq. (4.16), $\Delta E \cdot \Delta t \sim 1$, the above suggests $\delta\phi \sim H$, which is in agreement with Eq. (6.67).

In reality, things are more complicated. Perturbations of the field are, in general, uneven. But, being due to the uncertainty relation, their statistics are Gaussian, reflecting the inherent random nature of quantum fluctuations. Thus, they are characterised by a bell-shaped probability distribution, which peaks at the value given by the Hawking temperature. Formally, the above are quantified as follows.

Mirroring the discussion in Sec. 5.3.1, we switch to momentum space by Fourier expanding the perturbations as

$$\delta\phi(\boldsymbol{x}) = \frac{1}{(2\pi)^3}\int \delta\phi_{\mathbf{k}}e^{-i\boldsymbol{k}\cdot\boldsymbol{x}}d^3k, \tag{6.69}$$

where $\delta\phi_{\mathbf{k}} \equiv \delta\phi(\boldsymbol{k})$. Then, in analogy with Eq. (5.25), we define the two-point correlator, as

$$\langle\delta\phi_{\mathbf{k}}\delta\phi_{\mathbf{k}'}\rangle \equiv \delta^3(\boldsymbol{k}+\boldsymbol{k}')\frac{2\pi^2}{k^3}\mathcal{P}_{\delta\phi}(k), \tag{6.70}$$

where $k \equiv |\boldsymbol{k}|$, and $\mathcal{P}_{\delta\phi}(k)$ is called the power-spectrum of the field perturbations. As we show in Appendix C, studying the particle production rigorously results in

$$\sqrt{\mathcal{P}_{\delta\phi}} = \frac{H}{2\pi}, \tag{6.71}$$

which you might think of as a typical value of the field perturbations per Hubble time.

After becoming superhorizon, the perturbations evolve classically following the Klein-Gordon equation

$$\ddot{\delta\phi} + 3H\dot{\delta\phi} + V''\delta\phi = 0. \tag{6.72}$$

The above is easy to obtain from Eq. (6.8) by considering a perturbed field: $\phi = \bar\phi + \delta\phi$, where $\bar\phi(t)$ is the homogeneous field, which also satisfies Eq. (6.8) and we Taylor expand the slope of the potential as $V'(\bar\phi + \delta\phi) \simeq V'(\bar\phi) + V''(\bar\phi)\delta\phi$.

In view of the discussion after Eq. (6.26), we may set $V'' = m^2$, and considering quasi-de Sitter inflation with $H \simeq$ constant, we find the solution as follows. We consider first that the solution has the form $\delta\phi \propto e^{rt}$ with $r =$ constant. When putting this solution in Eq. (6.72) we find that $r = -\frac{3}{2}\left[1 \pm \sqrt{1 - (2m/3H)^2}\right]H$. Thus, the complete solution of Eq. (6.72) is

$$\delta\phi = C_1 \exp\left\{-\frac{3}{2}\left[1 - \sqrt{1 + \left(\frac{2m}{3H}\right)^2}\right]Ht\right\} + C_2 \exp\left\{-\frac{3}{2}\left[1 + \sqrt{1 + \left(\frac{2m}{3H}\right)^2}\right]Ht\right\},$$

(6.73)

where C_1 and C_2 are integration constants. From the above, we see that the solution is real (nonoscillatory) only if the field is light $m < \frac{3}{2}H$. In the limit when $m \ll H$ and using that in quasi-de Sitter inflation $a \propto e^{Ht}$ (cf. Eq. (4.23)), the solution in Eq. (6.73) becomes

$$\delta\phi \simeq \hat{C}_1 a^{-\eta} + \hat{C}_2 a^{-3},$$

(6.74)

where from Eq. (6.25) we have $\eta = \frac{1}{3}(m/H)^2$ and $\hat{C}_1$ and $\hat{C}_2$ are constants proportional to C_1 and C_2, respectively, which take the normalisation of $a(t)$ into account. Thus, we see that one mode decays exponentially with time because $a \propto e^{Ht}$, while the other mode remains roughly constant, because $|\eta| \ll 1$ in slow-roll. This means that $\delta\phi \simeq$ constant while superhorizon in quasi-de Sitter inflation. We say that the superhorizon perturbations are frozen. As demonstrated in Fig. 4.12, this implies that light scalar fields obtain a superhorizon spectrum of perturbations that is *scale-invariant*. The earlier a given fluctuation exits the horizon the larger the eventual lengthscale of the corresponding perturbation because the more it is enlarged by inflation, but its amplitude is roughly the same (given by $H/2\pi$). The lengthscale should not be confused with the amplitude of the perturbation in question.

6.5.2 From the field perturbations to the curvature perturbation

Our discussion in the previous section is valid for any light scalar field. In this section, however, we concentrate on the inflaton field, which is also light in quasi-de Sitter inflation, but in contrast to other light fields, determines the dynamics of the inflationary expansion and in particular terminates inflation when it reaches a critical value ϕ_{end}. Having perturbations in the value of the inflaton field ϕ means that the field reaches ϕ_{end} at different times at different points in space. As a result, inflation will continue a bit more in some places than others. This is what produces the perturbation in the curvature of space (see Fig. 6.4).

This can be understood as follows. After inflation, according to Eq. (2.41) we have $\rho \propto t^{-2}$. This means that the time difference of the end of inflation results in the generation of a density perturbation, with contrast:

$$\frac{\delta\rho}{\rho} \sim \frac{\delta t}{t} \sim \frac{\delta\phi}{\dot{\phi}t} \sim \frac{H^2}{\dot{\phi}},$$

(6.75)

where we used Eq. (6.67) and also that, at the end of inflation and afterwards $H \sim 1/t$ according to Eq. (2.40). This is a heuristic proof, which serves to give you an idea that a perturbation in the inflaton is a perturbation in the time that inflation ends and ultimately the Hot Big Bang begins. This perturbation in time generates a perturbation in density, or equivalently curvature.

The accurate value of the amplitude of the density perturbations is given by [LL00]

$$\frac{\delta\rho}{\rho} = \frac{H^2}{5\pi\dot{\phi}} = \frac{H}{5\pi\sqrt{2\epsilon}\,m_P},$$

(6.76)

where the above scale-dependent quantities are to be evaluated when the pivot scale exits the horizon and we used Eqs. (6.10) and (6.15), which result in $\dot{\phi} = \sqrt{2\epsilon}\,Hm_P$. The superhorizon

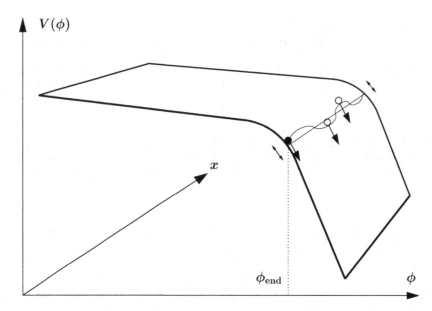

FIGURE 6.4 Plot of the inflationary potential $V(\phi)$. Inflation ends when the potential becomes steep and curved at a critical value ϕ_{end}. The figure also depicts the spatial direction x, assuming it is one-dimensional for illustrative purposes. Spatial perturbations of the inflaton field ϕ imply that, while it rolls down its potential, the inflaton reaches the value ϕ_{end} at different times at different locations (different values of x). This implies that inflation continues a bit more in some locations than in others, which results in the generation of the curvature perturbation.

curvature perturbation remains frozen. Assuming now slow-roll inflation and using Eqs. (6.19) and (6.20), the above is recast as

$$\frac{\delta\rho}{\rho} \simeq \frac{1}{5\sqrt{3}\pi} \frac{V^{3/2}}{m_P^3 |V'|}, \tag{6.77}$$

where the quantities are evaluated at horizon exit. We can relate the above with the curvature perturbation using Eq. (5.32), which suggests $\delta\rho/\rho = \frac{2}{5}\sqrt{\mathcal{P}_\zeta}$, so that we obtain

$$\sqrt{\mathcal{P}_\zeta} = \frac{H}{2\pi\sqrt{2\epsilon}\,m_P} \simeq \frac{1}{2\sqrt{3}\pi} \frac{V^{3/2}}{m_P^3 |V'|}. \tag{6.78}$$

This equation is evaluated when the pivot scale exits the horizon and the value of the left-hand-side is provided by the observations as shown in Eq. (5.42). Thus, we find that once the system assumes the slow-roll inflationary attractor, the amplitude of the curvature perturbation due to particle production of the inflaton field is determined solely by the characteristics of the inflaton scalar potential.

Employing Eq. (6.21), we can recast Eq. (6.78) as

$$V = 24\pi^2 \mathcal{P}_\zeta \varepsilon m_P^4. \tag{6.79}$$

Then, using the central value of the observations in Eq. (5.42) we obtain

$$V^{1/4} \simeq 0.0265 \times \varepsilon^{1/4} m_P. \tag{6.80}$$

The above suggests that, provided ε is not extremely small, the inflationary energy scale is comparable to the scale of grand unification (GUT scale) $V^{1/4} \sim 10^{16}$ GeV. Equation (6.80) is called the *COBE constraint*, because the amplitude of the primordial curvature perturbation (in fact, its very existence!), which determines the inflationary energy scale was first observed by the historic Cosmic Background Explorer (COBE) mission in 1989.

6.5.3 Scalar spectral index in the inflationary paradigm

As we have already discussed, the curvature perturbation is not exactly scale invariant, because H does mildly evolve during inflation so that the Hawking temperature is not exactly constant (so that the amplitude of the field perturbations depends slightly on exactly when scales exit the horizon) and also because the perturbations of the inflaton field do evolve slightly when superhorizon, since η is small but not zero in Eq. (6.74). A tilt in the spectrum of the curvature perturbation is parametrised by the spectral index n_s as discussed in Sec. 5.3.2. Since this is also an observable quantity, we can use information, obtained from the CMB for example, to further constrain the inflaton potential if we relate the spectral index to $V(\phi)$.

From Eq. (6.78) we have

$$\mathcal{P}_\zeta = \frac{1}{12\pi^2}\frac{V^3}{m_P^6(V')^2} \quad \Rightarrow \quad \ln \mathcal{P}_\zeta = \ln[V^3/(V')^2 m_P^6] - \ln(12\pi^2)$$

$$\Rightarrow \quad \frac{d\ln\mathcal{P}_\zeta}{d\phi} = 3\frac{V'}{V} - 2\frac{V''}{V'}\,. \tag{6.81}$$

We can also write

$$\frac{d\ln\mathcal{P}_\zeta}{d\phi} = \frac{1}{\dot\phi}\frac{d\ln\mathcal{P}_\zeta}{dt} = -\frac{3H^2}{V'}\frac{d\ln\mathcal{P}_\zeta}{H\,dt}\,, \tag{6.82}$$

where we used Eq. (6.19). Then, combining the above and using that $d\ln k = d\ln(aH) = H dt$,[*] the definition of the spectral index in Eq. (5.35) results in

$$n_s - 1 = \frac{d\ln\mathcal{P}_\zeta}{d\ln k} = \frac{d\ln\mathcal{P}_\zeta}{H\,dt} = -\frac{V'}{3H^2}\left(3\frac{V'}{V} - 2\frac{V''}{V'}\right)$$

$$\Rightarrow \quad n_s - 1 = 2\eta - 6\varepsilon\,, \tag{6.83}$$

where we considered Eqs. (6.20), (6.21) and (6.25). Hence, for slow roll inflation when $\varepsilon, |\eta| \ll 1$ we see that $n_s \approx 1$, i.e., the spectrum is indeed approximately scale invariant. However, the spectrum coming from inflation is expected not to be exactly scale-invariant, which is what is confirmed by observations in Eq. (5.41) as $n_s = 1$ is ruled out at 7-σ. Thus, there is clear observational support for inflation.

In a similar manner, for the running of the spectral index (defined in Eq. (5.36)) we have

$$n_s' = -16\varepsilon\eta + 24\varepsilon^2 + 2\xi\,, \tag{6.84}$$

where ξ here is a higher-order slow-roll parameter given by

$$\xi \equiv m_P^4 \frac{V'V'''}{V^2}\,. \tag{6.85}$$

Now, as evidenced from their definitions in Eqs. (6.21), (6.23), and (6.85) the slow-roll parameters ε, η and ξ are directly related to the inflaton potential, so observation of the spectral index and its running can directly constrain $V(\phi)$.

6.5.4 Tensors in the inflationary paradigm

Another by-product of inflation is the generation of gravitational waves. The latter can be represented by two scalar degrees of freedom φ_+, $\varphi_\times$ corresponding to the two possible polarisations

---

[*]At horizon exit during inflation the comoving lengthscale a/k equals the size of the horizon H^{-1}, so that we have $k = aH$.

of a plane gravitational wave. Each one of these behaves as a massless scalar field, very much like the inflaton field ϕ itself (but exactly massless). As a result, gravitational waves are also generated during inflation, due to the same reasons that perturbations of the inflaton $\delta\phi$ are generated. Therefore, their spectrum is also approximately scale invariant, with amplitude given by (c.f. Eq. (6.71))

$$\sqrt{\mathcal{P}_h} = 2\sqrt{16\pi G}\left(\frac{H}{2\pi}\right) \tag{6.86}$$

where $h_{+,\times} = \frac{1}{2}\sqrt{16\pi G}\,\varphi_{+,\times}$ are the components of the perturbation of the spacetime metric.

Using the above and Eqs. (5.39) and (6.10), the spectral index for the gravitational waves generated during inflation is

$$\mathcal{P}_h = \frac{16}{\pi}GH^2 \quad\Rightarrow\quad \ln\mathcal{P}_h = \ln(16/\pi) + \ln(GH^2)$$

$$\Rightarrow\quad n_t \equiv \frac{d\mathcal{P}_h}{d\ln k} = \frac{d\ln(GH^2)}{H\,dt} = 2\frac{\dot{H}}{H^2} = -2\epsilon. \tag{6.87}$$

Gravitational waves distort the CMB radiation similarly to the density perturbations. In fact, their effect is comparable and provides a consistency relation for single-field slow-roll inflation. Indeed, the ratio of the amplitude of the CMB temperature anisotropy due to gravity waves and density perturbations is given by

$$r \equiv \frac{\mathcal{P}_h}{\mathcal{P}_\zeta} = 16\varepsilon, \tag{6.88}$$

where we employed Eqs. (6.78) and (6.86), we used Eq. (6.20) and (6.21) and considered that $8\pi G = m_P^{-2}$ in natural units.

The tensor-to-scalar ratio r is constrained from the observations as shown in Eq. (5.41). It turns out that this can provide a bound on the excursion of the inflaton field while cosmological scales exit the horizon. From Eq. (6.21), using Eqs. (6.19) and (6.20) we find

$$\varepsilon = \frac{1}{2m_P^2}\left(\frac{\dot{\phi}}{H}\right)^2 \quad\Rightarrow\quad \left(\frac{d\phi}{dN}\right)^2 = 2\varepsilon m_P^2, \tag{6.89}$$

where we also used Eq. (4.24). Assuming that ε remains roughly constant, while cosmological scales exit the horizon (which is quite reasonable in slow-roll well before the end of inflation) we can integrate the above and find

$$\frac{\Delta\phi}{m_P} = \sqrt{2\varepsilon}\,\Delta N = \frac{1}{4}\sqrt{2r}\,\Delta N, \tag{6.90}$$

where we used $r = 16\varepsilon$, cf. Eq (6.88). and ΔN is the elapsing e-folds. The cosmological scales correspond to the observable range of multipoles in the CMB and in total amount to about $\Delta N \simeq 10$. Using the bound from observations in Eq. (5.41) $r < 0.056$ we find that, while the cosmological scales exit the horizon during inflation $\Delta\phi \lesssim m_P$, i.e., the inflaton excursion cannot be super-Planckian. The above is called the *Lyth bound*. In his original work in 1997 [Lyt97], David H. Lyth considered $\Delta N \simeq 4$, which corresponds to CMB multipoles in the region $1 < \ell < 100$, that were possible to probe at the time. As we have discussed in Sec. 5.1.3, we can in principle probe up to $\ell \sim 2000$ or so, which enlarges ΔN accordingly.

Finally, we can use the relation $\varepsilon = r/16$ in Eq. (6.80) which is rendered

$$V^{1/4} \simeq 0.0133 \times r^{1/4} m_P. \tag{6.91}$$

Then, the observational bound $r < 0.056$ suggests that the inflationary energy scale is bounded as $V^{1/4} < 1.57 \times 10^{16}$ GeV.

We should note here that the consistency relation in Eq. (6.88) is valid only in slow-roll and only if the inflaton, which drives the dynamics of inflation, is also responsible for the generation of the primordial curvature perturbation ζ. Even though this is the most minimal hypothesis, by all means it does not have to be so. In fact, several other proposals have been put forward where the observed ζ is not due to the inflaton field. We briefly discuss some of them next.

6.5.5 Beyond the inflaton paradigm

In the above, we focused on the minimal scenario, called the inflaton paradigm, where there is only one degree of freedom, namely the inflaton field, which controls all aspects of inflation, meaning it solves the flatness and horizon problems of the Hot Big Bang as well as produces the correct curvature perturbation spectrum. All these requirements may result in the need to tune the inflaton scalar potential in order to comply with observations. Such tuning is by far less than the extraordinary fine-tuning of the horizon and flatness problems (which inflation accounts for), but it is still undesirable. Inflation model-building may therefore be liberated if the task to generate the curvature perturbation was assigned to a field other than the inflaton.

This proposal may also result in producing distinct observational signatures, such as sizeable non-Gaussianity in the curvature perturbation, which minimal inflation cannot generate. Thus, the first decade of the 21st century saw an outburst of activity exploring scenarios beyond the inflaton paradigm, which would have been testable by imminent observations. Well, they were indeed tested and found wanting! The high precision observations, notably of the CMB, concluded that Nature seems to prefer to stay minimal, since all the exotic possibilities like significant non-Gaussianity signatures, were not found. This however, does not mean that mechanisms for the generation of the curvature perturbation ζ beyond the inflaton paradigm cannot be operational. It simply means that what they produce is hard to discern from the minimal case already outlined. Still, because fundamental theories upon which inflation model-building is based, contain a large multitude of degrees of freedom, scalar or otherwise, it seems natural to expect that such other fields also contribute to the generation of the curvature perturbation. However, because observations do not favour these more complicated scenarios, the principle of parsimony (Occam's razor) advocates that not much emphasis should be put to them at the moment, unless observations change their tune and require the departure from the minimal inflaton paradigm. This is why we will only briefly comment on two such possibilities, the *curvaton hypothesis* and the *modulated reheating* scenario.

In order to contribute to the curvature perturbation, a mechanism needs to affect the Universe expansion. For, as we have explained, the curvature perturbation is really a perturbation in the time when the Hot Big Bang begins. So, if an agent disturbs the Universe expansion and this agent is characterised by its own superhorizon perturbations, its disturbance would depend on location, which means that the history of the Universe will become modified by different degrees at different places. As we have seen, the simplest (but by no means the only) possibility is to consider a scalar field which is light during inflation as this agent. This scalar field is *not* the inflaton field and its behaviour does not in any way affect the dynamics or the history of inflation, including the end of inflation. This is why, such a field is usually referred to as a *spectator* field. In order to contribute to the curvature perturbation however, the once spectator field needs to affect the Universe expansion at some stage, so the field cannot remain a spectator for ever. How it does so depends on the scenario considered.

6.5.5.1 The curvaton hypothesis

This scenario is outlined in the graph in Fig. 6.5 and was originally put forward by David H. Lyth and David Wands in 2002 [LW02]. Consider a scalar field σ, which is not coupled to the inflaton but it is light during inflation, meaning its mass is $m_\sigma < H$. As we have discussed in

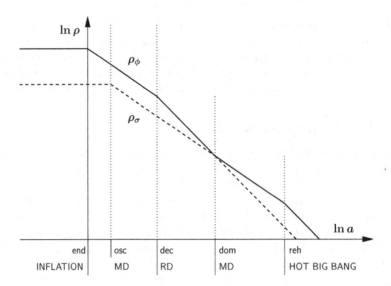

FIGURE 6.5   Log-log plot of the evolution of the densities of the inflaton ρ_ϕ and the curvaton ρ_σ and their decay products. The dominant density is depicted by the solid line. During inflation $\rho_\phi, \rho_\sigma \simeq$ constant. "end": denotes the end of inflation after which the inflaton oscillates around its VEV with $\rho_\phi \propto a^{-3}$ and $H(t)$ falls as $H = 2/3t$. At $m_\sigma \sim H$ the curvaton unfreezes and begins oscillating around its VEV, at a moment denoted by "osc." The inflaton decays into relativistic particles which scale as $\propto a^{-4}$ at the moment denoted by "dec." Thus, the curvaton's density fraction increases until the curvaton dominates the Universe at "dom." Eventually, the curvaton decays too, thereby reheating the Universe. This moment is the onset of the Hot Big Bang denoted by "reh."

Sec. 6.5.1, a light scalar field obtains a superhorizon spectrum of perturbations of typical amplitude $\delta\sigma = H/2\pi$. Such perturbations displace the field from the minimum of its potential, so it ends up having a nonzero potential density $V(\sigma)$, which however, is much smaller that the inflaton's potential density during inflation $V(\sigma) \ll V(\phi)$, for otherwise the field would affect inflationary expansion and it would not be a spectator. As we discuss in Chapter 9, a light spectator field remains frozen, with $\sigma \simeq$ constant, which means that its density is $\rho_\sigma \simeq V(\sigma) \simeq$ constant. This is so, even after the end of inflation as long as the field remains light.

After the end of inflation, the rate of expansion decreases $H(t) \propto 1/t$ [cf. Eq. (2.40)]. At some point, the field ceases to be light because now $m_\sigma > H(t)$. When this happens, the field unfreezes and begins coherent oscillations around its VEV (barring its small perturbations, the field has been homogenised by inflation) very much like the inflaton does before perturbative reheating. In general (says Taylor), the scalar potential is approximately of quadratic form near the VEV $V(\sigma) \sim m_\sigma^2 \sigma^2$, which means that the oscillations are quasi-harmonic and the density of the field is diluted as matter by the Universe expansion $\rho_\sigma \propto a^{-3}$, exactly as we discussed in Sec. 6.3.1 for the oscillating inflaton condensate [cf. Eq. (6.35)].

The density of the spectator field remains subdominant after inflation at least until it becomes heavy and its own oscillations begin. The Universe is dominated by the inflaton field, which after inflation presumably oscillates around its VEV as well. We assume now that the inflaton field decays sooner than the spectator, into relativistic degrees of freedom. Therefore, the Universe is now filled with radiation (which is a product of the inflaton decay) and the oscillating spectator field. Originally, the density of radiation is expected to be larger that that of the spectator field because the inflaton density was larger. However, radiation dilutes with the Universe expansion faster than matter, $\rho_r \propto a^{-4}$, which means that the density parameter of the spectator field increases with time $\Omega_\sigma \simeq \rho_\sigma/\rho_r \propto a(t)$, where we assumed $\rho_\sigma \ll \rho_r$. Provided the spectator field does not decay beforehand, it eventually comes to dominate the Universe.

This is the crucial moment. Because the value of the spectator field is perturbed on superhorizon scales, its potential density when frozen is similarly perturbed. As a result, when oscillating, its density is also perturbed $\delta\rho_\sigma/\rho_\sigma \neq 0$. This means that it dominates the Universe at slightly different moments in different locations and the Universe expansion changes from that of radiation dominated $a \propto t^{1/2}$ to that of matter dominated $a \propto t^{2/3}$ at slightly different times at different locations. The result of this is that the spectator field is responsible for a new contribution to the curvature perturbation, additional to the curvature perturbation due to the inflaton field, which we discussed in the previous subsection. This is why the spectator field in this case is called the *curvaton* field. It is also possible that the curvaton's contribution to ζ is the dominant one, which means that it is the curvaton which accounts for the CMB observations and the only requirement on the inflaton field is not to produce excessive perturbations. Needless to say that this massively liberates inflation model-building.

After imposing its curvature perturbation onto the Universe the curvaton is free to decay into the thermal bath of the Hot Big Bang. In principle, the curvaton could have decayed just before domination and could still account for the observed ζ (because $\mathcal{P}_\zeta \ll 1$). However, this possibility would result in sizeable non-Gaussianity which is excluded.* In contrast, the possibility that the curvaton decays after dominating the Universe, is not excluded.

6.5.5.2 Modulated reheating

Another way that a light spectator field can contribute to the curvature perturbation is by affecting the inflaton decay rate. This possibility was suggested by Gia Dvali, Andrei Gruzinov, and Matias Zaldarriaga in 2004 [DGZ04]. Unlike the curvaton hypothesis, in this case the inflaton is not uncorrelated with the spectator field. It is assumed that the coupling constant which determines the inflaton decay rate in Eq. (6.41) depends on the value of the spectator field $h = h(\sigma)$. Being light, the spectator scalar field obtains a spectrum of superhorizon perturbations as we explained in the curvaton scenario. This means that the inflaton decay rate is also perturbed on superhorizon scales $\delta\Gamma/\Gamma \neq 0$. As a result the oscillating inflaton condensate, after inflation has finished, decays into the Hot Big Bang radiation in slightly different times at different locations. Thus, the Universe switches from matter-dominated (dominated by inflaton particles) to radiation-dominated at different moments at different places, which is what gives rise to a contribution to the curvature perturbation, additional to the one due to the inflaton field. As with the curvaton mechanism, modulated reheating can be solely responsible for the observed curvature perturbations, which liberates inflation model-building.

We should note that, even though some alternative mechanism can contribute to the curvature perturbation ζ, even in a dominant way, they cannot negate it. Indeed, as we explained, the curvature perturbation amounts to a perturbation in the time when events in the history of the Universe happen. This perturbs also the onset of any alternative mechanism, which means that the effect of the inflaton's (or any other) previous contribution to ζ is carried over and cannot be erased by a subsequent mechanism for the generation of a contribution to ζ.

6.6 Inflationary e-folds revisited

As we have seen, in slow-roll inflation, inflationary observables such as A_s, n_s, and r are functions of the scalar potential, which also determines H during inflation through Eq. (6.20). The characteristics of the inflaton potential are parametrised by the slow-roll parameters ε and η. It

---

*Many variants of the story also have been put forward, for example the possibility that the curvaton does not oscillate even after domination, leading to a late inflation period (inflating curvaton).

turns out that they are both functions of the remaining number of e-folds of inflation when the cosmological scales exit the horizon, N_*. Thus, the calculation of N_* is of paramount importance, due to the delicate dependence of the inflationary observables on it. We look into this issue in this section.

The characteristics of the curvature perturbation (e.g., the amplitude A_s or the value of the scalar spectral index n_s) are computed on the *pivot* comoving scale k_*, which is somewhat larger than the comoving scale of the horizon at equality: $k_*/a_0 = 0.05\,\mathrm{Mpc}^{-1}$ for Planck 2018 so $k_*/a_0 \simeq 10^4 t_{\mathrm{eq}}$, where $t_{\mathrm{eq}} = 2 \times 10^{-3}\,\mathrm{Mpc}^{-1}$.* This means that it corresponds to the horizon scale some time after equality, in the matter dominated era. Note that the subscript '*' signifies the moment which the pivot scale left the horizon during inflation, which is N_* e-folds before the end of inflation.

The value of N_* is estimated by noting that the comoving scale of the horizon, at the horizon entry of the pivot scale, $R_k(t)$ scales like all lengthscales do: $R_k \propto a$. Therefore,

$$R_k(t_k) = \left(\frac{a_k}{a_*}\right) R_k(N_*) \Rightarrow$$

$$ct_k \simeq \left(\frac{a_{\mathrm{end}}}{a_*}\right)\left(\frac{a_{\mathrm{reh}}}{a_{\mathrm{end}}}\right)\left(\frac{a_{\mathrm{eq}}}{a_{\mathrm{reh}}}\right)\left(\frac{a_k}{a_{\mathrm{eq}}}\right) cH_*^{-1} \Rightarrow$$

$$\exp(N_*) \equiv \frac{a_{\mathrm{end}}}{a_*} \simeq H_* t_k \left(\frac{\rho_{\mathrm{reh}}}{\rho_{\mathrm{end}}}\right)^{1/3(1+w)} \frac{T_{\mathrm{eq}}}{T_{\mathrm{reh}}}\left(\frac{t_{\mathrm{eq}}}{t_k}\right)^{2/3}$$

$$\simeq t_{\mathrm{eq}}\left(\frac{t_k}{t_{\mathrm{eq}}}\right)^{1/3}\left(\frac{\pi^2 g_*}{30}\right)^{1/3(1+w)}\left(\frac{T_{\mathrm{reh}}}{V_{\mathrm{end}}^{1/4}}\right)^{4/3(1+w)}\frac{T_{\mathrm{eq}}}{T_{\mathrm{reh}}}\frac{\sqrt{V_*}}{\sqrt{3}m_P}$$

$$\simeq \frac{t_{\mathrm{eq}}T_{\mathrm{eq}}}{\sqrt{3}}\left(\frac{t_k}{t_{\mathrm{eq}}}\right)^{1/3}\left(\frac{\pi^2 g_*}{30}\right)^{1/3(1+w)}\left(\frac{T_{\mathrm{reh}}}{V_{\mathrm{end}}^{1/4}}\right)^{\frac{1-3w}{3(1+w)}}\sqrt{\frac{V_*}{V_{\mathrm{end}}}}\frac{V_{\mathrm{end}}^{1/4}}{m_P}, \quad (6.92)$$

where $a_k \equiv a(t_k)$ is the scale factor at the time t_k when the pivot scale reenters the horizon and we used that $R_k(t_k) \equiv D_H(t_k) \simeq ct_k$ by the definition of R_k and also, $R_k(N_*) \equiv D_H(N_*) \simeq cH_*^{-1}$ (cf. Eq. (2.57)) from the definition of N_* (in the second equation above we reinstated lightspeed c for clarity). In the above, the subscript 'eq' denotes the moment of matter and radiation equality, when the matter dominated era begins. We also used Eq. (6.1) (with g_* being the effective relativistic degrees of freedom at reheating), Eq. (6.20) with the approximation $\rho_{\mathrm{end}} \simeq V_{\mathrm{end}}$, $a \propto 1/T$ after reheating (cf. Eq. (3.11)), $a \propto t^{2/3}$ in the matter era of the Hot Big Bang and $\rho \propto a^{-3(1+w)}$ during the coherent inflaton oscillations (cf. Eq. (2.17)), where w is the barotropic parameter of the coherently oscillating inflaton condensate. For coherent oscillations in an approximately quadratic minimum of the scalar potential we have $w = 0$.[†]

---

*k is the wavenumber of a given scale ($k = 2\pi/\ell$, where ℓ is the comoving lengthscale).

[†]For oscillations in a quartic potential $V \propto \phi^4$ we have $w = \frac{1}{3}$ (see Sec. 8.5), which means that the dependence of N_* on T_{reh} disappears. This is because the density of a coherently oscillating scalar field condensate in a quartic potential scales on average as radiation $\propto a^{-4}$, which means that the Universe expansion is not changed when reheating occurs. Thus, in this case, the Universe behaves as if reheating is prompt and $V_{\mathrm{end}}^{1/4} \simeq T_{\mathrm{reh}}$.

Taking the time of equality as $t_{\rm eq} = 4.7 \times 10^4 \, {\rm y}$ and the temperature at equality as $T_{\rm eq} = 2.6 \, {\rm eV}$ (where $1 \, {\rm GeV}^{-1} = 6.5822 \times 10^{-25} \, {\rm sec}$ in natural units), Eq. (6.92) becomes

$$
\begin{aligned}
N_* & = 65.8 - \ln\left(\frac{k_*}{a_0 H_0}\right) + \frac{1}{3(1+w)} \ln\left(\frac{\pi^2 g_*}{30}\right) \\
& \quad - \frac{1-3w}{3(1+w)} \ln\left(\frac{V_{\rm end}^{1/4}}{T_{\rm reh}}\right) + \frac{1}{2}\ln\left(\frac{V_*}{V_{\rm end}}\right) + \ln\left(\frac{V_{\rm end}^{1/4}}{10^{16}\,{\rm GeV}}\right),
\end{aligned}
\tag{6.93}
$$

where we used that

$$
\left(\frac{t_k}{t_{\rm eq}}\right)^{1/3} \simeq \left(\frac{t_0}{t_{\rm eq}}\right)^{1/3} \frac{a_0 H_0}{k_*},
\tag{6.94}
$$

where the subscript "0" denotes the present time and we considered $t_0 = 13.8\,{\rm Gy}$. The above is obtained by considering that at the horizon entry of the pivot scale we have $k_*/a_k \simeq H(t_k) \simeq t_k^{-1}$ so that $k_*/a_0 \simeq (a_k/a_0)t_k^{-1}$, which means

$$
\frac{k_*}{a_0 H_0} \simeq \frac{a_k}{a_0}\frac{t_0}{t_k} \simeq \left(\frac{t_0}{t_k}\right)^{1/3}
\tag{6.95}
$$

and we used $t_0 \simeq H_0^{-1}.^*$ Assuming prompt reheating where $T_{\rm reh} \simeq V_{\rm end}^{1/4}$, slow-roll GUT-scale inflation where $V_* \simeq V_{\rm end}$ and $V_{\rm end}^{1/4} \simeq 10^{16}\,{\rm GeV}$ we find $N_* \simeq 60$, where we took $g_* = 106.75$ (this is the number of effective relativistic degrees of freedom at high energies in the standard model of particle physics), $H_0 = 67.8\,{\rm km/sec\,Mpc}$ and $k_*/a_0 = 0.05\,{\rm Mpc}^{-1}$ (so that $k_*/a_0 H_0 \simeq 222$). This value is a little smaller than the lower bound in Eq. (4.26) ($N_{\rm tot} \geq 62$), which corresponds to the solution of the horizon problem, the reason being that, in order to solve the horizon problem we need the scale of the observable Universe to exit the horizon during inflation. This scale is the size of the present horizon, which of course is larger than the comoving scale of the horizon when the pivot scale enters a_0/k_* ($a_0/k_* = 20\,{\rm Mpc}$ for $k_*/a_0 = 0.05\,{\rm Mpc}^{-1}$). A larger scale corresponds to a scale which exits the horizon earlier during inflation $N_{\rm tot} > N_*.^\dagger$ If we consider late reheating with $T_{\rm reh} \sim 1\,{\rm TeV}$ (such that electroweak baryogenesis is allowed during the electroweak phase transition) and we assume that, on average $w = 0$ during the coherent inflaton oscillations (i.e., the potential is approximately quadratic around the inflaton VEV), then we find $N_* \simeq 50$.

In this chapter, we have developed the arsenal needed to scrutinise inflation models and contrast them with observations. We take a look at some major proposals in the next chapter.

---

*Note that $t_0 = \frac{2}{3}H_0^{-1}$ for pure matter domination but $t_0 \simeq 2H_0^{-1}$ for vacuum dark energy, when the barotropic parameter of the Universe is $w = p_\Lambda/\rho = -\Omega_\Lambda = -0.7$.

†Recall that N counts e-folds before the end of inflation, so it decreases with time.

EXERCISES

1. (a) Reverse engineer a slow-roll inflationary scalar potential such that the scalar spectral index n_s of the generated curvature perturbation is constant, i.e., n_s does not change as the inflaton field ϕ slow-rolls down the potential.

    [Hint: A useful ansatz is $V \propto e^{\lambda\phi/m_P}$, where λ is some constant.]

   (b) Estimate the tensor-to-scalar ratio r for such a potential. Briefly discuss whether this choice would satisfy the observations.

2. The potential of *shaft* inflation is

$$V(\phi) = M^4 \phi^{2n-2}(\phi^n + m^n)^{\frac{2}{n}-2},$$

   where M and m are mass scales and $n \geq 2$ is an integer. From the above we see that the scalar potential approaches a constant for $\phi \gg m$, while for $\phi \ll m$ the potential becomes monomial, with $V \propto \phi^{2(n-1)}$.

   (a) Sketch the scalar potential, assuming a symmetry $\phi \to -\phi$ for $n = 2, 4, 8$, and 16.

   (b) Find the slow-roll parameters ε and η and the scalar spectral index n_s as functions of ϕ. By comparing ε and η show that inflation is terminated when $|\eta| \simeq 1$.

   (c) Assume that, throughout inflation, $\phi > m$ so that the potential deviates from a monomial.

    Calculate the value of the inflaton field when inflation is terminated. Then, calculate the dependence of the inflaton $\phi = \phi(N)$ on the remaining number of inflation e-folds N.

    Hence, find the scalar spectral index n_s and the tensor to scalar ratio r as functions of N.

   (d) Consider the limiting cases $n = 2$ and $n \gg 2$ and compare your results with the observations.

    Also, use the observations to establish the value of M.

   (e) As an order of magnitude, estimate the value of the running of the spectral index n_s' and check whether it is substantial.

7

Models of Inflation

7.1 Action principle 137
7.2 Archetype inflation models 139
     Chaotic inflation • Hilltop inflation • Natural inflation
     • Hybrid inflation
7.3 Modern inflation models 152
     Inflation of the Starobinsky type • Modified gravity •
     Higgs inflation • R^2 inflation • α-attractors
7.4 k-inflation 161
7.5 Inflation with heavy fields 162
7.6 Warm inflation 164

In this chapter, we discuss inflation model-building. We present some of the main slow-roll inflation models; some of the most successful ones. We begin by briefly introducing the action principle, upon which the entire edifice of modern physics is based.

7.1 Action principle

At school, the teacher used to tell us that a body on top of a hill rolls down the hill slope "to minimise its potential energy" V. Of course this is half the story. Indeed, a body at the base of the hill with a lot of kinetic energy rides up the hill slope to minimise its kinetic energy X (and increase along the way its potential energy V). The total energy $E = X + V$ remains the same,* so why does the body do anything at all? Dynamics has to do not with the sum but with the difference between the kinetic and the potential energy of a system. This difference is called *the Lagrangian* $L = X - V$. It seems that the world is keeping a delicate balance between X and V. A way to quantify this requirement is by defining *the action* of a system S, which is the integral of the Lagrangian over time, and consider that Nature prefers to keep S least varied.[†] This is called the *action principle*. Let's sketch how this requirement determines the dynamics of a system.

In terms of the Lagrangian L, the action is defined as

$$S = \int_{-\infty}^{+\infty} L(x, \dot{x}) dt, \qquad (7.1)$$

---

*Ignoring friction.

[†]In fact, the action is minimised, so its variation is zero, similar to the derivative of a function being zero at the minimum of the function.

where x denotes a spatial coordinate. The Lagrangian has dimensions of energy $[L] = E$, while in natural units time has dimensions of inverse energy $[t] = E^{-1}$. This means that the action S is dimensionless. The action principle requires that the action is least varied $\delta S = 0$. This implies

$$
\begin{aligned}
\delta S &= \int_{-\infty}^{+\infty} \left(\frac{\partial L}{\partial x} \delta x + \frac{\partial L}{\partial \dot{x}} \delta \dot{x} \right) \\
&= \left[\frac{\partial L}{\partial \dot{x}} \delta x \right]_{-\infty}^{+\infty} + \int_{-\infty}^{+\infty} \left[\frac{\partial L}{\partial x} - \frac{d}{dt} \left(\frac{\partial L}{\partial \dot{x}} \right) \right] \delta x \, dt = 0 \,,
\end{aligned}
\tag{7.2}
$$

where to get to the second line we have integrated by parts. The action principle requires that the system chooses the path which leaves its action invariant. The displacement δx considers variations of this path. The boundary term (total derivative) in the above is taken to be zero because we assume that at infinity δx vanishes (there is no variation at the end points of the path). Thus, from Eq. (7.2) we see that the action principle demands

$$
\frac{d}{dt} \left(\frac{\partial L}{\partial \dot{x}} \right) = \frac{\partial L}{\partial x} \,,
\tag{7.3}
$$

which is called the *Euler-Lagrange* equation. It produces the equation of motion that determines the dynamics of the system.

We can illustrate the above with a simple example. Consider Hook's law, of a spring, which states that the force F is proportional to the displacement x, i.e., $F = -Cx$, where $C > 0$ is a constant of proportionality. Because $F = -\partial_x V$, we have $V = \frac{1}{2}Cx^2$, where $\partial_x \equiv \partial/\partial x$. The kinetic energy is $X = \frac{1}{2}m\dot{x}^2$, where m is the mass at the end of the spring. The Lagrangian of the system is $L = X - V = \frac{1}{2}(m\dot{x}^2 - Cx^2)$. Using this we find $\partial L/\partial x = -Cx$ and $\partial L/\partial \dot{x} = m\dot{x}$. Then, Eq. (7.3) results in the equation of motion $m\ddot{x} = -Cx = F$, which is Newton's Second Law.

In four-dimensional spacetime the action principle is generalised by considering the *Lagrangian density* $\mathcal{L}$ and integrating it over all dimensions as

$$
S = \int d^4x \, \sqrt{-g} \, \mathcal{L} \,,
\tag{7.4}
$$

where $d\mathcal{V} = d^4x \, \sqrt{-g}$ is the invariant volume element, with g being the determinant of the spacetime metric (see Appendix A). Because S is dimensionless and $[x] = E^{-1}$, we see that $[\mathcal{L}] = E^4$. Then, the Euler-Lagrange equations are obtained by varying the action over a given field instead of the coordinates. As such, one obtains the equations of motion for the relevant fields. For example, the Euler-Lagrange equation takes the form

$$
\sum_{\mu} \nabla_{\mu} \left[\frac{\partial \mathcal{L}}{\partial (\nabla_{\mu} \phi)} \right] = \frac{\partial \mathcal{L}}{\partial \phi} \,,
\tag{7.5}
$$

when varying the action with respect to the scalar field ϕ, where ∇_{μ} denotes the *covariant* derivative, which also takes the curvature of spacetime into account on top of the variation of the differentiated quantity. In the above, μ parametrises all four spacetime coordinates.

The Lagrangian density of a scalar field is

$$
\mathcal{L} = \frac{1}{2}(\partial\phi)^2 - V(\phi) \,,
\tag{7.6}
$$

where the potential energy density is $V(\phi)$ and the kinetic energy density is given by $\rho_{\text{kin}} = \frac{1}{2}(\partial\phi)^2$. In flat FLRW spacetime, $(\partial\phi)^2 = \dot{\phi}^2 - a^{-2}(\nabla\phi)^2$, with $\nabla\phi$ being the comoving spatial gradient of the scalar field.* If the field is homogeneous, its spatial derivatives are zero and so $\rho_{\text{kin}} = \frac{1}{2}\dot{\phi}^2$, as in Eq. (6.6).

---

*The derivatives in $(\partial\phi)^2$ go over all the four dimensions, see Appendix D.

The equation of motion of a scalar field in a flat FLRW spacetime, as given by the Euler-Lagrange equation (7.6) is (see Appendix D, Eq. (D.7))

$$\ddot{\phi} + 3H\dot{\phi} - a^{-2}\nabla^2\phi + V'(\phi) = 0 \,, \tag{7.7}$$

where ∇^2 denotes the Laplacian operator in comoving coordinates, so that $a^{-2}\nabla^2$ is the Laplacian in physical coordinates.* Needless to say, if the field is homogeneous then $\nabla^2\phi = 0$ and the above equation reduces to the Klein-Gerdon equation (6.8). Finally, the Lagrangian density also specifies the energy density and pressure of a homogeneous scalar field in Eq. (6.6) (see Appendix D).

The theory is fully determined when the Lagrangian density is given. Thus, $\mathcal{L}$ can be thought of as a kind of Platonic ideal form, from which everything else follows. For a minimally coupled free scalar field (this is explained in Sec. 7.3.3) Eq. (7.6) suggests that this amounts to specifying $V(\phi)$. It is worth noting that a given theory (for example a ToE, such as string theory) is characterised by a single Lagrangian density, which contains all the degrees of freedom (fields). When dealing with a scalar field, we ignore the rest of the "master" Lagrangian density assuming that there is no relevant dynamics due to it, i.e., either the fields other than our scalar field are all at fixed expectation values (which may or may not be their VEVs) or their dynamics are negligible, as with spectator fields, for example. Therefore, inflation model-building in most cases is all about the form and origin of $V(\phi)$. As we have seen, observations put tight constraints on the scalar potential. We show now how these constraints are realised in the context of specific models.

7.2 Archetype inflation models

It is frequently said that inflation is a theory in search of a model. This does not mean that there are no models around. On the contrary, the abundance of models is so large that inflation is criticised for having no predictability. However, every specific model is definitely predictive and can be tested by contrast with the observations. In fact, the dramatic increase in the precision of the latest observations has resulted in ruling out entire classes of inflation models. Moreover, it is really unfair to talk about gazillions of inflationary models because most are just variants of a handful of basic models, which I call "archetype" inflation models. In this section, we briefly review the main ones.

7.2.1 Chaotic inflation

Probably the simplest inflation model is *chaotic* inflation with $V \propto \phi^q$, where q is some positive power. If q is an integer, this family of models is called *monomial* inflation. The name "chaotic" is due to the initial conditions, which may be random, as we discuss in Sec. 8.9.1. Chaotic inflation was introduced by Andrei D. Linde in 1983 [Lin83]. In chaotic inflation, the potential V is steep but the density scale of inflation is very high so that large Hubble friction (because $H^2 = \rho/3m_P^2$ from the Friedmann equation (6.2)) ensures slow-roll (see Fig. 6.1).

The potential can be written as

$$V(\phi) = V_0 \left(\frac{\phi}{m_P}\right)^q \,, \tag{7.8}$$

---

*Derivatives with respect to physical coordinates are $\partial/\partial r = a^{-1}\partial/\partial x$, where $x = r/a$ are comoving coordinates cf. Eq. (2.5).

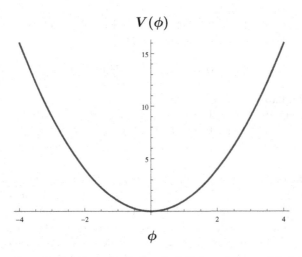

FIGURE 7.1   The scalar potential in quadratic chaotic inflation. Axes are labeled in fiducial units.

where V_0 is some constant density scale. From Eqs. (6.21) and (6.23) it is straightforward to find the slow-roll parameters

$$\varepsilon = \frac{q^2}{2}\left(\frac{m_P}{\phi}\right)^2 \quad \text{and} \quad \eta = q(q-1)\left(\frac{m_P}{\phi}\right)^2 . \tag{7.9}$$

From the above we see that slow-roll is possible only when $\phi \gtrsim m_P$. This is why chaotic inflation belongs to *large-field* inflation models. Using Eq. (6.22), we readily obtain $\phi^2 = \phi_{\text{end}}^2 + 2qm_P^2N$. The value of the inflaton at the end of slow-roll inflation is obtained by $\varepsilon(\phi_{\text{end}}) = 1$, which suggests $\phi_{\text{end}} = (q/\sqrt{2})m_P$. Using this, we express the expectation value of the inflaton field as a function of remaining e-folds of inflation as

$$\phi(N) = \sqrt{2q}\, m_P \sqrt{N + \frac{q}{4}} . \tag{7.10}$$

Combining this with Eqs. (6.83) and (7.9) we get

$$n_s = 1 - \frac{q+2}{2\left(N + \frac{q}{4}\right)} . \tag{7.11}$$

Similarly, combining Eqs. (6.88) and (7.9) we obtain

$$r = \frac{4q}{N + \frac{q}{4}} . \tag{7.12}$$

Finally, using Eqs. (6.78), (7.8), and (7.10) we find

$$\frac{V_0}{m_P^4} = 12\pi^2 q^2 \mathcal{P}_\zeta \left[2q\left(N + \frac{q}{4}\right)\right]^{-\frac{q+2}{2}} . \tag{7.13}$$

Historically, the most celebrated models would consider $q = 2$ or $q = 4$. For $q = 4$ (quartic chaotic inflation), we have $V = \frac{1}{4}\lambda\phi^4$, where $\lambda = 4V_0/m_P^4$. This could correspond to a free and massless scalar field. Employing Eq. (7.13) we find that $\lambda = \frac{6\pi^2 \mathcal{P}_\zeta}{(N+1)^3}$. Evaluating this at horizon crossing of the pivot scale, we have $\sqrt{\mathcal{P}_\zeta} \simeq 4.579 \times 10^{-5}$ (cf. Eq. (5.42)) and taking $N_* \simeq 50 - 60$ we find $\lambda \sim 10^{-12}$. The natural expectation for the value of a dimensionless coupling is of order unity. Therefore, the required value of λ is fine-tuned but still much better than the level of fine-tuning of the horizon and flatness problems which inflation tackles. For the spectral index,

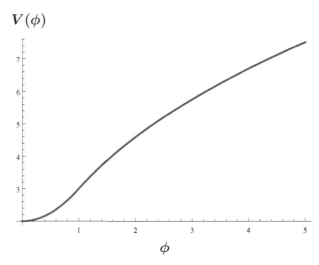

FIGURE 7.2   The scalar potential in chaotic monodromy inflation with $V \propto \phi^{2/5}$, when $\phi \gg 0$. Near the origin the potential becomes quadratic $V \propto \phi^2$. Axes are labeled in fiducial units.

we find from Eq. (7.11) $n_s = 1 - \frac{3}{N+1}$. Taking $N_* \simeq 60$ we get $n_s \simeq 0.951$, which is clearly observationally excluded (cf. Eq. (5.41)). In this model, late reheating would not affect the value of N_* because the effective barotropic parameter of an oscillating scalar field in a quartic potential is that of radiation (see Sec. 8.5). In any case, N_* would need to grow to satisfy the observations. With $n_s \simeq 0.965$, we find $N_* \simeq 85$ which is impossible (we obtain an upper bound on N_* in Sec. 9.7.2).

Similarly, for $q = 2$ (quadratic chaotic inflation) we have $V = \frac{1}{2}m^2\phi^2$ with $m = \sqrt{2V_0}/m_P$. The potential is shown in Fig. 7.1. This corresponds to a free massive field, with mass m. This time, Eq. (7.13) suggests $m/m_P = \frac{\sqrt{6}\pi\sqrt{\mathcal{P}_\zeta}}{N+\frac{1}{2}}$. Evaluating again when the cosmological scales exit the horizon (and taking $N_* \simeq 50 - 60$) we obtain $m \sim 10^{13}$ GeV, which is a reasonable intermediate scale. Equation (7.11) suggests that the spectral index is $n_s = 1 - \frac{4}{2N+1}$. For $N_* = 60(50)$ this gives $n_s \simeq 0.967(0.960)$, which is spot on (cf. Eq. (5.41)). However, things are not as great when one calculates the tensor-to-scalar ratio. Indeed, Eq. (7.11) gives $r = \frac{16}{2N+1}$. For $N_* = 60(50)$ this gives $r = 0.13(0.16)$, which fails badly in comparison to the observations in Eq. (5.41), which provide the bound $r < 0.056$. Thus, quadratic chaotic inflation is excluded.

Because the tensor bound seems to be the toughest constraint for chaotic inflation we use it to find which models are still allowed. Combining Eqs. (7.11) and (7.12), we find the relation

$$r = \frac{8q}{q+2}(1-n_s).\qquad(7.14)$$

Using the above and setting $n_s \simeq 0.965$, the bound $r < 0.056$ results in the requirement $q < 1/2$. Fractional powers like this bound are found to have some theoretical justification. One particular example is *axion monodromy* inflation [SW08].* In this theoretical framework, the choice $q = 2/5$ is possible (see Fig. 7.2). With this value, Eqs. (7.11) and (7.12) suggest $n_s = 1 - \frac{12}{10N+1}$ and

---

*Acually, in axion monodromy inflation a sinusoidal potential is superimposed on the power-law potential, so that the total potential is of the form:

$$V = V_0|\phi/m_P|^q + \hat{V}_0[\cos(\phi/f) - 1],$$

where $0 < q \leq 1$ and f is called the axion decay constant. This potential is a combination of the chaotic and the natural inflation models; see Sec. 7.2.3. However, observations demand that $\hat{V}_0 \ll V_0$, which implies that the sinusoidal axionic term is negligible and the potential is basically of monomial form.

$r = \frac{16}{10N+1}$, respectively. Demanding $n_s \simeq 0.965$ we obtain $N_* \simeq 34$ and $r \simeq 0.047$, which is on the verge of being observed/excluded. The value of N_* is rather small but this can be accounted for if one envisages a subsequent stage of thermal inflation, as we discuss in Sec. 8.4. The models with fractional powers discussed above approximate quadratic models after the end of inflation so reheating proceeds normally. Predictions of some chaotic models for $N_* = 60(50)$ are depicted in Fig. 5.5 for $q = 2, 4/3, 1, 2/3$ (see legend). For low values of q, we see that n_s is not red enough, but this can be remedied if the value of N_* is reduced, which in effect moves the predicted points in the graph towards a redder spectrum (to the left of the graph).

Quadratic and quartic chaotic inflation are examples of models which were originally deemed very successful to the point that, even though excluded, they are still used as benchmark models when discussing other early Universe research, under the assumption that the actual inflation model would not deviate from these that much. However, as we continue our tour of inflation models it becomes apparent that this assumption is not justified, because inflation can be very different from chaotic. Still, it is a demonstration of the power of the recent precision observations that such popular models are now excluded.

7.2.2   Hilltop inflation

Chaotic inflation is a large-field model because, as we have seen, during inflation $\phi > \phi_{\rm end} \gtrsim m_P$. The vacuum expectation value (VEV), however, is $\langle \phi \rangle = 0$, which means that the translation of the inflaton is more than m_P in field space. It turns out that the background theory cannot be trusted over such variations of the field. Why is this?

You may recall that the Planck mass corresponds to the energy scale of gravity because $m_P^{-2} = 8\pi G$. Now, all matter/energy is minimally coupled gravitationally to all other matter/energy through the factor of $\sqrt{-g}$ in the action in Eq. (7.4). This factor multiplies all terms in the Lagrangian density $\mathcal{L}$, which includes all the dynamical degrees of freedom (fields) of the theory. This is why matter/energy attracts (or repels, in the case of dark energy) other matter/energy.[*] This implies that there are couplings of the inflaton which are suppressed by m_P. If we Taylor expand $V(\phi)$ around a particular value (e.g., its VEV), $V(\phi)$ reduces to a series of monomial terms proportional to the factor $\sim (\Delta\phi/m_P)^q$, where q is an integer and for the field variation we have $\Delta\phi \ll m_P$ for the expansion to be valid and the terms of the form $\delta\mathcal{L} \sim (\Delta\phi)^{4+q}/m_P^q$ to be negligible for large values of q. Thus, we would be entitled to consider a *perturbative* (meaning Taylor expanded) potential of the form

$$V(\phi) \sim m^2\phi^2 + \lambda\phi^4 \,, \tag{7.15}$$

where, without loss of generality, we have set the origin in field space at the VEV (the minimum of V), so that the odd powers in the above expansion are missing. Higher orders in the above series are suppressed by factors of m_P, e.g., $\sim \phi^6/m_P^2$ or $\sim \phi^8/m_P^4$. The monomial potential could correspond to the possibility that one of the terms in the series is dominant. However, if $\Delta\phi > m_P$ then the Taylor expansion does not hold true as all the factors proportional to $(\Delta\phi/m_P)^q$ blow up. Unable to Taylor expand, we have to consider the true form of the scalar potential, which is unlikely to be polynomial. The motivation behind the monomial potential is thus lost if $\phi \gg m_P$, when the VEV is taken at zero.

This is why a different type of potential with $\phi \ll m_P$ has been put forward, where the inflaton is taken to roll down from near the top of a hill in the scalar potential. This type of inflation is called *hilltop* inflation [BL05]. The name was coined by David H. Lyth and Lotfi

---

[*]There are rare exceptions to this rule. These are called topological terms in $\mathcal{L}$ (e.g., the Chern-Simons term), which are proportional to $1/\sqrt{-g}$.

$$V(\phi)$$

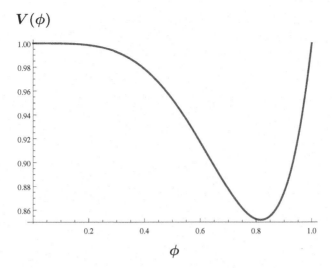

$$\phi$$

FIGURE 7.3 The scalar potential in hilltop inflation. The potential density is labeled as a fraction of V_0, while the inflaton field is measured in units of $m_P/\lambda^{1/q}$.

Boubekeur in 2005, but versions of the model preexisted for many years, dating back to *new inflation*, which was introduced by Andrei D. Linde in 1982 [Lin82a]. The scalar potential is

$$V(\phi) = V_0 \left[1 - \lambda \left(\frac{\phi}{m_P} \right)^q \right] + \cdots , \tag{7.16}$$

where V_0 is a constant density scale, $q > 0$ is a parameter and λ is a dimensionless constant, due to the fact that ϕ is not weighted by m_P, but by the mass-scale $m_P/\lambda^{1/q}$. The ellipsis in Eq. (7.16) denotes terms which stabilise the potential, meaning they do not allow V to become negative. This is motivated by the fact that $V < 0$ would have similar effects to a negative cosmological constant, which would lead the Universe to collapse. There are other destructive instabilities in theories with negative potential density, which have no relation to cosmology. Thus, we assume that stabilising terms create a minimum for $V(\phi)$ and a corresponding VEV $\langle \phi \rangle$ for the inflaton. During inflation however, the field is rolling down the potential hill (see Fig. 7.3) and its value is much smaller than the VEV. Thus, the effect of the stabilising terms is negligible because they become important only near the VEV, i.e., for $\phi \sim \langle \phi \rangle$. For this reason, we ignore these terms from now on.

Let us study slow-roll inflation. From Eqs. (6.21) and (6.23), the slow-roll parameters are

$$\varepsilon = \frac{(\lambda q)^2}{2} \left(\frac{\phi}{m_P} \right)^{2(q-1)} \quad \text{and} \quad \eta = -\lambda q(q-1) \left(\frac{\phi}{m_P} \right)^{q-2} , \tag{7.17}$$

where we have considered that $V(\phi) \simeq V_0$ during inflation. The reason for this approximation is that, during inflation while the inflaton slow-rolls down the potential hill, we have $\phi < \langle \phi \rangle$. From Eq. (7.16), the inflaton VEV is $\langle \phi \rangle \sim m_P/\lambda^{1/q}$, because $V(\langle \phi \rangle) \sim 0$. Therefore, if we assume $\phi \ll \langle \phi \rangle$, the factor in angular brackets in Eq. (7.16) is close to unity and $V(\phi) \simeq V_0$. From the above, we see that slow-roll is possible only when $\phi \lesssim m_P$. This is why hilltop inflation belongs to *small-field* inflation models.

Taking the ratio of the slow-roll parameters we get

$$\frac{\varepsilon}{|\eta|} = \frac{\lambda q}{2(q-1)} \left(\frac{\phi}{m_P} \right)^q . \tag{7.18}$$

Equation (7.18) shows that $\varepsilon \ll |\eta|$ during inflation because $\phi \ll \langle\phi\rangle \sim m_P/\lambda^{1/q}$ (unless $q \approx 1$).

Thus, under the assumption $\phi \ll \langle\phi\rangle$, slow-roll inflation ends when $|\eta| = 1$, with less than an e-fold of inflation remaining until ϵ grows to unity and the expansion ceases to be accelerated.[*] Taking $|\eta(\phi_{\text{end}})| = 1$ we find

$$\left(\frac{\phi_{\text{end}}}{m_P}\right)^{2-q} = \lambda q(q-1) . \tag{7.19}$$

Using this and Eq. (6.22), we express the expectation value of the inflaton field as a function of remaining e-folds of inflation as

$$\frac{\phi(N)}{m_P} = \left[\lambda q(q-2)\left(N + \frac{q-1}{q-2}\right)\right]^{\frac{1}{2-q}} , \tag{7.20}$$

where we assumed that $q \neq 2$. Combining this with Eqs. (6.83) and (7.17), we get

$$n_s = 1 - 2\frac{q-1}{q-2}\left(N + \frac{q-1}{q-2}\right)^{-1} . \tag{7.21}$$

Similarly, combining Eqs. (6.88) and (7.17) we obtain

$$r = 8(\lambda q)^2 \left[\lambda q(q-2)\left(N + \frac{q-1}{q-2}\right)\right]^{-2\left(\frac{q-1}{q-2}\right)} . \tag{7.22}$$

Similarly, using Eqs. (6.78), (7.16), and (7.20) we find

$$\frac{V_0}{m_P^4} = 12\pi^2(\lambda q)^2 \mathcal{P}_\zeta \left[\lambda q(q-2)\left(N + \frac{q-1}{q-2}\right)\right]^{-2\left(\frac{q-1}{q-2}\right)} . \tag{7.23}$$

Let us now use the above to find some predictions. Assume $q = 4$ (quartic hilltop inflation). Then, Eq. (7.21) reduces to $n_s = 1 - \frac{6}{2N+3}$. Using $N_* = 60$ we obtain $n_s = 0.951$ which is too low compared with the observations in Eq. (5.41), which suggest $n_s \simeq 0.965$. To approach this value would require unacceptably large N_* (about $N_* \simeq 84$). We need to consider a larger value of q to counteract this. To find out how large q would need to be we rewrite Eq. (7.21) as

$$N = \frac{q-1}{q-2}\left(\frac{2}{1-n_s} - 1\right) . \tag{7.24}$$

Demanding that $N_* \lesssim 60$ and taking $n_s \simeq 0.965$ we find the bound $q \gtrsim 16$. This is a very large and rather unrealistic value. In the limit $q \to \infty$ Eq. (7.21) becomes $n_s = 1 - \frac{2}{N+1}$. Taking $N_* = 60$ we get $n_s \simeq 0.967$, which is excellent but requires $q \gg 1$.

In the above we considered $\varepsilon \ll |\eta|$. What if it is the other way around and $\varepsilon \gg |\eta|$? In this case, Eq. (6.83) reduces to $n_s \simeq 1 - 6\varepsilon = 1 - \frac{3}{8}r$, where we used Eq. (6.88), which suggests $r = 16\varepsilon$. From the observational bound in Eq. (5.41), we have $r < 0.056$, which gives $n_s > 0.979$, which is too large. Thus, we find that when $\varepsilon \gg |\eta|$ the spectrum is not red enough, while when $\varepsilon \ll |\eta|$ the spectrum is too red (unless $q \gg 1$). You might expect that, if we tune $\varepsilon \sim |\eta|$, we may get it exactly right then.[†] Indeed, it turns out that, provided $10^{-8} < \lambda < 10^{-4}$, quartic hilltop

---

[*]When $q = 1$ then Eq. (7.17) suggests that $\varepsilon = \text{constant}$, which means that inflation also ends when $|\eta| = 1$ (assuming $\varepsilon < 1$ to have inflation in the first place).

[†]This amounts to $\phi \lesssim \langle\phi\rangle$ rather than $\phi \ll \langle\phi\rangle$ during inflation.

inflation ($q = 4$) can work fine, as depicted in Fig. 5.5. Note however, that the VEV is super-Planckian since $\langle \phi \rangle \sim m_P/\lambda^{1/4}$ with $\lambda \ll 1$, which means that the validity of the perturbative origin of the hilltop potential is questionable.

Before concluding this section, we also take a look at the case $q = 2$, which is different from the above. With $q = 2$ the potential can be written as

$$V(\phi) = V_0 \left[1 - \lambda \left(\frac{\phi}{m_P} \right)^2 \right] + \cdots = V_0 - \frac{1}{2} m^2 \phi^2 + \cdots , \tag{7.25}$$

where $m^2 = 2\lambda V_0/m_P^2$. This can be thought of as a perturbative potential. The top of the hill can be an *enhanced symmetry point* (ESP), which means that thermal corrections due to a thermal bath preexisting inflation (subsequently inflated away) can arrange to bring the field to the top with $\phi = 0$ (see Sec. 6.3.4 for more details). In this case, the model is called *new inflation*.

From Eqs. (6.21) and (6.23), the slow-roll parameters are

$$\varepsilon = 2\lambda^2 \left(\frac{\phi}{m_P} \right)^2 \quad \text{and} \quad \eta = -2\lambda = \text{constant} . \tag{7.26}$$

To have slow-roll inflation, we need $|\eta| < 1$, which means $\lambda < 1/2$. Because $\eta = $ constant, inflation ends this time when $\varepsilon(\phi_{\text{end}}) = 1$, which results in $\phi_{\text{end}} = m_P/\sqrt{2}\lambda$. Now, Eq. (6.22) suggests $\ln(\frac{\phi}{\phi_{\text{end}}}) = -2\lambda N$. Combining these findings we get

$$\phi(N) = \frac{m_P}{\sqrt{2}\lambda} e^{-2\lambda N} . \tag{7.27}$$

Combining the above with Eq. (7.26) we obtain

$$\varepsilon = e^{-4\lambda N} . \tag{7.28}$$

Thus, Eq. (6.83) suggests

$$n_s = 1 - 4\lambda - 6e^{-4\lambda N} . \tag{7.29}$$

It is easy to check that the above is impossible to reconcile with $n_s \simeq 0.965$ and $N_* \lesssim 60$ whatever the value of λ. Therefore, this version of hilltop inflation (called *quadratic hilltop* inflation) is excluded.

7.2.3 Natural inflation

The next effort attempts to combine the two previous models by featuring both a hilltop and a quadratic minimum, in the hope of retaining the advantages of both approaches and escaping their drawbacks. Is this effort successful? Let us see.

Natural inflation was introduced by Katherine Freese, Joshua A. Frieman, and Angela V. Olinto in 1990 [FFO90]. The potential can be written as

$$V(\phi) = V_0[1 - \cos(\phi/qm_P)] , \tag{7.30}$$

where again V_0 is a constant density scale and q is a parameter.[*] The potential is shown in Fig. 7.4. The rationale behind this potential is as follows.

---

[*] We can change the origin in field space and write the potential as $V(\phi) = V_0[1 + \cos(\phi/qm_P)]$ by taking the transformation $\phi \to \pi q m_P + \phi$. The physics remains the same.

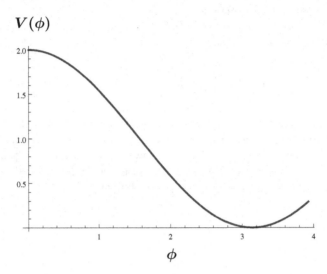

FIGURE 7.4 The scalar potential in natural inflation. The potential density is labeled as a fraction of V_0, while the inflaton field is measured in units of qm_P.

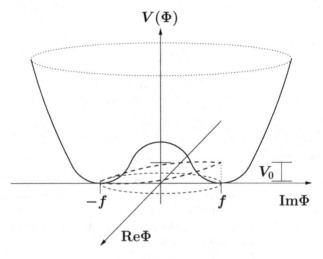

FIGURE 7.5 The appearance of a sinusoidal potential because of the controlled breakdown of the shift symmetry, associated with the angular direction of a complex scalar field. The circle of degenerate minima (vacuum manifold), depicted with a thin dashed line, corresponds to zero potential density and satisfies a shift symmetry $\phi \to \phi + C$, C being an arbitrary constant. A phase transition "tilts" the vacuum manifold, now depicted by the thick dashed line, such that the angular direction ϕ is now characterised by sinusoidal potential density, while a remnant of the original shift symmetry is preserved as $\phi \to \phi + 2\pi f$.

One of the basic problems of inflation model-building is that a relatively flat scalar potential is needed, which is difficult to maintain against radiative corrections, that is contributions to $V(\phi)$ from higher loops in the theory. A way out of this problem is protecting the inflaton direction in field space with some symmetry. One such possibility is a shift-symmetry, which amounts to the Lagrangian density being invariant under the shift $\phi \to \phi + C$, with C some constant. How to create such a situation? Well, consider a complex scalar field $\Phi = \sigma e^{i\theta}$, where both σ and θ are real. Then, suppose the scalar potential contains a term of the form $\Delta V = \kappa(|\Phi|^2 - f^2/2)^2$, where f is some large mass scale (called *axion decay constant*) and κ is a dimensionless coupling. This generates a "Mexican hat" potential,* which sends the radial field σ to its minimum at $\sigma = f/\sqrt{2}$ where $\Delta V = 0$. The VEV of the complex field can now be written as $\langle\Phi\rangle = (f/\sqrt{2})e^{i\phi/f}$, where we have defined $\phi \equiv f\theta$. Such a field is called a *Goldstone boson*. The circle at the bottom of the potential (see Fig. 7.5) is called the *vacuum manifold*. As it is evident, the potential density is constant (zero) along the ϕ-direction, meaning there is a shift symmetry $\phi \to \phi + C$ as we go around the vacuum manifold. But one cannot have inflation with zero $V(\phi)$. Indeed, there is more to come. At some lower energy scale a phase transition "tilts" the vacuum manifold by a small degree, generating thus a sinusoidal variation in $V(\phi)$, with amplitude $V_0 \sim m^2\langle\phi\rangle^2 \sim (mf)^2$, where m is the mass of ϕ. Now the field is called a *pseudo Nambu-Goldstone boson* (PNGB) or equivalently, an *axion*. In Eq. (7.30), we have defined $f \equiv qm_P$ because it turns out that $f \gtrsim m_P$. Thus, the flatness of the inflaton direction is preserved by the remnant of the broken shift-symmetry.

Let us study the potential in Eq. (7.30). The VEV (the minimum of V) is at zero. Taylor expanding around the VEV, when $\phi \ll qm_P$, we have

$$V \simeq \frac{1}{2}m^2\phi^2 + \cdots, \tag{7.31}$$

where $m^2 = V_0/(qm_P)^2$. Thus, in the limit $q \to \infty$ the potential approximates quadratic monomial inflation. Far away from the VEV however, the potential looks more like hilltop inflation (see Fig. 7.3). It is true that when ϕ/qm_P becomes large all the higher-order terms in the Taylor expansion blow up. But their alternate signs[†] make sure that $V(\phi)$ does not diverge since $V(\phi) \le 2V_0$.

Following the same procedure as before, using Eqs. (6.21) and (6.23) we find the slow-roll parameters

$$\varepsilon = \frac{1}{2q^2}\frac{1 + \cos\left(\frac{\phi}{qm_P}\right)}{1 - \cos\left(\frac{\phi}{qm_P}\right)} \quad \text{and} \quad \eta = \frac{1}{q^2}\frac{\cos\left(\frac{\phi}{qm_P}\right)}{1 - \cos\left(\frac{\phi}{qm_P}\right)}. \tag{7.32}$$

From the above, we obtain the relation

$$\eta = \varepsilon - \frac{1}{2q^2}. \tag{7.33}$$

Using this in Eq. (6.83), we find

$$n_s = 1 - \frac{1}{q^2} - 4\varepsilon. \tag{7.34}$$

---

*Sometimes also called "bottom of the bottle" potential, depending on whether one lives in a place with many Mexicans or not.

[†]Recall that

$$\cos x = \sum_{n=0}^{\infty}\frac{(-1)^n}{(2n)!}x^{2n} = 1 - \frac{x^2}{2!} + \frac{x^4}{4!} - \frac{x^6}{6!} + \cdots.$$

Similarly, in view of Eq. (5.40), we get

$$r = 16\varepsilon = 4\left(1 - n_s - \frac{1}{q^2}\right), \qquad (7.35)$$

where we also employed Eq. (7.34). Using that $n_s \simeq 0.965$ and $r < 0.056$ from Eq. (5.41), we obtain the bound $q < 6.9.$*

Now, from the requirement that $\varepsilon(\phi_{\text{end}}) = 1$ we obtain that $\cos^2(\frac{\phi_{\text{end}}}{2qm_P}) = \frac{2q^2}{2q^2+1}$. Then, Eq. (6.22) suggests that[†]

$$\cos^2\left(\frac{\phi(N)}{2qm_P}\right) = \frac{2q^2}{2q^2+1}e^{-N/q^2}. \qquad (7.36)$$

Using this, Eq. (7.32) suggests

$$\varepsilon = \frac{1}{(2q^2+1)e^{N/q^2} - 2q^2}. \qquad (7.37)$$

Eqs. (7.34) and (7.37) allow the exploration of the parameter space of the model. It turns out that the value of n_s is too small unless one considers large values of q. However, as we have shown, $q < 6.9$. As an indicative value, we choose $q = 6.0$ and $N_* \simeq 60$. Then Eqs. (7.34) and (7.37) suggest $n_s \simeq 0.960$ and Eq. (7.35) results in $r \simeq 0.051$, which is at the verge of observability. If the bound on r tightens, it would force smaller values of q however, which would result in low values of n_s, that might be unacceptable. Thus, natural inflation seems to work for now, but it may become excluded (or observed!) in the very near future. The model predictions are contrasted with the Planck results in Fig. 5.5 (see legend), where it is evident that the model is at the verge of being excluded.

Finally, using Eqs. (6.78), (7.30), and (7.36) we find

$$\frac{V_0}{m_P^4} = 12\pi^2(2q^2+1)\mathcal{P}_\zeta e^{N/q^2}[(2q^2+1)e^{N/q^2} - 2q^2]^{-2}. \qquad (7.38)$$

Putting the numbers in the above, using $N_* = 60$ and $q = 6.0$ with $\sqrt{\mathcal{P}_\zeta} \simeq 4.579 \times 10^{-5}$ (cf. Eq. (5.42)), we find $V_0^{1/4} \simeq 1.36 \times 10^{16}$ GeV, which is at the GUT scale.

So how natural is natural inflation? Well, the model does manage to address the observations (for now) with little fine-tuning if any. The theoretical setup is rather generic in fundamental theories. In particular, the prime candidate for the inflaton field is a string axion, especially since the value of f is so large. Such string axions are abundant in the various incarnations of string theory. However, as we have seen, $f = qm_P > m_P$ and accommodating a super-Planckian f is problematic for string theory. This is why we may choose to climb down to merely GUT theories, where theorists feel much more comfortable. Next, we look into such a possibility.

7.2.4 Hybrid inflation

Hybrid inflation was originally introduced by Andrei D. Linde in 1994 [Lin94] to overcome the problem of super-Planckian expectation values of the inflaton field in monomial chaotic inflation.

---

*Note that Eq. (7.14) suggests that $r = 4(1 - n_s)$ for quadratic chaotic inflation, which is identical to Eq. (7.35) in the limit $q \to \infty$.

[†]Note the integral

$$\int \frac{1 - \cos x}{\sin x}dx = -2\ln[\cos(x/2)] + C,$$

where C is an integration constant.

The idea is the following. In chaotic inflation the need to go super-Planckian is because only then can the inflaton field ϕ climb high enough up the scalar potential $V(\phi)$ so that Hubble friction can become large ($H^2 \simeq V/3m_P^2$, cf. Eq. (6.20)) and lead to slow-roll. What if we had a lot of potential density regardless of the value of ϕ? For example, consider $V(\phi) = V_0 + \frac{1}{2}m^2\phi^2$, with V_0 being a large constant density scale. Then the inflaton would not have to have a large value to lead to slow-roll because the contribution of the quadratic term might be subdominant to V_0, governing only the dynamics of inflation, i.e., the roll of ϕ due to V' in the Klein-Gordon Eq. (6.8). This sounds swell, but there is one problem; inflation would never end because V_0 is a constant that we cannot get rid of. Or can we? Linde's idea was to incorporate a second scalar field χ in the model, whose task is only to eliminate V_0 when the time is right, meaning after enough inflation has taken place. How does χ manage this? This is the hybrid mechanism, which is sketched below.

The scalar potential is (see Fig. 7.6)

$$
\begin{aligned}
V(\phi, \chi) &= V_0 - \frac{1}{2}m_0^2\chi^2 + \frac{1}{4}\lambda\chi^4 + \frac{1}{2}g^2\phi^2\chi^2 + U(\phi) \\
&= \frac{1}{4}\lambda(M^2 - \chi^2)^2 + \frac{1}{2}g^2\phi^2\chi^2 + U(\phi),
\end{aligned}
\tag{7.39}
$$

where $V_0 \equiv \frac{1}{4}\lambda M^4$ and $m_0^2 \equiv \lambda M^2$ with λ, g being coupling constants and M being a large mass scale. In the above, $U(\phi)$ serves only to enable the roll of ϕ towards smaller values. In our example above, $U(\phi) = \frac{1}{2}m^2\phi^2$.

The effective mass-squared of the χ-field is

$$
\frac{\partial^2 V}{\partial\chi^2} = g^2\phi^2 - m_0^2 + 3\lambda\chi^3.
\tag{7.40}
$$

This is positive for large values of the inflaton field ϕ. Thus, if ϕ is originally large, the χ-field is driven to zero, and its effective mass-squared becomes

$$
\left.\frac{\partial^2 V}{\partial\chi^2}\right|_{\chi=0} = g^2\phi^2 - m_0^2.
\tag{7.41}
$$

The above shows that the effective mass-squared of χ is positive if $\phi > \phi_c$ and negative if $\phi < \phi_c$, where

$$
\phi_c \equiv \frac{m_0}{g} = \frac{\sqrt{\lambda}}{g}M.
\tag{7.42}
$$

Hence, when $\phi > \phi_c$, then $\chi \to 0$ and the scalar potential becomes $V = V_0 + U$, while when $\phi < \phi_c$, then $\chi \to M$, $\phi \to 0$ and the scalar potential becomes $V = 0$, assuming $U(0) = 0$. Therefore, when the inflaton field, as it decreases in time, reaches the value $\phi_{\text{end}} \simeq \phi_c$, a phase transition sends χ off from the origin down to its VEV, thereby cancelling V_0 in the potential. This rapid roll of χ to its VEV, triggered when ϕ reaches ϕ_c is the reason χ is the called the *waterfall* field. Even though the hybrid mechanism requires two fields, because $\chi = 0$ (and heavy) during inflation, we are effectively talking about a single-field model, where all the dynamics during inflation are determined by $U(\phi)$.

Soon it was realised that there was a problem when trying to "hybridise" quadratic or quartic chaotic inflation. Because the potential density during hybrid inflation is dominated by the constant V_0, the scalar potential is very flat, and so the ε slow-roll parameter is extremely small since $\varepsilon \propto V'$ as $V \simeq V_0$ in this case. This means that the spectral index in Eq. (6.83) is given by $n_s \simeq 1 + 2\eta$. The potential for quadratic or quartic inflation is curved upwards, meaning $V'' > 0$. Consequently, $\eta > 0$ which would result in a blue spectrum of scalar curvature perturbations $n_s > 1$. However, the observations conclusively find a red spectrum instead, with

$$V(\phi, \chi)$$

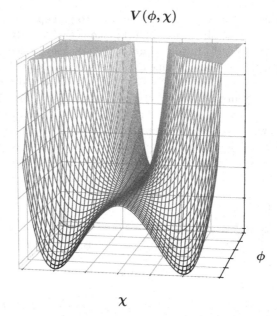

χ

FIGURE 7.6   The scalar potential in hybrid inflation.

$n_s < 1$ (cf. Eq. (5.41)). Thus, for $U(\phi)$ we should opt for a concave potential instead, which would lead to $\eta < 0$ and $n_s < 1$.

A promising possibility is a Coleman-Weinberg one-loop radiative correction along the inflaton direction. In supersymmetric theories this has the following form [LR97]

$$U(\phi) = qV_0 \ln\left(\frac{\phi}{Q}\right), \qquad (7.43)$$

where Q is some large mass scale (called a renormalisation scale) and $q = y^2/16\pi^2$, with $y \leq 1$ being a perturbative coupling.

The dynamics of slow-roll are solely determined by U because they involve derivatives of the scalar potential. Using Eqs. (6.21) and (6.23), we find the slow-roll parameters

$$\varepsilon = \frac{q^2}{2}\left(\frac{m_P}{\phi}\right)^2 \quad \text{and} \quad \eta = -q\left(\frac{m_P}{\phi}\right)^2. \qquad (7.44)$$

Hybrid inflation does not end by the violation of the slow-roll condition but by the phase transition at $\phi = \phi_c$. In view of this, Eq. (6.22) suggests

$$\begin{aligned} \phi^2 &= \phi_c^2 + 2qm_P^2 N \\ \Rightarrow \quad \phi(N) &\simeq \sqrt{2qN}\, m_P, \end{aligned} \qquad (7.45)$$

where in the last equation we have taken $\phi(N) > \phi_c$. The above and Eq. (6.83) give

$$n_s = 1 - \left(1 + \frac{3}{2}q\right)\frac{1}{N} \simeq 1 - \frac{1}{N}, \qquad (7.46)$$

where we used that $q \ll 1$. Similarly, Eq. (5.40) suggests

$$r = 16\varepsilon = \frac{4q}{N}. \qquad (7.47)$$

Considering $N_* = 60$, Eq. (7.46) results in $n_s \simeq 0.983$, which is too large. Using $n_s \simeq 0.965$ as suggested by the observations in Eq. (5.41), we find $N_* \simeq 29$. This is very small, but it can be accommodated if there is a period of late thermal inflation (see Sec. 8.4). Using $N_* \simeq 29$ and in view of the bound $r < 0.056$ in Eq. (5.41) we obtain the bound $q < 0.4$ or so, which is rather mild. In fact, we expect q to be much smaller than this.

The predictions of the model are contrasted with the Planck observations in Fig. 5.5. The model is denoted as SB SUSY, which stands for symmetry breaking (SB) (referring to the end of hybrid inflation) in supersymmetry (SUSY) (so that the Coleman-Weinberg potential is the one shown in Eq. (7.43)). The predictions of the model fall well out of the Planck contours for $N_* = 50 - 60$. As argued, this can be remedied if N_* is decreased down to $N_* \simeq 30$, which would mean moving the points in the plot to the left, enough to satisfy the observations.

The introduction of another degree of freedom, namely the waterfall field, can be seen as a weakness of the hybrid scenario. Indeed, if one involves hundreds of new and unobservable fields then one can dispense with any fine-tuning. This, however, is overcome if the existence of the waterfall field is expected anyway. Indeed, noticing that hybrid inflation is at the GUT scale, as most inflation models are, it was assumed that χ is in fact the GUT Higgs field, that is the GUT equivalent to the electroweak Higgs field, which was observed in CERN in 2012. In this case, the phase transition that ends hybrid inflation is the breaking of grand unification. Under this consideration, the mass-scale M in Eq. (7.39) is the GUT scale $M \sim 10^{16}$ GeV. Then, if the GUT-Higgs field self-coupling λ is not too small (the self-coupling of the electroweak Higgs field is 0.13 or so), $V_0^{1/4} \sim 10^{16}$ GeV too. Using Eq. (7.39), Eq. (6.78) suggests that

$$\frac{V_0}{m_P^4} = \frac{6\pi^2 q \mathcal{P}_\zeta}{N} . \tag{7.48}$$

Using that $\sqrt{\mathcal{P}_\zeta} \simeq 4.579 \times 10^{-5}$ (cf. Eq. (5.42)) and $N_* \simeq 29$, with $V_0/m_P^4 \lesssim 10^{-8}$ we find $q \sim 0.1$ or so. In view of Eq. (7.47), we find $r \sim 10^{-2}$, which will be observable in the near future.

After the phase transition which ends inflation both the coupled scalar fields ϕ and χ oscillate around their VEVs. This results in parametric resonance as described in Sec. 6.3.3. Preheating may lead to prompt reheating especially if the waterfall field is the GUT Higgs, in which case it may also decay into the *super-massive bosons* of the GUT.* The phase transition may also lead to copious production of primordial black holes (PBHs), which can be a significant fraction (if not the entirety) of dark matter Needless to say, hybrid inflation can have rich phenomenology. Moreover, the hybrid mechanism is generic and does not depend much on the choice of $U(\phi)$ in Eq. (7.43).

This concludes our brief tour of some of the most basic inflation models. Many variants of such models exist, going by names such as *supernatural* inflation or *mutated* hybrid inflation. Such models might score better in comparison with observations but they are usually more complicated and seem artificial. The observations in Eq. (5.41) have seriously affected inflation model building, rendering entire families of inflation models (such as monomial inflation) excluded. However, they also pointed towards new directions and resulted in an explosion of novel inflation models, which outperform the "archetype" models and supersede traditional thinking. Ironically, one model that is tremendously successful, is the oldest inflation model ever, introduced by Alexei A. Starobinsky in 1980 and used to be partly forgotten in the decades that followed. Not anymore. Much of the plethora of new models are successful exactly because they mimic Starobinsky inflation. In the next section, we discuss some of the main proposals.

---

*Also called X and Y bosons.

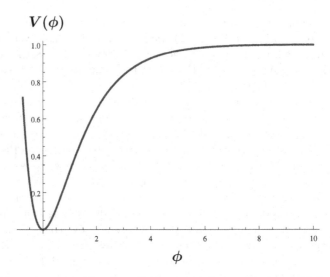

FIGURE 7.7 The scalar potential in Starobinsky type inflation. The potential density is labeled as a fraction of V_0, while the inflaton field is measured in units of m_P/q.

7.3   Modern inflation models

In this section, we take a look at more recent inflation models, which are better suited to the observational findings. These models feature an inflationary plateau exactly because the observations seem to prefer it. First we analyse the generic form of these proposals, and then we briefly discuss some of the main candidates, including the original Starobinsky model itself.

7.3.1   Inflation of the Starobinsky type

The scalar potential is of the form

$$V = V_0 \left(1 - e^{-q\phi/m_P}\right)^2 , \tag{7.49}$$

where V_0 is a constant density scale and $q \sim 1$ is a parameter. The potential is shown in Fig. 7.7.

Employing Eqs. (6.21) and (6.23), we find the slow-roll parameters

$$\varepsilon = \frac{2q^2}{(e^{q\phi/m_P} - 1)^2} \quad \text{and} \quad \eta = -2q^2 \frac{e^{q\phi/m_P} - 2}{(e^{q\phi/m_P} - 1)^2} . \tag{7.50}$$

We see that the slow-roll parameters are small when $\phi > m_P/q$. Thus, this is a large-field inflation model, as with chaotic inflation.

Taking $\varepsilon(\phi_{\text{end}}) = 1$ we obtain

$$e^{q\phi_{\text{end}}/m_P} = \sqrt{2}q + 1 . \tag{7.51}$$

Then, from Eq. (6.22) we get

$$
\begin{aligned}
N &= \frac{1}{2q^2} \left[(e^{q\phi/m_P} - e^{q\phi_{\text{end}}/m_P}) - \left(\frac{q\phi}{m_P} - \frac{q\phi_{\text{end}}}{m_P} \right) \right] \\
&= \frac{1}{2q^2} \left[e^{q\phi/m_P} - \left(1 + \frac{q\phi}{m_P}\right) - \sqrt{2}q + \ln(\sqrt{2}q + 1) \right] \\
\Rightarrow e^{q\phi(N)/m_P} &\simeq 2q^2 N + \sqrt{2}q - \ln(\sqrt{2}q + 1) \simeq 2q^2 N ,
\end{aligned}
\tag{7.52}
$$

where we used Eq. (7.51) and that $e^x > 1 + x$ for $x > 1$.* Using this and Eq. (7.50), it is straight-forward to obtain

$$\varepsilon \simeq \frac{1}{2q^2 N^2} \quad \text{and} \quad \eta \simeq -\frac{1}{N} \, . \tag{7.53}$$

Inserting the above into Eqs. (6.83) and (5.40), we find

$$n_s = 1 - \frac{2}{N} \quad \text{and} \quad r = \frac{8}{q^2 N^2} \, . \tag{7.54}$$

Using $N_* = 60(50)$, we find $n_s \simeq 0.967(0.960)$ and $r = 2.2(3.2) \times 10^{-3}/q^2$, which is in excellent agreement with the observations in Eq. (5.41) for q not too small ($q > 0.2$ or so).

From Eq. (6.78) we get

$$\frac{V_0}{m_P^4} \simeq \frac{12\pi^2 \mathcal{P}_\zeta}{q^2 N^2} \, . \tag{7.55}$$

Thus, using Eq. (5.42), for the inflation energy scale $V_0^{1/4}$ we get

$$V_0^{1/4} = \frac{0.022 \, m_P}{\sqrt{qN}} = \frac{5.4 \times 10^{16} \text{GeV}}{\sqrt{qN}} \, , \tag{7.56}$$

which suggests that $V_0^{1/4} \lesssim 10^{16}$ GeV for $q \sim 1$ and inflation is at the GUT-scale as expected.

The above demonstrates the undeniable success of the model in Eq. (7.49). This is why many attempts aim to end up with a scalar potential of this form. In what follows, we present some of the most prominent of them. But first we need to briefly discuss modified gravity because it is a popular way to construct inflation models of the Starobinsky type, including the original proposal by Starobinsky himself.

7.3.2 Modified gravity

The action principle is also employed in general relativity (GR). In fact the Lagrangian density of GR is deceptively simple, given by

$$\mathcal{L} = \frac{R}{16\pi G} = \frac{1}{2} m_P^2 R \, , \tag{7.57}$$

where R is the *scalar curvature*[†] comprised of second derivatives of the metric and of quadratic products of derivatives of the metric (see Appendix A). The spacetime integral of the above is called the *Einstein-Hilbert* action.[‡]

However, for a long time (almost immediately after its inception) there have been numerous efforts to go beyond and generalise GR. One of the most prominent attempts to do this is *scalar-tensor* theory in which the gravitational action is written as

$$S = \int d^4x \sqrt{-g} \left[\frac{1}{2} m_P^2 f(R, \chi) + \frac{1}{2} j(\chi)(\partial \chi)^2 - U(\chi) \right] , \tag{7.58}$$

where f, j, U are real functions and χ is a scalar field. As you can see, this proposal mixes gravity with a scalar field, which is in fact part of the matter content of the theory. The original scalar-tensor theory is the *Brans-Dicke* theory [BD61] in which $f = R\chi$ and $j = \omega/\chi$, where ω is called the Brans-Dicke parameter.

---

*Recall that $e^x = 1 + x + \frac{1}{2}x^2 + \frac{1}{6}x^3 + \cdots$.

[†]Also called the Ricci scalar.

[‡]We have ignored the contribution of a nonzero cosmological constant. Incorporating it leads to the Lagrangian density $\mathcal{L} = \frac{1}{2}m_P^2(R - 2\Lambda)$.

We may be able to formulate the theory in Eq. (7.58) in such a way that the Einstein-Hilbert action reemerges and the theory appears as normal GR. This is done through a *conformal transformation* of the metric $g_{\mu\nu} \to \Omega^2 g_{\mu\nu}$, where the conformal factor Ω can be a function of both the location in spacetime and also of the field χ. To yield the desired GR action the conformal factor must be chosen to be

$$\Omega^2 = \frac{\partial f}{\partial R} \, . \tag{7.59}$$

Of particular interest are theories where Ω^2 is determined only by a single degree of freedom, either due to the field χ or due to higher-order terms in the geometry $\Omega^2 = \Omega^2(R)$. Now, the kinetic term of the scalar field χ is in general noncanonical, meaning it is not of the form $\frac{1}{2}(\partial\chi)^2$. In the case when Ω^2 is determined only by a single degree of freedom, we can generate the canonically normalised scalar field ϕ through the expression

$$\phi = \int d\chi \sqrt{\frac{3}{2}m_P^2 \left(\frac{\partial \ln \Omega^2}{\partial \chi}\right)^2 + \frac{j(\chi)}{\Omega^2}} \, , \tag{7.60}$$

so that we obtain the familiar action

$$S = \int d^4x \sqrt{-g} \left[\frac{1}{2}m_P^2 R + \frac{1}{2}(\partial\phi)^2 - V(\phi)\right] , \tag{7.61}$$

with

$$V = \Omega^{-4} U \tag{7.62}$$

because the conformal transformation implies that $\sqrt{-g} \to \Omega^{-4}\sqrt{-g}$.

The above can be better understood when put in action, as we do below. Before this we need to briefly discuss the meaning of the conformal transformation. This is not a coordinate transformation, because GR is invariant under those. It is a transformation which mixes the gravitational field with the matter content of spacetime, meaning it redefines the gravitational and matter degrees of freedom. A particular definition of the gravitational degrees of freedom is called a (conformal) *frame*.[*] If gravity is described by GR and the gravitational action is the Einstein-Hilbert action then we say we are in the *Einstein frame*. In contrast, if gravity is not GR, as in Eq. (7.58) with $f \neq R$, then we say that we are in the *Jordan frame*, named after Ernst Pascual Jordan.[†] Being a mathematical field redefinition the conformal transformation would not change the physics. However, there are subtleties that make this difficult, having to do with the choice of a cutoff (M_P in case of GR) or of a "smooth" gravitational background in quantum field theory in curved spacetime. When differences do appear between frames, it is still debatable, which is the "true" frame, the Einstein or the Jordan one.

7.3.3 Higgs inflation

In recent years, it has increasingly been claimed that, in curved spacetime, scalar fields are expected to develop a direct coupling to gravity of the form shown below[‡]

$$S = \int d^4x \sqrt{-g} \left[\frac{1}{2}m_P^2 R + \frac{1}{2}\xi R\chi^2 + \frac{1}{2}(\partial\chi)^2 - U(\chi)\right] , \tag{7.63}$$

---

[*]The use of the term "frame" here should not be confused with a choice of coordinate system, such as an "inertial frame."

[†]Who would probably have won a Nobel prize if he had not joined the Nazi party.

[‡]This coupling arises naturally when considering quantum corrections in curved spacetime.

where ξ is a non-perturbative coupling constant, meaning it can assume large values $\xi \gg 1$. As we have said, all fields are minimally coupled to gravity because of the factor $\sqrt{-g}$ in the above. Thus, the extra coupling $\xi R \chi^2$ of χ to gravity (to the scalar curvature R) is called "nonminimal". The gravitational Lagrangian density is therefore $\mathcal{L} = \frac{1}{2}(m_P^2 + \xi \chi^2)R$, which is not the one in Eq. (7.57). As a result we are not in GR but in a scalar-tensor theory.

We now attempt to reformulate the above theory in the Einstein frame. Comparing Eq. (7.63) with Eq. (7.58), we see that $j(\chi) = 1$ and

$$f = \left(1 + \frac{\xi \chi^2}{m_P^2}\right)R.\tag{7.64}$$

Thus, in view of Eq. (7.59), the conformal factor is

$$\Omega^2 = \frac{df}{dR} = 1 + \frac{\xi \chi^2}{m_P^2} = \begin{cases} 1 & \text{when} \quad \xi \chi^2 \ll m_P^2, \\ \xi \chi^2 / m_P^2 \gg 1 & \text{when} \quad \xi \chi^2 \gg m_P^2. \end{cases}\tag{7.65}$$

Using Eq. (7.60), the canonically normalised field is obtained as follows

$$\left(\frac{d\phi}{d\chi}\right)^2 = \frac{\Omega^2 + 6\xi^2 \chi^2/m_P^2}{\Omega^4} = \begin{cases} 1 & \text{when} \quad \xi \chi^2 \ll m_P^2, \\ \frac{\Omega^2 + 6\xi\Omega^2}{\Omega^4} = \frac{1 + 6\xi}{\Omega^2} & \text{when} \quad \xi \chi^2 \gg m_P^2, \end{cases}\tag{7.66}$$

where we used $\Omega^2 \simeq \xi \chi^2/m_P^2$ when $\xi \chi^2 \gg m_P^2$.

The above shows that when $\xi \chi^2 \ll m_P^2$, we have $\Omega \simeq 1$ and $\phi \simeq \chi$, that is, the χ field is canonically normalised. Indeed, in this limit the nonminimal term in Eq. (7.63) becomes negligible and we are back in GR. Thus, when the value of χ is small it behaves as any other scalar field would and, in particular, it can be the electroweak Higgs field, for which

$$U(\chi) = \frac{1}{4}\lambda(\chi^2 - v^2)^2,\tag{7.67}$$

where $v \simeq 246\,\text{GeV}$ is the VEV of the electroweak Higgs field and its self-coupling is $\lambda = \frac{1}{2}(m_H/v)^2 \simeq \frac{1}{2}(\frac{125}{246})^2 \simeq 0.13$, with $m_H \simeq 125\,\text{GeV}$ being the mass of the Higgs particle.* The idea of employing a nonminimally-coupled electroweak Higgs field as the inflaton was put forward by Fedor L. Bezrukov and Mikhail Shaposhnikov in 2008 [BS08], who called the model *Higgs inflation*.

However, in the limit $\xi \chi^2 \gg m_P^2$ things are different. As we have seen $\Omega^2 \simeq \xi \chi^2/m_P^2$, which means $d\Omega \simeq (\sqrt{\xi}/m_P)\,d\chi$. Thus we can write

$$\left(\frac{d\phi}{d\chi}\right)^2 \simeq \frac{\xi}{m_P^2}\left(\frac{d\phi}{d\Omega}\right)^2 \simeq \frac{1 + 6\xi}{\Omega^2},\tag{7.68}$$

where we used Eq. (7.66). Integrating the above and using $\xi \gg 1/6$, we find

$$\phi \simeq \frac{\sqrt{6}}{2}m_P \ln \Omega^2 \;\Rightarrow\; \Omega^2 = e^{\sqrt{\frac{2}{3}}\phi/m_P}.\tag{7.69}$$

For large values of χ, we can ignore the VEV and write Eq. (7.67) as $U \simeq \frac{1}{4}\lambda\chi^4$. Then the scalar potential of the canonically normalised field ϕ is given by

$$V(\phi) = \Omega^{-4}\frac{1}{4}\lambda\frac{m_P^4}{\xi^2}(\Omega^2 - 1)^2 = \frac{\lambda m_P^4}{4\xi^2}\left(1 - e^{-\sqrt{\frac{2}{3}}\phi/m_P}\right)^2,\tag{7.70}$$

---

*We have $m_H^2 = V''(v) = 2\lambda v^2$, with the prime denoting differentiation with respect to χ.

where we used Eqs. (7.62), (7.65), and (7.69). The above is of the form of Eq. (7.49) with $q = \sqrt{\frac{2}{3}}$ and $V_0 = \lambda m_P^4/4\xi^2$. From Eq. (7.54), we have $N_* = 2/(1 - n_s) \simeq 57$, where $n_s \simeq 0965$ as suggested by Eq. (5.41). Then, Eq. (7.56) suggests that $\xi \simeq 47000\sqrt{\lambda}$. If χ is indeed the electroweak Higgs field with $\lambda = 0.13$ we find $\xi \simeq 17000$. In this case, Eq. (7.54) suggests $r = 12/N^2$. Using $N_* \simeq 57$ we obtain $r = 3.7 \times 10^{-3}$, which will be within reach of observations in the near future.

Once the inflaton field slides over the plateau, it begins its coherent oscillations around the minimum of its potential. As we have discussed in Sec. 6.3.1, these oscillations correspond to massive particles. If the inflaton is the electroweak Higgs field these are Higgs particles. Higgs inflation has the unique property of leading to fully determined reheating, because the decay rates of the Higgs field to standard model particles are known. The only unknown parameter in the model is the nonminimal coupling ξ, which is fully determined by the observations.

Still, χ does not have to be the electroweak Higgs field. The nonminimal coupling $\xi R\chi^2$ is generic and is expected to appear for all scalar fields. As we have seen, the effect of such nonminimal coupling is to "flatten" an otherwise steep potential (for without the coupling the potential is $\propto \chi^4$) and give rise to the desired inflationary plateau. Thus, it would seem that the inflationary plateau is also a generic feature if there are scalar fields in the theory.

It turns out that there are other ways to achieve this "flattening" of the scalar potential at large field values as we discuss below when we mention α-attractors. Before this, however, we consider another modified gravity model, that of R^2-gravity.

7.3.4 R^2 inflation

Another subset of scalar-tensor theories is called $f(R)$ gravity. In this case, the dependence of the action in Eq. (7.58) on the scalar field χ is eliminated, such that the action can be written as

$$S = \int d^4x \sqrt{-g}\, \frac{1}{2} m_P^2 f(R)\,. \tag{7.71}$$

The above can be readily recast as

$$S = \int d^4x \sqrt{-g} \left(\frac{1}{2} m_P^2 \Omega^2 R - U \right)\,, \tag{7.72}$$

where U is defined as

$$U = \frac{1}{2} m_P^2 (R\Omega^2 - f)\,. \tag{7.73}$$

This is because the conformal transformation $g_{\mu\nu} \rightarrow \Omega^2 g_{\mu\nu}$ results in

$$\int d^4x \sqrt{-g}\, \frac{1}{2} m_P^2 \Omega^2 R \rightarrow \int d^4x \sqrt{-g}\, \frac{1}{2} m_P^2 [R + 6(\partial \ln \Omega)^2]\,, \tag{7.74}$$

where we used that $\sqrt{-g} \rightarrow \Omega^{-4}\sqrt{-g}$ and we disregarded a total derivative (see Sec. 7.1). Thus, the conformal transformation brings the theory in the Einstein frame as it is meant to do.

Now, in view of Eq. (7.62), the scalar potential in the Einstein frame is

$$V = \frac{m_P^2 (R\Omega^2 - f)}{2\Omega^4}\,, \tag{7.75}$$

where we used Eq. (7.73). Let us now consider a particular $f(R)$ theory.

In his seminal 1980 paper [Sta80], which was among the very first on inflation (the name had not even been coined yet), Alexei Alexandrovich Starobinsky introduced the following Lagrangian density

$$\mathcal{L} = \frac{1}{2} m_P^2 R + \beta R^2\,, \tag{7.76}$$

where β is a dimensionless non-perturbative coupling (meaning one can consider $\beta > 1$). The term proportional to R^2 introduces an additional degree of freedom. When the theory is reformulated in the Einstein frame, this degree of freedom becomes a canonically normalised scalar field ϕ called the *scalaron*, because its origin is due to the scalar curvature R. In the Einstein frame this theory ends up with a potential of the form in Eq. (7.49). Let us sketch how.

For this theory one readily obtains

$$f(R) = R + \frac{2\beta}{m_P^2}R^2 \,. \tag{7.77}$$

Hence, the conformal factor is obtained by Eq. (7.59) as

$$\Omega^2 = \frac{df}{dR} = 1 + \frac{4\beta}{m_P^2}R \;\Rightarrow\; R = \frac{m_P^2}{4\beta}(\Omega^2 - 1) \,. \tag{7.78}$$

Inserting Eqs. (7.77) and (7.78) into (7.75) we find

$$V = \frac{m_P^4}{16\beta}(1 - \Omega^{-2})^2 \,. \tag{7.79}$$

As in Eq. (7.69) (obtained directly from Eq. (7.60) with $j(\chi) = 0$), we write

$$\Omega^2 = e^{\sqrt{\frac{2}{3}}\phi/m_P} \;\Rightarrow\; \phi/m_P = \sqrt{6}\ln\Omega \,. \tag{7.80}$$

Using this, the potential in Eq. (7.79) becomes

$$V(\phi) = \frac{m_P^4}{16\beta}\left(1 - e^{-\sqrt{\frac{2}{3}}\phi/m_P}\right)^2 \,, \tag{7.81}$$

which is of the form in Eq. (7.49) with $q = \sqrt{\frac{2}{3}}$ and $V_0 = m_P^4/16\beta$. From Eq. (7.80), we also have $6(\partial\ln\Omega)^2 = (\partial\phi)^2/m_P^2$. Thus, in view of Eq. (7.74), we end up with the theory in Eq. (7.61) with $V(\phi)$ given by Eq. (7.81). As we did with Higgs inflation, we consider $N_* \simeq 57$ and $q = \sqrt{\frac{2}{3}}$ in Eq. (7.56) and obtain $V_0^{1/4} \simeq 7.9 \times 10^{15}$ GeV. Thus, we find $\beta = m_P^4/16V_0 \simeq 5.5225 \times 10^8$. As with Higgs inflation $r = 12/N^2$, which leads to $r = 3.7 \times 10^{-3}$ when $N_* \simeq 57$. Contrasting this model with the CMB observations is depicted in Fig. 5.5 (see legend). As shown, the model with $N_* \simeq 60$ hits the sweet spot of the Planck data.

The similarity between Starobinsky inflation and Higgs inflation is remarkable because in both cases $q = \sqrt{\frac{2}{3}}$. In fact, the two models are equivalent during inflation. This can be understood as follows. From Eq. (7.63) the Lagrangian density of Higgs inflation during inflation can be written as

$$\mathcal{L} \simeq \frac{1}{2}m_P^2 R + \frac{1}{2}\xi R\chi^2 - \frac{1}{4}\lambda\chi^4 \,, \tag{7.82}$$

where the contribution of the kinetic density term $\frac{1}{2}(\partial\chi)^2$ is negligible during inflation (because of potential domination) and we have considered that $U(\chi) \simeq \frac{1}{4}\lambda\chi^4$ during inflation when the value of χ is large. If we vary the action $S = \int d^4x\sqrt{-g}\,\mathcal{L}$ (cf. Eq. (7.1)) with respect to the field χ we obtain

$$\frac{\delta S}{\delta\chi} = \int d^4x\sqrt{-g}\left(\xi R\chi - \lambda\chi^3\right) \,. \tag{7.83}$$

From the least action principle, requiring $\delta S = 0$ implies

$$\chi^2 = \frac{\xi}{\lambda}R \,. \tag{7.84}$$

The reason for this is that omitting the kinetic term $\frac{1}{2}(\partial\chi)^2$ in Eq. (7.82) reduces χ to an *auxiliary* field, which is not dynamical and in place of its equation of motion, we have a constraint given by Eq. (7.84). Inserting this constraint into the Lagrangian density in Eq. (7.82) we obtain

$$\mathcal{L} = \frac{1}{2}m_P^2 R + \frac{\xi^2}{4\lambda}R^2 . \tag{7.85}$$

The above is the same as Eq. (7.76) if we identify $\beta = \xi^2/4\lambda$. Indeed, inserting the value $\xi = 47{,}000\sqrt{\lambda}$, which the observations suggest, we readily obtain $\beta = 5.5225 \times 10^8$.

However, Starobinsky inflation is not identical to Higgs inflation, because the two models do differ after inflation has ended and may lead to distinct reheating histories since the decay of the scalaron can be much different to the decay of the Higgs (or any other nonminimally-coupled scalar field χ). Thus the actual value of N_* would not necessarily be the same in both cases, which means the inflationary observables n_s and r can also differ.[*]

7.3.5 α-attractors

There are other ways to get to the Starobinsky-type inflation discussed in Sec. 7.3.1 apart from modified gravity. Indeed, in 2013, Renata Kallosh, Andrei Linde, and Diederik Roest [KLR13] showed that in supergravity, considering a nontrivial geometry not in physical spacetime but in field-space (also called configuration space) may introduce poles in the kinetic term of the Lagrangian density of a minimally-coupled (to gravity) scalar field φ, which can be written as

$$\mathcal{L}_{\text{kin}} = \frac{\frac{1}{2}(\partial\varphi)^2}{\left[1 - \frac{1}{6\alpha}\left(\frac{\varphi}{m_P}\right)^2\right]^2} , \tag{7.86}$$

where α is a constant parameter. Soon it was realised that one does not have to rely on super-gravity to obtain the above kinetic poles. Other theoretical motivations, e.g., from conformal field theory can also result in Eq. (7.86).

The form of the kinetic term in Eq. (7.86) suggests that one can switch to a canonically normalised scalar field ϕ with $\mathcal{L}_{\text{kin}} = \frac{1}{2}(\partial\phi)^2$ through the transformation

$$\frac{\mathrm{d}\varphi}{1 - \frac{1}{6\alpha}\left(\frac{\varphi}{m_P}\right)^2} = \mathrm{d}\phi$$

$$\Rightarrow \quad \varphi = \sqrt{6\alpha}\,m_P \tanh\left(\frac{1}{\sqrt{6\alpha}}\frac{\phi}{m_P}\right) . \tag{7.87}$$

From the above, we see that the poles at $\varphi = \pm\sqrt{6\alpha}\,m_P$ have been transposed to infinity $\phi \to \pm\infty$ by virtue of $\tanh(\pm\infty) = \pm1$. In effect, this "stretches" the potential of the canonical field ϕ near the poles of the $V(\varphi)$. Such "flattening" of the potential, gives rise to the desired inflationary plateau. The canonical field ϕ cannot run beyond infinity, which means that the noncanonical field φ cannot cross the poles. Indeed, the noncanonical field φ varies slower and slower as it approaches the poles.

---

[*]Higgs and R^2 inflation are special cases where the function $f(R,\chi)$ in Eq. (7.58) can be written as $f = FR$, where F is determined by a single degree of freedom $F = F(\chi)$, or $F = F(R)$. As we have explained, it is possible to reformulate the theory in the Einstein frame, where this degree of freedom gives rise to the canonical scalar field ϕ via Eq. (7.60). However, with a generic $f(R,\chi)$ we end up with two degrees of freedom, one due to the original χ-field and the other due to the scalaron.

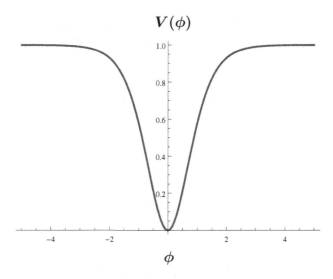

$$V(\phi)$$

FIGURE 7.8 The scalar potential in T-model inflation. The potential density is labeled as a fraction of V_0, while the inflaton field is measured in units of m_P/q.

The Starobinsky-type model is retrieved by considering that the potential of the original noncanonical field is

$$V(\varphi) = \frac{\frac{1}{2}m^2\varphi^2}{\left(1 + \frac{1}{\sqrt{6\alpha}}\frac{\varphi}{m_P}\right)^2},\tag{7.88}$$

where $m = $ constant. In terms of the canonical field, this becomes

$$V(\phi) = \frac{3}{4}\alpha m^2 m_P^2 \left(1 - e^{-\sqrt{\frac{2}{3\alpha}}\phi/m_P}\right)^2,\tag{7.89}$$

where we used Eq. (7.87). This potential is the same as the one in Eq. (7.49) with $V_0 = \frac{3}{4}\alpha m^2 m_P^2$ and $q = \sqrt{\frac{2}{3\alpha}}$ so that $\alpha = 1$ corresponds to the actual Starobinsky model in Eq. (7.81). In view of Eq. (7.54), we have

$$n_s = 1 - \frac{2}{N} \quad \text{and} \quad r = \frac{12\alpha}{N^2}.\tag{7.90}$$

Thus, we see that the value of the spectral index is independent of α. In the $(n_s\text{-}r)$-plot of Fig. 5.5, these models correspond to vertical lines because r grows proportionally to α, while n_s remains constant. As one can easily see, if α is not too large, agreement with the observations is excellent.

The model in Eq. (7.88) is called *E-model*. However, this is not the only possibility for this method of forming the inflationary plateau. Indeed, a simpler choice of $V(\varphi)$ is

$$V(\varphi) = \frac{1}{2}m^2\varphi^2,\tag{7.91}$$

with $m = $ constant. Again, employing Eq. (7.87) we arrive at the potential of the canonical field

$$V(\phi) = 3\alpha m^2 m_P^2 \tanh^2\left(\frac{1}{\sqrt{6\alpha}}\frac{\phi}{m_P}\right).\tag{7.92}$$

The form of the potential is shown in Fig. 7.8. Because of its shape it is called *T-model*.

Let us quickly study this model, which we can write as

$$V = V_0 \tanh^2 x,\tag{7.93}$$

where $V_0 = 3\alpha m^2 m_P^2$ and we define $x \equiv \frac{1}{\sqrt{6\alpha}} \phi/m_P$. Then, it is straightforward to obtain*

$$V' = \frac{2V_0}{\sqrt{6\alpha}\,m_P} \frac{\tanh x}{\cosh^2 x} \quad \text{and} \quad V'' = \frac{V_0}{3\alpha m_P^2} \frac{1 - 2\sinh^2 x}{\cosh^4 x}. \tag{7.94}$$

Using Eqs. (6.21) and (6.23), we find the slow-roll parameters

$$\varepsilon = \frac{1}{3\alpha} \frac{1}{\sinh^2 x(\sinh^2 x + 1)} \quad \text{and} \quad \eta = \frac{1}{3\alpha} \frac{1 - 2\sinh^2 x}{\sinh^2 x(\sinh^2 x + 1)}. \tag{7.95}$$

Putting the above in Eq. (6.83), we obtain

$$n_s = 1 - \frac{4}{3\alpha} \frac{1}{\sinh^2 x}. \tag{7.96}$$

Next, we find out when inflation ends, by setting $\varepsilon(x_{\rm end}) = 1$. Using Eq. (7.95), we find

$$\sinh^2 x_{\rm end} = \frac{1}{2}\left(\sqrt{1 + \frac{4}{3\alpha}} - 1\right), \tag{7.97}$$

where $x_{\rm end} \equiv \frac{1}{\sqrt{6\alpha}} \phi_{\rm end}/m_P$. From this and Eq. (6.22), we get

$$\sinh^2 x(N) = \frac{2N}{3\alpha} + \frac{1}{2}\left(\sqrt{1 + \frac{4}{3\alpha}} - 1\right). \tag{7.98}$$

Inserting the above into Eq. (7.96), we arrive at

$$n_s = 1 - \frac{2}{N + \frac{3\alpha}{4}\left(\sqrt{1 + \frac{4}{3\alpha}} - 1\right)} \simeq 1 - \frac{2}{N}. \tag{7.99}$$

Similarly, using Eq. (7.95), we find

$$r = 16\varepsilon = \frac{16}{\frac{4}{3\alpha}N^2 + 2N\sqrt{1 + \frac{4}{3\alpha}} + 1} \simeq \frac{12\alpha}{N^2}, \tag{7.100}$$

where in the last equation in Eqs. (7.99) and (7.100) we took $\alpha < 2N/3 \simeq 40$. The above findings are the same with the E-model case in Eq. (7.90) despite the fact that the T-model potential in Eq. (7.92) is very different from the Starobinsky potential. Indeed, it turns out that the form of the kinetic poles in Eq. (7.86) produces the same expressions for inflationary observables (the ones in Eq. (7.90)) regardless of the form of the scalar potential $V(\varphi)$. This is why these models are called "attractors."

Poles in the kinetic term of the scalar field Lagrangian density do not have to be of the same form as in Eq. (7.86). As a result, switching to the canonical field ϕ does not have to reproduce the α-attractors results, but the general rule is that kinetic poles "stretch" the scalar potential of the canonical field around the poles and produce thereby the inflationary plateau. Such inflation models are collectively called *pole inflation*. Depending of the order of the pole, when we turn to the canonical field we may arrive at Starobinsky inflation, T-model inflation but also hilltop inflation or inverted hilltop inflation and so on. If the noncanonical scalar potential $V(\varphi)$ also features the same pole, then the inflationary potential for the canonical field may not feature an inflationary plateau, depending on the order of the pole in the kinetic and the potential term.

---

*Note that the prime still denotes derivative with respect to ϕ.

7.4   k-inflation

In the last section we showed that it is possible to realise inflation not by means of a suitable scalar potential but by a suitable kinetic term for the inflaton field instead. This has been known for some time, and it is called *kinetically driven inflation* or k-inflation for short. It was first proposed by Christian Armendariz-Picon, Thibault Damour, and Viatcheslav F. Mukhanov at the end of last century, in 1999 [APDM99]. The idea is as follows.

Consider that the Universe is dominated by a single scalar field with Lagrangian density

$$\mathcal{L} = p(\varphi, X)\,, \tag{7.101}$$

where we defined

$$X \equiv \frac{1}{2}(\partial\varphi)^2\,. \tag{7.102}$$

If the field is homogeneous the above becomes $X = \frac{1}{2}\dot{\varphi}^2$, and the field is equivalent to a perfect fluid with pressure and density given, respectively, by

$$p = \mathcal{L} \quad \text{and} \quad \rho = 2X\partial_X p - p\,, \tag{7.103}$$

where $\partial_X \equiv \partial/\partial X$. In the familiar case of a minimally coupled, single scalar field we have $\mathcal{L} = X - V(\varphi)$ and the above reduce to Eq. (6.6).

Combining Eqs. (6.2) and (6.4) we obtain

$$\dot{\rho} = -\sqrt{3\rho}\,(\rho + p)/m_P\,. \tag{7.104}$$

Using this relation we can see that inflation can be a fixed point of the system. To this end, consider first that $p = p(X)$, i.e., p is a function of X only. The extrema of this function correspond to $\partial_X p = 0$. Thus, Eq. (7.103) suggests that $p = -\rho$ at the extrema, which is the equation of state of quasi-de Sitter inflation. Indeed, Eq. (7.104) (or Eq. (6.4)) suggests that, when $p = -\rho$ we have $\dot{\rho} = 0$ and from the Friedmann equation (6.2) we find $H = $ constant, which implies $a \propto e^{Ht}$. These are fixed points in phase space, and it can be shown that they are attractors. Indeed, Eq. (7.104) suggests that when $\rho > -p$ ($\rho < -p$), we have $\dot{\rho} < 0$ ($\dot{\rho} > 0$) so that the system is driven to the condition of inflation $\rho = -p$ and $\dot{\rho} = 0$. Crucially, however, we need to have a nontrivial kinetic term such that at the attractor $\partial_X p = 0$ we still have $X \neq 0$ (meaning $\dot{\varphi} \neq 0$). Otherwise, the "inflation" fixed point corresponds to $p = -\rho = 0$.

Such is the situation when p only depends on X, $p = p(X)$. However, exactly because inflation is an attractor of the system, this inflation would never end. To achieve "graceful exit" from inflation we need to introduce some explicit φ-dependence into p, exactly as in the α-attractors case, where $p(\varphi, X) = X[1 - \frac{1}{6\alpha}(\varphi/m_P)^2]^{-2}$. The Klein-Gordon equation is obtained using the continuity equation (6.4) in conjunction with Eq. (7.103). Employing that $\dot{X} = \dot{\varphi}\ddot{\varphi}$, we readily get

$$\ddot{\varphi}[2 + \dot{\varphi}^2\partial_X(\ln\partial_X p)] + 3H\dot{\varphi} + \dot{\varphi}^2\,\partial_\varphi(\ln\partial_X p) - \frac{\partial_\varphi p}{\partial_X p} = 0\,, \tag{7.105}$$

where we used $\frac{\mathrm{d}}{\mathrm{d}t}\partial_X p = \dot{\varphi}\partial_\varphi\partial_X p + \dot{X}\partial_X^2 p$ with $\partial_\varphi \equiv \partial/\partial\varphi$ and $\partial_X^2 p = \partial_X(\partial_X p)$. The logarithmic terms are negligible during inflation.*

For the scalar perturbations, we have

$$\mathcal{P}_\zeta = \frac{H^2}{8\pi^2 m_P^2 c_s \epsilon}\,, \tag{7.106}$$

---

*These terms are exactly zero in the standard case $\mathcal{L} = X - V = p$, when $\partial_X p = 1$. In this case, noting that $\partial_\varphi p = \ddot{\varphi} - V'$ ($\partial_\varphi X = \ddot{\varphi}$), it is straightforward to see that Eq. (7.105) reduces to Eq. (6.8).

where ϵ is given by Eq. (6.10) and the sound speed is (cf. Eq. (2.16))

$$c_s^2 \equiv \frac{\partial p}{\partial \rho} = \frac{\partial_X p}{\partial_X \rho} = \frac{\rho + p}{2X \partial_X \rho} = \frac{\partial_X p}{\partial_X p + 2X \partial_X^2 p} \,, \tag{7.107}$$

where $\partial_X^2 p \equiv \partial_X(\partial_X p)$, and we used Eq. (7.103). Setting $c_s \to c = 1$, Eq. (7.106) is readily reduced to Eq. (6.78), under the slow-roll approximation when $\epsilon \simeq \varepsilon$ given by Eq. (6.21) and the Friedmann equation is approximated with Eq. (6.20).*

For the scalar spectral index we have

$$n_s - 1 = -2\epsilon - \mathcal{E} - s \,, \tag{7.108}$$

where

$$\mathcal{E} \equiv -\frac{d \ln \epsilon}{dN} = \frac{\dot{\epsilon}}{\epsilon H} \,, \tag{7.109}$$

and

$$s \equiv -\frac{d \ln c_s}{dN} \,, \tag{7.110}$$

with $dN = -H dt$ as usual. Under the slow-roll approximation, we have $\epsilon \simeq \varepsilon$ and it is easy to show that $\mathcal{E} \simeq 4\varepsilon - 2\eta$ (cf. Eq. (6.24)) so that Eq. (7.108) reduces to Eq. (6.83) when $s = 0$.

For the tensor perturbation spectrum we have

$$\mathcal{P}_h = \frac{2H^2}{\pi^2 m_P^2} \,, \tag{7.111}$$

which is the same as Eq. (6.86). Dividing the above with Eq. (7.106), we find the tensor to scalar ratio

$$r = 16 \, c_s \epsilon \,. \tag{7.112}$$

Comparing this with the standard consistency relation in Eq. (6.88), $r = 16\varepsilon$, reveals the appeal of the k-inflation setup. By arranging such that $c_s < 1$, one may relax the observational constraint e.g., on the models of chaotic inflation which produce too much gravitational radiation. Thus, a suitable modification of the kinetic term in the Lagrangian density might render quadratic or quartic chaotic inflation acceptable. However, $c_s \ll 1$ results in substantial non-Gaussianity in the curvature perturbation, which has not been observed. Indeed, the observations set a lower bound on c_s, which allows the range:

$$0.024 < c_s \leq 1 \,. \tag{7.113}$$

Nevertheless, the above means that quadratic chaotic k-inflation may still be allowed.

Finally, k-inflation also incorporates the *Dirac-Born-Infeld* (DBI) inflation [AST04] arising in string theory, for which $p(\varphi, X) = -T(\varphi)\sqrt{1 - 2X/T(\varphi)} + T(\varphi) - V(\varphi)$, where $T(\varphi)$ is the "warp function" determined by the geometry of the extra dimensions. Obviously, when $2X \ll T(\varphi)$, the DBI Lagrangian density reduces to the standard $\mathcal{L} = p = X - V$.

7.5   Inflation with heavy fields

Modifying the kinetic term in the Lagrangian density, as in k-inflation, is not the only way to achieve a low value of the speed of sound c_s. In 2011, Ana Achúcarro, Jinn-Ouk Gong, Sjoerd

---

*In contrast to Eq. (6.78) though, the scalar perturbation spectrum in Eq. (7.106) is calculated at the exit of the *sound horizon* c_s/H.

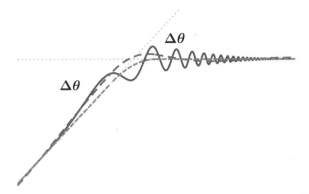

FIGURE 7.9 The inflaton's "trajectory" in field space when traversing a bend of the potential valley. The bottom of the valley is depicted with the short-dashed line. If the "velocity" of the inflaton in field space is moderate we expect that the field moveas away from the bottom of the valley and rises up the valley slope as it takes the turn. This case is depicted with the long-dashed line. If, however, the inflaton "velocity" is large then it oscillates around the bottom of the valley as depicted by the solid line. The total change in direction is parametrised by the angle $\Delta\theta$ between the original and the final directions, shown with dotted lines.

Hardeman, Gonzalo A. Palma, and Subodh P. Patil [AGH+11] showed that another way is considering an inflaton trajectory which is undergoing a turn in field space. Then there is a mixing of the light inflaton mode, which corresponds to the flat valley in which the system is rolling, with the heavy mode that corresponds to the perpendicular direction to the valley of minima. What this means is that the rolling system departs from the minimum of the value and climbs up the valley walls. In fact, depending on the steepness of the walls and the strength of the turn, the rolling system can begin oscillating around the minimum of the valley, as shown in Fig. 7.9. As a result, there is a reduction to the speed of sound, with which the curvature perturbations propagate. This reduction is given by

$$\frac{1}{c_s^2} = 1 + \frac{4\varpi}{M^2} = 1 + \frac{4\dot{\varphi}^2}{\kappa^2 M^2} = 1 + \left(\frac{\Delta\phi}{\kappa}\right)^2 \left(\frac{t_m}{t_\perp}\right)^2, \qquad (7.114)$$

where κ is the radius of curvature of the turn in field space,[*] and the turn has been parametrised by the angle θ, which corresponds to the change in the direction of roll $\Delta\varphi = \kappa\Delta\theta$, where $\varpi \equiv \dot{\theta} = \dot{\varphi}/\kappa$ is the angular velocity in field space. In the above equation, M is the effective mass, which is given by

$$M^2 = m^2 - \varpi^2, \qquad (7.115)$$

where m is the bare mass of the heavy mode. In Eq. (7.114), we have defined $t_m \equiv 1/M$, which is the period of oscillations and $t_\perp \equiv \Delta\varphi/\dot{\varphi} = \kappa\Delta\theta/\dot{\varphi}$, which is the time taken to complete the turn, where $\Delta\theta$ corresponds to the total change in the direction of roll.

The adiabaticity condition is

$$\left|\frac{\dot{\varpi}}{\varpi}\right| = \left|\frac{d}{dt}\ln(c_s^{-2}-1)\right| \ll M, \qquad (7.116)$$

which can be shown to be equivalent to $t_m \ll t_\perp$, and it means that the duration of the turn allows many oscillations to take place. Violation of adiabaticity results in particle production which backreacts on the rolling inflaton (the adiabatic mode).

---

[*]Note that in field space $[\kappa] = E$.

The scenario is very probable to realise in the context of fundamental theories, which include a multiple of scalar fields and where the scalar potential characterising these fields is complicated, as is the case of the string landscape for example (see Sec. 8.8.3). Turns in field space can be strong, where there is a substantial reduction of c_s and they can also be sudden when the variation of the speed of sound $\dot{c}_s$ is large (and $s > 1$ in Eq. (7.110)). However, a strong or a sudden turn does not necessarily lead to the violation of slow-roll.

As we have said, the speed of sound has to lie in the range shown in Eq. (7.113). Equations (7.106), (7.108), and (7.112) are still valid, although there can be oscillations in the power spectrum $\mathcal{P}_\zeta$ superimposed on the value given by Eq. (7.106). Such oscillations can be a characteristic signature of a turn in field space when the cosmological scales exit the horizon during inflation.

7.6 Warm inflation

In the previous sections, we saw that a suitable modification of the kinetic term can force the slow-roll of the inflaton field even if the scalar potential is steep. In the case of α-attractors (or in general pole inflation), we demonstrated that a pole in the kinetic term can halt the roll of the field because the latter cannot cross the pole. However, there are other ways to affect the roll of the inflaton and make sure that slow-roll is achieved regardless of the steepness of the scalar potential. One popular such way is to introduce an extra friction force to dampen the inflaton's variation, on top of the usual Hubble friction in the Klein-Gordon equation of motion (6.8). This extra friction is due to dissipation of the inflaton's energy density.

During inflation the inflaton field does not have a particle interpretation (because it is not oscillating) so it cannot directly decay into other fields a la perturbative reheating. However, suitable interactions between the inflaton field and other fields may allow it to excite other degrees of freedom, thereby dissipating some of its kinetic energy density and slowing-down its roll. In turn, the excited field(s) can produce a subdominant thermal bath, which does not destabilise inflation but can have a profound effect on the production of the curvature perturbation. This is because, apart from its quantum fluctuations, the inflaton field is now subject to thermal fluctuations as well. If the effect of thermal fluctuations is dominant over the quantum fluctuations then we depart from the usual, cold inflation paradigm with temperature $T \to 0$ that we have been discussing so far. This new version of inflation is called *warm inflation*, introduced by Arjun Berera in 1995 [Ber95].

In warm inflation the Klein-Gordon of the homogeneous scalar field is

$$\ddot{\phi} + (3H + \Upsilon)\dot{\phi} + V' = 0, \qquad (7.117)$$

where Υ is called the dissipation coefficient and it serves to dampen the roll of the inflaton field down the slope of the scalar potential, in a similar manner to Γ in Eq. (6.36). Its origin, however, is very different, as Γ is the perturbative decay rate which is inapplicable during inflation.

Dissipation generates a subdominant thermal bath of temperature T.* When $T < H$, then this thermal bath and the associated dissipation are negligible, and we are back, in effect, to the usual cold inflation. However, when $T > H$ the thermal fluctuations of the field cannot be ignored and inflation is now warm. There is ample range for T such that the radiation bath is subdominant with $\rho_\phi > \rho_r$, but inflation is warm with $T > H$. This range is

$$\sqrt{V}/m_P \sim H < T < V^{1/4}, \qquad (7.118)$$

where we used the slow-roll Friedmann equation (6.20), $V \sim (m_P H)^2$, and considered that during inflation $\rho_\phi \simeq V$ and $T \sim \rho_r^{1/4}$ as suggested by Eq. (6.1).

---

*We assume the generated radiation thermalises faster than a Hubble time.

The continuity equation (6.4) is valid for the total density $\rho = \rho_\phi + \rho_r$. However, the inflaton field and the radiation bath are not independent fluids because the inflaton net-decays to radiation. Thus, they are characterised by the equations*

$$\dot{\rho}_\phi + 3H(\rho_\phi + p_\phi) = -\Upsilon(\rho_\phi + p_\phi)\,, \tag{7.119}$$

$$\dot{\rho}_r + 3H(\rho_r + p_r) = \Upsilon(\rho_\phi + p_\phi)\,. \tag{7.120}$$

Using that $\rho_\phi = \frac{1}{2}\dot{\phi}^2 + V$ and $p_\phi = \frac{1}{2}\dot{\phi}^2 - V$ according to Eq. (6.6), we have $\rho_\phi + p_\phi = \dot{\phi}^2$. So Eq. (7.119) becomes Eq. (7.117) while Eq. (7.120) becomes

$$\dot{\rho}_r + 4H\rho_r = \Upsilon\dot{\phi}^2\,, \tag{7.121}$$

where we used the radiation equation of state $p_r = \frac{1}{3}\rho_r$ (cf. Sec. 2.3.4).

The generated radiation bath is inflated away but it is constantly replenished.[†] In fact, the system soon approaches the slow-roll attractor in which the acceleration term $\ddot{\phi}$ is negligible in the Klein-Gordon Eq. (7.117) and also $\dot{\rho}_r \ll 4H\rho_r$ in Eq. (7.121). Thus, the slow-roll equations for warm inflation are

$$3H(1+Q)\dot{\phi} \simeq -V' \tag{7.122}$$

$$\text{and} \qquad \rho_r \simeq \frac{3}{4}Q\dot{\phi}^2\,, \tag{7.123}$$

where we have defined

$$Q \equiv \frac{\Upsilon}{3H}\,. \tag{7.124}$$

The slow-roll conditions are now

$$\varepsilon, |\eta|, |\beta|, |\sigma| < 1 + Q\,, \tag{7.125}$$

where ε and η are given by Eqs (6.21) and (6.23), respectively, while β and σ are new slow-roll parameters defined as

$$\beta \equiv m_P^2 \frac{\Upsilon'V'}{\Upsilon V} \quad \text{and} \quad \sigma \equiv m_P^2 \frac{V'}{\phi V}\,. \tag{7.126}$$

Additionally, we need to satisfy the condition $\left|\frac{\mathrm{d}\ln\Upsilon}{\mathrm{d}\ln T}\right| < 4$. From Eq. (7.125), slow-roll requires $|\eta| < 1 + Q$. Provided $Q > 1$, we could have inflation even with $|\eta| \sim 1$, i.e., $m \sim H$ for the inflaton field (cf. Eq. (6.25)). Thus, the infamous η-problem of inflation model-building in supergravity and string theory is alleviated (see Sec. 6.2.4).

Combining Eqs. (7.122) with (7.123) and using Eq. (6.20), it is straightforward to show that

$$\rho_r \simeq \frac{\varepsilon Q V}{2(1+Q)^2}\,. \tag{7.127}$$

The above implies that the subdominant thermal bath can reheat the Universe after the end of inflation without the need of any additional decay mechanism. Indeed, at the end of inflation when $\varepsilon \simeq 1 + Q$ the temperature of the thermal bath is $T \sim (\frac{Q}{1+Q}V)^{1/4}$, where we used $\rho_r \sim T^4$, which suffices to reheat the Universe.

Considering that $dN = -Hdt = -(H/\dot{\phi})d\phi$, it can be readily shown that the number of remaining e-folds of inflation are

$$N = \frac{1}{m_P^2} \int_{\text{end}} \frac{V(1+Q)}{V'} d\phi\,, \tag{7.128}$$

where we employed also Eq. (7.122). This is the equivalent to Eq. (6.22) in warm inflation.

---

*Which when summed together reproduce Eq. (6.4).

[†]This is similar to the situation in gravitational reheating, as we discussed in Sec. 6.3.4.

The curvature perturbation spectrum is[*]

$$\mathcal{P}_\zeta = \left(\frac{H}{\dot{\phi}}\right)^2 \mathcal{P}_{\delta\phi}, \tag{7.129}$$

which, in cold inflation, can readily be shown to result in Eq. (6.78) provided that the inflaton perturbation spectrum is taken as $\mathcal{P}_{\delta\phi} = (H/2\pi)^2$ as in Eq. (6.71). However, in warm inflation the spectrum of the inflaton perturbations is

$$\mathcal{P}_{\delta\phi} = \left(\frac{H}{2\pi}\right)^2 \mathcal{F}, \quad \text{where} \quad \mathcal{F} \equiv \left(1 + 2\mathcal{N}_* + \frac{T}{H}\frac{2\pi Q}{\sqrt{1+\frac{4\pi}{3}Q}}\right), \tag{7.130}$$

where $\mathcal{N}_* = (e^{H/T} - 1)^{-1}$, so that $1 + 2\mathcal{N}_* = \coth(H/2T)$.[†] Note that in cold inflation when $T \to 0$ we have $\mathcal{N}_* = 0$ and $\mathcal{F} = 1$, and the above suggests that $\mathcal{P}_{\delta\phi} = T_H^2$, where $T_H = H/2\pi$ is the Hawking temperature of de Sitter space (cf. Eq. (4.27)). However, in the limit $T \gg H$ we have $\mathcal{N}_* \simeq T/H \gg 1$.

Eq. (7.122) suggests that $H/\dot{\phi} = -3H^2(1+Q)/V'$. Using this, Eqs. (7.129) and (7.130) give

$$\mathcal{P}_\zeta = \frac{H^2(1+Q)^2\mathcal{F}}{8\pi^2\varepsilon m_P^2}, \tag{7.131}$$

where ε is given by Eq. (6.21). The spectrum of the tensor perturbations is unaffected by warming up inflation, and it is still given by Eq. (7.111) (also Eq. (6.86)). Thus, we find the tensor to scalar ratio

$$r = \frac{\mathcal{P}_h}{\mathcal{P}_\zeta} = \frac{16\varepsilon}{(1+Q)^2\mathcal{F}}. \tag{7.132}$$

Because $\mathcal{F} \geq 1$ and $Q \geq 0$ we see that $\mathcal{P}_\zeta$ is enhanced while r is suppressed compared to the cold inflation case in Eqs. (6.78) and (6.88), respectively. This is great news for all these chaotic inflation models, which were very popular but produce excessive gravitational radiation as we have seen. Indeed, quadratic and quartic chaotic inflation become viable if "warmed up," i.e., if dissipation effects are nonnegligible. Moreover, the extra friction due to the dissipation effects implies that inflation continues even for sub-Planckian field values, in contrast to Eq. (7.10). If $\phi < m_P$, then the Planck-suppressed higher-order terms in the scalar potential are under control and the chaotic models are better justified theoretically (see the discussion in Sec. 7.2.2).

Finally, for the scalar spectral index, in the case of warm inflation we have

$$n_s - 1 = -\frac{17+9Q}{4(1+Q)^2}\varepsilon + \frac{3}{2(1+Q)}\eta - \frac{1+9Q}{4(1+Q)^2}\beta. \tag{7.133}$$

It is evident that the crucial quantity determining the dynamics but also the observables of warm inflation is the value of Q. If $Q \ll 1$ and $T < H$ dissipation is negligible and we are in fact in the usual cold inflation, with $\mathcal{F} = 1$. If, however, $T > H$ inflation is warm. Then there are two possibilities.

---

[*]This relation suggests

$$\zeta = \frac{H}{\dot{\phi}}\delta\phi = H\delta t = \delta N,$$

which directly follows from the definition of ζ in Eq. (5.23).

[†]There is also a minor correction to $\mathcal{F}$ when $Q \gtrsim 1$.

If $Q < 1$ then the dissipation does not affect the dynamics of inflation because Eq. (7.122) reduces to the standard slow-roll equation (6.19). However, because $T > H$, thermal fluctuations dominate quantum fluctuations and $\mathcal{F} \simeq T/H > 1$. This means that the spectrum of inflaton perturbations is intensified as $\sqrt{\mathcal{P}_{\delta\phi}} \sim \sqrt{T/H}(H/2\pi)$. Consequently, the curvature perturbation is enhanced accordingly. Matching to observations allows inflation to take place at lower energy. This is called the *weak dissipative regime*.

If we have $Q > 1$, the dynamics of inflation is now affected additionally to the perturbations. Extra friction in the Klein-Gordon equation allows inflation even with a steep scalar potential, e.g., even when the field is not very high up in the chaotic potential, which may achieve sub-Planckian chaotic inflation. The perturbation spectrum is further intensified $\sqrt{\mathcal{P}_{\delta\phi}} \sim Q^{1/4}\sqrt{T/H}(H/2\pi)$. This is called the *strong dissipative regime*. In summary we have:

$$
\begin{array}{llll}
\text{Cold Inflation} & (Q \ll 1, T < H): & \sqrt{\mathcal{P}_{\delta\phi}} = H/2\pi & r = 16\varepsilon \qquad (7.134) \\
\text{Weak Dissipation} & (Q < 1, T > H): & \sqrt{\mathcal{P}_{\delta\phi}} \simeq \sqrt{T/H}(H/2\pi) & r \simeq 16\varepsilon(H/T) \\
\text{Strong Dissipation} & (Q > 1, T > H): & \sqrt{\mathcal{P}_{\delta\phi}} \simeq Q^{1/4}\sqrt{T/H}(H/2\pi) & r \simeq 16\varepsilon(H/T)Q^{-5/2}
\end{array}
$$

The value of Q is determined by the value of the dissipation coefficient Υ, which is a product of the dissipation mechanism and it is highly model dependent. As we have discussed, the dissipation of the inflaton's kinetic energy density is indirect. Through a suitable coupling, the inflaton density is transferred to a mediator field χ, which in turn decays to light fermion fields that form the thermal bath. Depending on whether the mediator field is heavy or not determines two regimes.

In the *high-temperature* regime, we have $m_\chi < T$, where m_χ is the mass of the χ-field. Then, the dissipation coefficient is linear to temperature $\Upsilon \propto T$. In the *low-temperature* regime, however, $m_\chi > T$ and the mediator field is heavy. Then the dissipation coefficient is of the form $\Upsilon \propto T(T/m_\chi)^\alpha$, where the value of α depends on the nature of the χ-field. If the χ-field is a fermion then $\alpha = 4$. But this is subdominant to possible dissipation through a boson field, in which case $\alpha = 2$. Since $m_\chi \propto \phi$, we have that $\Upsilon \propto T^3/\phi^2$ in this case. The above are valid for supersymmetric theories. In non-supersymmetric theories $\Upsilon \propto 1/T$, while it is also possible that Υ is actually T-independent and $\Upsilon \propto \phi$. All these possibilities can be explored if we write

$$
\Upsilon = C \frac{T^m}{\phi^{m-1}}, \qquad (7.135)
$$

where C is a dimensionless model-dependent constant. We see that $m = 1$ for the high-T regime, while $m = 3$ for the low-T regime, but we also can have $m = 5$ for a fermion mediator field, or $m = -1$ for non-supersymmetric theories and $m = 0$ for no T-dependence of Υ. The more researched possibilities are $\Upsilon = C_T T$ in the high-T regime, where $C_T = 10^{-7} - 10^{-1}$ and $\Upsilon = C_\phi T^3/\phi^2$ in the low-T regime, where $C_\phi = 10^2 - 10^6$ or so.

The mechanism of warm inflation has been considered with a multitude of inflation models, mostly of the archetype kind, like chaotic, hilltop or hybrid inflation. It also has been combined with other ideas, such as k-inflation, which we have discussed in Sec. 7.4. For example, in the weak dissipative regime, warm k-inflation results in

$$
\mathcal{P}_\zeta \simeq \frac{1}{c_s} \frac{T}{H} \left(\frac{H}{\dot{\phi}}\right)^2 \left(\frac{H}{2\pi}\right)^2 \quad \text{and} \quad r \simeq \frac{H}{T} 16 c_s \varepsilon, \qquad (7.136)
$$

which can be obtained by Eqs. (7.106) and (7.112) by multiplying or dividing, respectively, with the factor $T/H > 1$. We see that, because $c_s \leq 1$, both the k-inflation and the warm inflation effects push the values of $\mathcal{P}_\zeta$ and r in the same directions, enhancing the curvature perturbation and suppressing the tensor-to-scalar ratio.

In this chapter, we have taken a brief look into how the inflationary paradigm accounts for cosmological observations. This is attained through the slow-roll attractor, which determines the dynamics and the form of successful inflation models. However, slow-roll is a necessary but not sufficient condition for this success because the precision of the observations is by now so selective as to rule out classes of models and put severe tension on others. Nevertheless, there is more to inflation than slow-roll. In the next chapter, we discuss a number of other possibilities, which correspond to how inflation may occur before or after the horizon exit of the cosmological scales. In fact, we mention ways that a dominant scalar field can lead to inflation beyond slow-roll. This provides insights on the behaviour of a scalar field condensate, when dominant and of the resulting dynamics of the Universe. Moreover, it might reveal more on the thorny issue of inflation's initial conditions, or inflation's end.

EXERCISES

1. In Dirac-Born-Infeld (DBI) inflation, the pressure is

$$\mathcal{L} = p = -T\sqrt{1 - 2X/T} + T - V\,,$$

where T is called the warp function and V is the scalar potential, with X standing for the kinetic density $X = \rho_{\text{kin}}$.
Find c_s^2, where c_s is the speed of sound for this fluid. Demonstrate that, if we demand $c_s^2 \approx 1$ then we must have $T \gg 2X$. Show that, in this limit we recover a minimally coupled scalar field with $\mathcal{L} = X - V$.

2. Study the model of quartic hilltop inflation without any assumptions apart from that the value of the inflaton field is smaller than its VEV, $\phi < \langle\phi\rangle$ (but not much smaller), such that higher order terms in the scalar potential are safely ignored. In particular:

   (a) Find the slow-roll parameters ε and η as functions of ϕ.

   (b) Show that the inflaton field as a function of the number N of remaining e-folds of inflation $\phi = \phi(N)$ is given by

$$\frac{\phi}{m_P} = 2\sqrt{\bar{N}[Z]}\,,$$

   where $[Z] \equiv 1 - \sqrt{1 - \frac{1}{Z}} > 0$, with $Z \equiv 16\lambda\bar{N}^2 > 0$. In the above, we have defined $\bar{N} = N + N_{\text{end}}$ with

$$N_{\text{end}} \equiv \frac{1}{8\lambda}\left[\left(\frac{\phi_{\text{end}}}{m_P}\right)^{-2} + \lambda\left(\frac{\phi_{\text{end}}}{m_P}\right)^2\right]\,,$$

   where inflation ends when $\phi = \phi_{\text{end}}$.

   (c) Show that the scalar spectral index is

$$n_s = 1 - \frac{3}{\bar{N}}\frac{Z^2[Z]^2}{(1 - Z[Z])^2}\,.$$

   The above reduces to $n_s = 1 - 3/\bar{N}$ in both limits $Z \gg 1$ and $Z \ll 1$.
   [Hint: Use that $Z[Z]^2 = 2Z[Z] - 1$.]

   (d) The expression for the spectral index can be rewritten as

$$\frac{Z[Z]}{|1 - Z[Z]|} = \sqrt{(1 - n_s)\bar{N}/3} \equiv X > 0\,.$$

   Taking $Z[Z] < 1$, solve the above and find the relation between Z and X. Then verify the condition $Z[Z] < 1$.

   [Hint: Define $\omega \equiv \sqrt{1 - \frac{1}{Z}}$, such that $[Z] = 1 - \omega$ and $\frac{1}{Z} = 1 - \omega^2$.]

   (e) Consider both cases when inflation ends because $|\eta| \simeq 1$ or $\varepsilon \simeq 1$. Show that, no matter which slow-roll parameter is taken to end inflation, we have

$$N_{\text{end}} \simeq \begin{cases} \frac{1}{4\sqrt{\lambda}} & \text{when} \quad \lambda \ll 1\,, \\[2mm] \lesssim 1 & \text{when} \quad \lambda \gg 1\,. \end{cases}$$

(f) When $\lambda \gg 1$ the contribution of N_{end} to $\bar{N}$ is negligible. This really means that the spectrum is too red when $N \approx 60$. However, this is not so when $\lambda \ll 1$, in which case $\bar{N} \simeq N + \frac{1}{4\sqrt{\lambda}}$ is large. Using this, find the following analytic expression for λ, which can generate any given value of n_s and N.

$$\sqrt{\lambda} = \frac{2(1 - n_s)N - 3}{4N[3 - (1 - n_s)N]} \, .$$

Choosing $n_s = 0.965$ and $N = 60$, find the value of λ.

(g) Show that the tensor to scalar ratio is

$$r = \frac{8}{3}(1 - n_s) \left\{ 1 - \frac{\sqrt{3[2(1 - n_s)N - 3]}}{(1 - n_s)N} \right\} \, .$$

Choosing $n_s = 0.965$ and $N = 60$, find the value of r.

8

Beyond Slow-Roll Inflation

8.1 Old inflation 171
8.2 Power-law inflation.................................. 173
8.3 Fast-roll inflation 175
8.4 Thermal inflation.................................... 177
8.5 Oscillating inflation 179
8.6 Ultra-slow-roll inflation............................. 181
8.7 Stochastic inflation.................................. 185
8.8 Eternal inflation 188
     The basic idea • Eternal inflation in action • The
     multiverse hypothesis
8.9 The beginning of time 195
     Chaotic initial conditions • A Universe from nothing •
     Sub-Planckian inflation • Homogenising the horizon
     before inflation • Inflation after the bounce

In the previous chapter, we have discussed ways to attain slow-roll inflation within the inflationary paradigm. We have seen that slow-roll inflation can be the result of a suitably flat scalar potential for the inflaton, or of a suitable nonminimal inflaton kinetic term, or a nonminimal direct coupling of the inflaton to gravity, or extra impedance in the inflaton's variation due to dissipative effects. Such proposals can be in agreement with observations provided the model parameters are tuned accordingly. In fact, each model has testable predictions which can be used to constrain it or even rule it out.

In contrast, in this chapter we lose contact with the observations and look into ways that a scalar field can make the Universe inflate without leading to slow-roll and therefore cannot be in agreement with the CMB. Why bother then? Well, the behaviour of a scalar field when it dominates the Universe is instructive regarding how exactly inflation is facilitated, what its initial conditions are (is inflation itself likely?), whether slow-roll can be destabilised and how and so on. Furthermore, it is possible that observations might be indirectly affected, for example by a period of thermal inflation, and this can resurrect some of the models reviewed in the previous chapter, which would otherwise be ruled out. Finally, accelerated expansion beyond slow-roll might also account for current dark energy, as we explore in Chapter 9.

Thus, we begin our tour of the bizarre and the exotic, with one of the oldest ideas.

8.1 Old inflation

In his 1981 seminal paper, where he coined the term "inflation," Alan H. Guth was the first to demonstrate that an inflationary period can overcome the horizon and flatness problems of the Hot Big Bang (see Secs. 4.1.1 and 4.1.2). To this end, he presented an inflation model which is now called *old inflation* [Gut81].

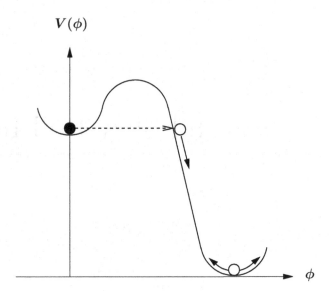

FIGURE 8.1 Schematic representation of the scalar potential in old inflation. The field is trapped in a metastable minimum, until it quantum-mechanically tunnels through the potential barrier and falls into its VEV.

The idea of old inflation is that the inflaton scalar field is not varying, but sitting instead at a metastable minimum in its scalar potential, see Fig. 8.1. While the field is located in its metastable minimum, the false vacuum constant potential density acts an effective cosmological constant and forces the Universe to inflate. Classically, this situation would remain unchanged forever and inflation would never end. Quantum mechanics though, suggests that the system eventually tunnels through the potential barrier and finds itself in the true vacuum. In physical space, this would mean that bubbles of the true vacuum phase would be nucleated here and there. The energy difference between the false and the true vacuum would make the bubbles expand with lightspeed. This is but a first-order phase transition, like that of the boiling of water, which completes when all space is filled with bubbles which coalesce and so, eventually the system is in the true vacuum everywhere. Only in old inflation, the bubbles cannot coalesce!

The reason is that between the bubbles, the Universe is still in the false vacuum and therefore undergoes inflation, which means that space in between bubbles is expanding superluminally. Thus, even though the bubble walls of neighbouring bubbles move with lightspeed they do not have a chance to meet each other so that the bubbles can coalesce. OK, you might say, forget the other bubbles. Suppose that we find ourselves inside one of these bubbles and its walls are still expanding with lightspeed but beyond our cosmological horizon today, so that we cannot see them. Sure, there is more of the Universe still undergoing inflation beyond the bubble walls but we cannot experience this either. Wouldn't this mean that old inflation can indeed create a homogeneous and isotropic universe? Bubble coalescence would not be needed for this.

It is true that such a Universe would comply with the cosmological principle, but it would be empty. This is because all the latent energy released by the switch from the false to the true vacuum is given to the bubble walls and propels their motion. If the walls of our bubble are beyond the horizon, this energy is lost to us and there is nothing left over from the inflationary past apart from gravitational radiation (i.e., no "us"). Thus, it is imperative for the bubbles to coalesce, because only such collisions can reheat the Universe. On the other hand, if one tries to tune the model such that the bubble nucleation rate is bigger than the Hubble rate, so that bubbles can coalesce before the Universe expansion inflates them apart, then the inflation era ends immediately and it is not enough to overcome the horizon and flatness problems. This problem is called the problem of *graceful exit*.

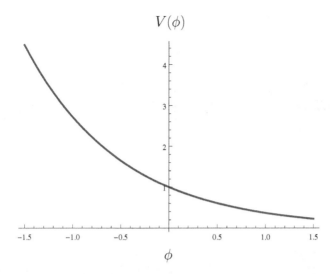

FIGURE 8.2 The scalar potential in power-law inflation. Fiducial units are used.

There is no solution. Old inflation is all but excluded (but a phase of old inflation could still be incorporated in inflation model-building).

8.2 Power-law inflation

Most processes in Nature are of two types. In one case, an effect has an outcome that counteracts its cause. This backreacts to the effect and reverses it, leading to oscillations, which are typically harmonic. In the other case, an effect has an outcome which reinforces its cause. This leads to further intensification of the effect, leading to a runaway explosive behaviour, which typically results in exponential growth. In fact, both possibilities are two sides of the same coin, since exponential explosion corresponds to a vicious (or virtuous) cycle, while the exponent of an imaginary number is harmonic, $e^{i\theta} = \cos\theta + i\sin\theta$. This is why a scalar potential of harmonic or exponential form is quite natural. We have already explored a harmonic inflaton scalar potential in the case of natural inflation in Sec. 7.2.3. In this section we consider an exponential potential. Such a nonperturbative potential can have many theoretical justifications.

Consider the potential

$$V(\phi) = V_0 \exp(-\lambda\phi/m_P), \qquad (8.1)$$

where V_0 is a constant density scale and λ is a constant. Without loss of generality we take $\lambda > 0$, noting that a negative λ can be turned positive if we switch the orientation of ϕ in field space: $\phi \to -\phi$.* The value of V_0 has no physical significance because if we choose to transpose the origin in field space $\phi \to \phi + \phi_0$ with $\phi_0 = $ constant, V_0 is modified accordingly $V_0 \to V_0 e^{-\lambda\phi_0/m_P}$. This is not so for λ, whose value has a profound impact on the dynamics as we discuss below. The potential is shown in Fig. 8.2.

For the potential in Eq. (8.1), the Klein-Gordon equation (6.8) of a homogeneous scalar field has the following exact solution:

$$V = \frac{2(6 - \lambda^2)}{\lambda^4} \left(\frac{m_P}{t}\right)^2 \quad \text{and} \quad \rho_{\text{kin}} \equiv \frac{1}{2}\dot{\phi}^2 = \frac{2}{\lambda^2} \left(\frac{m_P}{t}\right)^2. \qquad (8.2)$$

---

*The trivial case $\lambda = 0$ corresponds to a cosmological constant $\Lambda = V_0/m_P^2$.

Combining the above with Eq. (8.1) we find that

$$\phi(t) = -\frac{1}{\lambda}m_P \ln\left[\frac{2(6-\lambda^2)}{\lambda^4 V_0}\left(\frac{m_P}{t}\right)^2\right].$$ (8.3)

Stability analysis demonstrates that this is an attractor solution, similar to the slow-roll attractor.
    Using the flat Friedmann equation (6.2), we find

$$H(t) = \sqrt{\frac{\rho_{\rm kin} + V}{3m_P^2}} = \frac{2}{\lambda^2 t}.$$ (8.4)

Putting the above in Eq. (6.8), it can be readily shown that Eq. (8.2) is indeed a solution to the Klein-Gordon equation (6.8).* From Eq. (8.2), we obtain the barotropic parameter

$$w = \frac{\rho_{\rm kin} - V}{\rho_{\rm kin} + V} = \frac{\lambda^2}{3} - 1,$$ (8.5)

where we used Eq. (6.7). Since w is constant, we can employ the findings of Sec. 2.8. Indeed, Eqs. (2.40) and (8.5) readily confirm Eq. (8.4). Using Eq. (6.11), we obtain

$$\epsilon = -\dot{H}/H^2 = \lambda^2/2.$$ (8.6)

Then, from Eq. (2.39) we find

$$a \propto t^{2/\lambda^2} = t^{1/\epsilon},$$ (8.7)

while from Eq. (2.17) (valid for constant w) we get

$$\rho_\phi \propto a^{-3(1+w)} = a^{-\lambda^2} = a^{-2\epsilon}.$$ (8.8)

    The above findings demonstrate that we can have inflation when $\lambda < \sqrt{2}$ because then $\epsilon < 1$ (equivalently $w < -\frac{1}{3}$). However, this type of inflation is not quasi-de Sitter because H decreases as $H \propto 1/t$. Consequently, the scale factor does not grow exponentially with time, but instead as a power-law, as shown in Eq. (8.7). This is why this kind of inflation is called *power-law inflation*, a term coined by Francesco Lucchin and Sabino Matarrese, who introduced the model in 1985 [LM85]. Of course, long enough inflation successfully deals with the horizon and flatness problems like old inflation does. But, also like old inflation, it has a graceful exit problem. Indeed, if $\lambda < \sqrt{2}$ inflation does take place but never stops. To facilitate reheating, one would need to augment the model, e.g., by using a hybrid mechanism as discussed in Sec. 7.2.4,
    In power-law inflation, the cosmological horizon is an event horizon and we do have particle production, (cf. Sec. 4.3.4) because, for small λ, we have $m^2 \equiv V'' < H^2$ so that the inflaton field is light. If λ is very small the generated spectrum of perturbations is approximately scale invariant. Indeed, Eq. (8.6) suggests $\epsilon \ll 1$ for $\lambda \ll 1$. In this case, Eq. (6.13) implies that H is "robust", i.e., it remains largely unchanged while the cosmological scales exit the horizon. Then, we can calculate the slow-roll parameters using Eqs. (6.21) and (6.23), which give

$$\eta = 2\varepsilon = \lambda^2,$$ (8.9)

---

*To verify this, you may use

$$V' = \frac{\dot{V}}{\dot{\phi}} = \frac{2(6-\lambda^2)}{\lambda^3}\frac{m_P}{t^2}.$$

where comparison with Eq. (8.6) reveals that $\varepsilon = \epsilon$ exactly. The spectral index and the tensor-to-scalar ratio are

$$n_s = 1 - \lambda^2 \quad \text{and} \quad r = 6\lambda^2 , \tag{8.10}$$

where we used Eq. (6.83). Note that the inflationary observables do not depend on the number of remaining e-folds of inflation as is typically the case. From Eq. (5.41), we have that $n_s \simeq 0.965$, which suggests $\lambda^2 \simeq 0.035$. Then, the above equation implies that $r \simeq 0.21$, which is way too large compared to the upper bound in Eq. (5.41). Indeed, the bound $r < 0.056$ suggests that $n_s > 0.991$, which is too large. Therefore, power-law inflation cannot be used to explain the CMB primordial anisotropy or structure formation. The spectacular failure of the model is shown in Fig. 5.5, where its predictions are depicted by the dashed line (see legend). However, it can be a phase of inflation which does not correspond to the exit from the horizon of the cosmological scales, but is still part of inflation model-building, see Sec. 8.9.4.

8.3   Fast-roll inflation

Fast-roll inflation was introduced by Andrei Linde in 2001 [Lin01]. It involves the scalar potential of quadratic hilltop inflation in Eq. (7.25), repeated here

$$V(\phi) = V_0 - \frac{1}{2}m^2\phi^2 + \cdots , \tag{8.11}$$

but with the crucial difference that $|\eta| \sim 1$ so that the field does not undergo slow-roll but instead it fast-rolls down the scalar potential. Then, Eq. (6.25) suggests

$$|\eta| = \frac{1}{3}\left(\frac{m}{H}\right)^2 \sim 1 \Rightarrow m \sim H . \tag{8.12}$$

As we have already mentioned, this is natural in supergravity theories, and, in fact, it is the source of the infamous η-problem of inflation model-building.

Before the inflaton field reaches its VEV $\langle\phi\rangle$, the undetermined stabilising term in the potential (the one which is implied with the ellipsis in Eq. (8.11)) is not important. At the minimum, we presume that the residual potential density is negligible,* so that $V(\langle\phi\rangle) \approx 0$. This means that

$$m^2\langle\phi\rangle^2 \sim V_0 \simeq \rho = 3H^2m_P^2 \Rightarrow \langle\phi\rangle \sim m_P , \tag{8.13}$$

and we used the flat Friedmann equation (6.2). A Planckian VEV suggests that the inflaton field might be a string modulus, in the sense that such fields are typically flat directions in field space (flat means $V \simeq$ constant), whose scalar potential density is only expected to vary over at least Planckian distances in field space. This is why such a model is also called *modular inflation*.

However, do we indeed have inflation here? Let us look into it. When $\phi \ll \langle\phi\rangle$ we may ignore the undetermined stabilising term in the potential. Then the Klein-Gordon equation (6.8) is

$$\ddot{\phi} + 3H\dot{\phi} - m^2\phi \simeq 0 . \tag{8.14}$$

We assume that $H \simeq$ constant as implied by Eq. (8.12). In this case, the above equation admits solutions of the form $\phi \propto e^{\omega\Delta t}$, with Δt being the elapsing time and ω being

$$\omega = -\frac{3}{2}\left[1 \pm \sqrt{1 + \frac{4}{9}\left(\frac{m}{H}\right)^2}\right]H . \tag{8.15}$$

---

*This might allow a tiny cosmological constant, which is way smaller than the energies considered here.

Keeping only the growing mode we obtain that

$$\phi = \phi_0 \exp(F \Delta N) \,, \tag{8.16}$$

where $\Delta N = H \Delta t$ are the elapsing e-folds, ϕ_0 is some initial inflaton value and

$$F \equiv \frac{3}{2} \left(\sqrt{1 + \frac{4}{3}|\eta|} - 1 \right) \,, \tag{8.17}$$

where we also used Eq. (8.12). We see that $F \sim 1$ unless $|\eta|$ is very small.

Eq. (8.16) suggests that $\dot{\phi} = FH\phi$. Using this and also Eq. (6.15), we find

$$\epsilon = -\frac{\dot{H}}{H^2} = \frac{1}{2}F^2 \left(\frac{\phi}{m_P} \right)^2 \,. \tag{8.18}$$

Thus, as the inflaton fast-rolls down the potential hill, we have $\epsilon \ll 1$, because before reaching the VEV, $\phi \ll m_P$. This not only means that there is inflation but it also supports the "robustness of H" (cf. Eq. (6.13)), which substantiates our assumption of $H \simeq$ constant. The reason is that, when $\phi \ll \langle\phi\rangle \sim m_P$, we have $V_0 \gg m^2\phi^2$ so that $V(\phi) \simeq V_0 =$ constant and inflation is quasi-de Sitter.

The total e-folds of fast-roll inflation are easy to obtain using Eq. (8.16), which gives

$$N_{\text{FR}} = \frac{1}{F} \ln\left(\frac{\langle\phi\rangle}{\phi_0}\right) \simeq \frac{1}{F} \ln\left(\frac{m_P}{H}\right) \simeq \frac{1}{2F} \ln\left(\frac{m_P^4}{V_0}\right) \,, \tag{8.19}$$

where we have assumed that the inflaton is kicked off from the top of the potential hill by a quantum fluctuation, which implies that the initial value is $\phi_0 \sim \delta\phi \sim H \sim m$ (cf. Eq. (6.67)). We also considered Eq. (6.20), which suggests $H \sim \sqrt{V_0}/m_P$.

Fast-roll inflation cannot reproduce the observed perturbation spectrum because the spectrum it generates is too tilted (the spectral index is $n_s \simeq 1 + 2\eta$, with $|\eta| \sim 1$). However, if it occurs much later (meaning at a much lower energy scale) than primordial inflation, which is the one responsible for the CMB primordial anisotropy and structure formation, fast-roll inflation can be useful in diluting dangerous relics such as monopoles, gravitinos, or moduli particles if these appear after the end of primordial inflation (so that they are not diluted by it). For example, if primordial inflation were hybrid inflation and the phase transition which terminates it generated monopoles (or primordial black holes), then we would need some additional inflation period to get rid of them. The energy scale of fast-roll inflation is $V_0^{1/4} \sim \sqrt{m_P m} > 10^{10}$ GeV, where we assumed that the inflaton mass is $m > 100$ GeV for otherwise it could have already been seen in CERN, for example. Note that fast-roll inflation dilutes away any preexisting thermal bath due to primordial inflation, which means that it is the inflaton field of fast-roll inflation whose decay produces the thermal bath of the Hot Big Bang. Therefore, the fast-roll inflaton particle must be coupled to some standard model particles, and hence we cannot assume it is very light if this coupling is large enough to render it observable.

The existence of a period of fast-roll inflation at some intermediate energy scale $V_0^{1/4}$ means that the observable 60 or so e-folds of inflation do not correspond only to primordial inflation. In this case, the total number of e-folds is a sum of the fast-roll e-folds, N_{FR}, and the primordial ones, N_*. Hence, $N_* + N_{\text{FR}} \simeq 60$, which means that the value of N_* can be substantially smaller than 60. This can resurrect some inflation models, for example supergravity hybrid inflation (see Sec. 7.2.4). However, how long fast-roll inflation lasts not only depends on the model parameters, such as the mass of the inflaton field, it also depends on its initial value ϕ_0. How can we arrange to have the inflaton located up the potential hill in the first place? One way is to consider thermal effects, as we discuss next.

8.4 Thermal inflation

Thermal effects can have a profound impact on the dynamics of a scalar field coupled to the thermal bath. This was studied by David H. Lyth and Ewan D. Stewart, who proposed *thermal inflation* in 1996 [LS96]. The idea is the following.

Consider that the Universe, after primordial inflation, is dominated by a radiation thermal bath, but that there is also a scalar field present coupled to this radiation. Then, the density of the Universe is (c.f. Eq. (6.1))

$$\rho = \rho_r + \rho_\phi = \frac{\pi^2}{30}g_*T^4 + \frac{1}{2}(\partial\phi)^2 + V(\phi, T)\,, \tag{8.20}$$

with

$$V(\phi, T) = V_0 - \frac{1}{2}m^2\phi^2 + \lambda_n \frac{\phi^{2n+4}}{m_P^{2n}} + \frac{1}{2}g^2T^2\phi^2\,, \tag{8.21}$$

where $\lambda_n, g \lesssim 1$ are perturbative coupling constants and $n > 0$ is a parameter, determining the order of the term stabilising the potential. We have assumed that the quartic self-interaction term is absent and (for simplicity) we considered only even powers, i.e., we have imposed a mirror symmetry $\phi \to -\phi$.* A scalar field with no self-interaction is called a *flaton* field. It can be a flat direction in a supersymmetric theory, lifted by supersymmetry breaking effects, which introduce a "soft" mass $m \sim 1\,\mathrm{TeV}$ (tachyonic in our case).[†]

The VEV $\langle\phi\rangle$ is obtained by setting $V'(\phi, T = 0) = 0$, which gives

$$\langle\phi\rangle = \pm\left[\frac{m\,m_P^n}{\sqrt{2(n+2)\lambda_n}}\right]^{\frac{1}{n+1}} \sim \pm(m\,m_P^n)^{\frac{1}{n+1}}\,, \tag{8.22}$$

where in the last equality we assumed that λ_n is not too small. Requiring that there is no residual potential density means that $V(\langle\phi\rangle, T = 0) = 0$, which suggests

$$V_0 = \frac{1}{2}\left(\frac{n+3}{n+2}\right)m^2\langle\phi\rangle^2 \sim m^2\langle\phi\rangle^2\,. \tag{8.23}$$

Now, when the temperature is very high, the thermal correction term in Eq. (8.21) makes sure that the minimum of the scalar potential is at the origin $\phi = 0$. This means that the scalar field soon finds itself stabilised on top of the potential hill. Then, the density of the Universe is

$$\rho = \frac{\pi^2}{30}g_*T^4 + V_0\,. \tag{8.24}$$

The effective mass-squared on top of the hill is given by

$$m_{\mathrm{eff}}^2(T) = V''(\phi = 0, T) = g^2T^2 - m^2\,. \tag{8.25}$$

The above suggest that the field remains on top of the potential hill even if the temperature drops below

$$T_0 = \left(\frac{30}{\pi^2 g_*}\right)^{1/4}V_0^{1/4} \sim V_0^{1/4}\,, \tag{8.26}$$

---

*This is called Z_2-symmetry.

[†]The thermal correction term in Eq. (8.21) can be understood in comparison with the interaction term in Eq. (6.44), where for a thermalised field we have $\chi \sim T$.

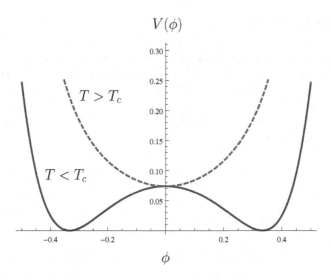

FIGURE 8.3 The scalar potential in thermal inflation. The origin is a global minimum only when $T > T_c$. Otherwise, two new minima at $\phi \neq 0$ develop, which approach $\pm|\langle\phi\rangle|$ when $T \to 0$.

as long as the effective mass-squared is positive, which remains so as long as the temperature is higher than

$$T_c = m/g\,. \tag{8.27}$$

Indeed, in the region $T_c < T < T_0$ we have $\rho \simeq V_0$ and $m_{\text{eff}}^2 > 0$. Thus, in this region of temperatures, the Universe is dominated by a constant potential density, which results in a period of inflation, called thermal inflation.* The scenario is shown in Fig. 8.3. In order to have thermal inflation, we need $T_0 > T_c$, which results in the bound

$$g > \left(\frac{\pi^2 g_*}{30}\right)^{1/4} \frac{m}{V_0^{1/4}} \sim \frac{m}{V_0^{1/4}}\,. \tag{8.28}$$

The total e-folds of thermal inflation are easy to find

$$N_T = \ln\left(\frac{a_{\text{end}}}{a_{\text{beg}}}\right) = \ln\left(\frac{T_0}{T_c}\right) = \ln\left[\left(\frac{30}{\pi^2 g_*}\right)^{1/4} \frac{g V_0^{1/4}}{m}\right] \simeq \ln\left(\frac{g V_0^{1/4}}{m}\right), \tag{8.29}$$

where "beg" and "end" denote the beginning and the end of thermal inflation, respectively, and we considered $a \propto 1/T$ (cf. Eq (3.11)). In view of Eqs. (8.22) and (8.23), the above can be written as

$$N_T \simeq \ln\left[g\left(\frac{m_P}{m}\right)^{\frac{n}{2(n+1)}}\right]. \tag{8.30}$$

The number of thermal inflation e-folds is maximised when $g \sim 1$ and $n \gg 1$. In this case, the above suggests $N_T^{\text{max}} \simeq \frac{1}{2}\ln(m_P/m)$. Taking $m \sim 1\,\text{TeV}$, we find that $N_T^{\text{max}} \simeq 17$. Thus, it is evident that thermal inflation cannot solve the horizon and flatness problems of the Hot Big Bang, because N_T is too small. However, it can inflate away unwanted relics, like monopoles, generated after (or at the end) of primordial inflation, as in the case of fast-roll inflation. Moreover, in the limit $n \gg 1$, Eq. (8.22) suggests that $|\langle\phi\rangle| \sim m_P$ and the $T = 0$ thermal inflation potential

---

*When $T > T_0$ the thermal inflaton field ϕ is confined at the origin, but its contribution to the density budget of the Universe is negligible because $T^4 \gg V_0$ (cf. Eq. (8.24)).

reduces to the (modular) fast-roll inflation potential in Eq. (8.11), with $m \sim H \sim \sqrt{V_0}/m_P$. In this case, thermal inflation can be followed by a period of fast-roll inflation when $T < T_{\text{end}}$ and the field is released from the origin. Then, the total e-folds of late inflation are $N_T + N_{\text{FR}}$, which implies that the number of remaining e-folds of primordial inflation when the cosmological scales exit the horizon is reduced even further $N_* \simeq 60 - (N_T + N_{\text{FR}})$. As we discussed in the previous section, this may resurrect inflation models, which would otherwise be excluded by observations.

Thermal inflation ends by the decay of the thermal inflaton field ϕ when oscillating around its VEV. The resulting radiation is the thermal bath of the Hot Big Bang, with temperature $T_{\text{reh}} \sim V_0^{1/4}$. Why doesn't this thermal bath send the field up the potential hill due to the thermal correction in Eq. (8.21)? The reason is that the new thermal bath is different from the one preexisting thermal inflation. To the new thermal bath the inflaton is coupled via a different coupling of the form $\frac{1}{2}\hat{g}^2 T^2 (\phi - \langle\phi\rangle)^2$, where $\hat{g} \neq g$. This coupling confines the field to its VEV $\langle\phi\rangle$. Thus, the preexisting thermal bath is not comprised of standard model degrees of freedom. This important subtlety supports the possibility that the thermal inflaton field is a modulus field, because we have two ESPs, one on top of the potential hill and one at the bottom (the VEV) which are $\sim m_P$ apart (because $|\langle\phi\rangle| \sim m_P$ in this case).

Thus, thermal inflation is one way to make sure that, initially, the inflaton field finds itself at the top of the potential hill. This was employed long ago in quadratic hilltop inflation, when it was called new inflation. However, this is not the only way to bring the inflaton to the top of the hill at the onset of inflation. Another possibility is to consider a hybrid setup, such that the role of the inflaton is played by the waterfall field (see Sec. 7.2.4) with the crucial difference that inflation continues after the phase transition of the hybrid mechanism and while the waterfall field rolls down towards its VEV. This roll can be slow-roll if the waterfall field is light with mass $m < H$ (in which case, we have *false vacuum inflation*) or fast-roll if its mass is $m \sim H$.

As we have seen, thermal inflation can substantially affect the predictions of primordial inflation in that it changes the number N_* of remaining e-folds of primordial inflation when the cosmological scales exit the horizon. This can have dramatic consequences for the choice of model of primordial inflation, excluding otherwise successful models or leading to success otherwise excluded models. In the following section, we briefly consider an even more extreme possibility, which does the same thing, i.e., adds inflationary e-folds such that N_* is diminished. In this case, slow-roll is so badly violated that the system undergoes oscillations, but still the Universe keeps inflating. Let us see how this is possible.

8.5   Oscillating inflation

Fast-roll inflation demonstrates that slow-roll is a sufficient but not necessary condition for inflation. In fact, inflation can be achieved even when the inflaton field is undergoing oscillations. This was first considered by Michael Turner in 1983 [Tur83]. He studied a scalar field with an even potential $V(\phi) = V(-\phi)$ oscillating around the minimum at $\phi = 0$. Assuming that the frequency of oscillations is much larger than the Hubble scale $\omega \gg H$ we may regard two timescales; the expansion timescale H^{-1} (Hubble time) and the period of oscillations $\mathcal{T} = 2\pi/\omega \ll H^{-1}$. Thus, considering Eqs. (6.5) and (6.6), we have

$$(1 + w_\phi)\rho_\phi = \rho_\phi + p_\phi = 2\rho_{\text{kin}} = \dot{\phi}^2 \,. \tag{8.31}$$

Taking the average over many oscillations the above suggests

$$1 + \bar{w}_\phi = \frac{\overline{\dot{\phi}^2}}{\bar{\rho}_\phi} \,. \tag{8.32}$$

For times $\mathcal{T} < \Delta t < H^{-1}$, we may ignore the Universe expansion and consider that $\rho_\phi = \frac{1}{2}\dot{\phi}^2 + V = V_{\text{max}} \simeq \text{constant}$, where $V_{\text{max}} = V(\phi_{\text{max}})$ with ϕ_{max} being the amplitude of the oscillations, such that $\rho_{\mathrm{kin}}(\phi_{\max}) = 0$. Then, we have $\dot{\phi}^2 = 2(V_{\max} - V)$, and we find the period of oscillations as

$$\frac{d\phi}{dt} = \sqrt{2(V_{\max} - V)} \quad \Rightarrow \quad \mathcal{T} = \int_0^{\mathcal{T}} dt = 4 \int_0^{\phi_{\max}} \frac{d\phi}{\sqrt{2(V_{\max} - V)}} \,. \tag{8.33}$$

Then, using Eq. (8.32), we have

$$1 + \bar{w}_\phi = \frac{1}{V_{\max}} \times \frac{1}{\mathcal{T}} \int_0^{\phi_{\max}} \dot{\phi}\, d\phi = 2 \frac{\int_0^{\phi_{\max}} (1 - V/V_{\max})^{1/2} d\phi}{\int_0^{\phi_{\max}} (1 - V/V_{\max})^{-1/2} d\phi} \,, \tag{8.34}$$

where we also used Eq. (8.33). Now, assume that the potential is of the form $V \propto |\phi|^q$, with $q > 0$. Then $V/V_{\max} = (|\phi|/\phi_{\max})^q$. Putting this into the above it can be shown that

$$\bar{w}_\phi = \frac{q-2}{q+2} \quad \Leftrightarrow \quad q = 2\frac{1 + \bar{w}_\phi}{1 - \bar{w}_\phi} \,. \tag{8.35}$$

Thus, we have

$$a \propto t^{\frac{2}{3(1+\bar{w}_\phi)}} = t^{\frac{q+2}{3q}} \,, \tag{8.36}$$

$$\rho_\phi \propto a^{-3(1+\bar{w}_\phi)} = a^{-\frac{6q}{q+2}} \propto t^{-2} \,, \tag{8.37}$$

$$\rho_\phi \propto \phi_{\max}^q \Rightarrow \phi_{\max} \propto a^{-\frac{6}{q+2}} \,, \tag{8.38}$$

where we employed Eqs. (2.17) and (2.39) and we considered that the scalar field dominates the Universe. If $q = 2$ then the field oscillates in a quadratic potential, we have $\bar{w}_\phi = 0$ and the oscillating condensate behaves as matter, as we discussed in Sec. 6.3.1. If $q = 4$ then the field oscillates in a quartic potential, $\bar{w}_\phi = \frac{1}{3}$ and the oscillating condensate behaves as radiation. When $q \gg 1$ we have $\bar{w}_\phi \approx 1$ and the field remains kinetically dominated, bouncing on vertical potential walls.

From the above, we also see that the Universe can undergo inflation if the potential is concave

$$\bar{w}_\phi < -\frac{1}{3} \quad \Leftrightarrow \quad q < 1 \,. \tag{8.39}$$

Such inflation is called *oscillating inflation*, a named coined by Andrew R. Liddle and Anupam Mazumdar in 1998 [LM98]. The potential in oscillating inflation is shown in Fig. 8.4.

One might think that a potential of the form $V \propto |\phi|^q$ with $q < 1$ is difficult to explain in particle physics. However, one can obtain a potential of this form for a canonically normalised field, starting with a Lagrangian density with a noncanonical kinetic term. Indeed, considering $\mathcal{L} = \frac{1}{2}(\partial\phi)^2 - V(\phi)$ we can always find a reasonable noncanonical Lagrangian density (cf. Sec. 7.4)

$$\mathcal{L} = \frac{1}{2}K(\varphi)(\partial\varphi)^2 - \frac{1}{2}m^2\varphi^2 \,, \tag{8.40}$$

where m is a mass scale. The form of $K(\varphi)$ is obtained as follows. Using that $\phi^q \propto \varphi^2$ we have $d\phi \propto \varphi^{\frac{2}{q}-1} d\varphi$ so that $(\partial\phi)^2 \propto \varphi^{2(\frac{2}{q}-1)}(\partial\varphi)^2$, i.e., $K(\phi) \propto \varphi^{2(\frac{2}{q}-1)}$. Demanding $q < 1$, implies $K \propto \varphi^n$, with $n > 2$.

There is a discontinuity at $\phi = 0$, where perturbative degrees of freedom (small oscillations) are not well defined. We can "regularise" this discontinuity by setting $\phi \to \sqrt{\phi^2 + \phi_c^2}$ near the VEV, where ϕ_c is some mass scale. In this spirit, Thibault Damour and Viatcheslav F. Mukhanov in 1998 [DM98] introduced the phenomenological potential for oscillating inflation

$$V(\phi) = \frac{1}{q}M^4 \left[\left(\frac{\phi^2}{\phi_c^2} + 1 \right)^{q/2} - 1 \right] \,, \tag{8.41}$$

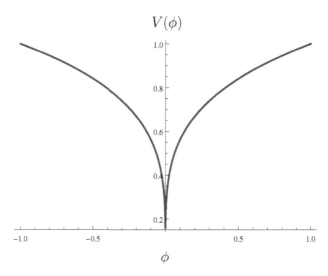

$$V(\phi)$$

FIGURE 8.4 The scalar potential in oscillating inflation. Fiducial units are used.

where M is a mass scale. The above potential becomes $V \simeq \frac{1}{q}M^4(|\phi|/\phi_c)^q \propto |\phi|^q$ when $|\phi| \gg \phi_c$. When $0 < q \ll 1$ the potential in Eq. (8.41) becomes $V \simeq \frac{1}{2}M^4 \ln[1 + (\phi/\phi_c)^2]$. If both limits are applied, i.e., $|\phi| \ll \phi_c$ and $0 < q \ll 1$, we have $V \simeq M^4 \ln(|\phi|/\phi_c)$, which is similar the the Coleman-Weinberg potential in supergravity (cf. Eq. (7.43)). The potential near the VEV (when $|\phi| \ll \phi_c$) is $V \simeq \frac{1}{2}M^4(\phi/\phi_c)^2$. Thus, the form of the potential is a parabola at the bottom, opening up, as one moves away from the VEV, into shallow wings. Despite oscillating around a parabolic minimum, the inflaton field ϕ spends most of the time on the upper parts of the potential, where $\rho_{\rm kin} < V$. Consequently, the main contribution to $\bar{w}_\phi$ comes from the potential density, which leads to inflation. As time passes, the amplitude of the oscillations $\phi_{\rm max}$ decreases (cf. Eq. (8.38)) approaching ϕ_c, the latter determining the core quadratic region. At some point, when $\phi_{\rm fin} \simeq (1-q)^{-1/q}\phi_c$, oscillating inflation ends. Thus, oscillating inflation lasts as long as $\phi_{\rm fin} < \phi_{\rm max}(t) < \phi_{\rm end}$, where $\phi_{\rm end}$ denotes the end of slow-roll inflation (as the field rolls down the $V \propto |\phi|^q$ potential), which precedes the oscillations.

Calculating the number of e-folds of oscillating inflation $N_{\rm osc}$ is rather challenging. However, numerical simulations have shown that they cannot exceed $N_{\rm osc} \lesssim 10$ or so.

8.6 Ultra-slow-roll inflation

So far we have seen that the potential may become too steep or curved to support slow-roll and yet we may still have inflation, such as fast-roll or oscillating inflation. In this section we consider the opposite case, when slow-roll does not occur because the potential is too flat.

To see how this comes about, let us consider that, in some region in field space, the inflaton scalar potential becomes extremely flat $V' \to 0$ so that the Friedmann and the Klein-Gordon equations (6.2) and (6.8) become

$$3m_P^2 H^2 = \frac{1}{2}\dot{\phi}^2 + V_0 \,, \tag{8.42}$$

and

$$\ddot{\phi} + 3H\dot{\phi} = 0 \,, \tag{8.43}$$

respectively, where $V_0 = \text{constant}$. Multiplying Eq. (8.43) with $\dot{\phi}$ we immediately obtain $\dot{\rho}_{\text{kin}} = -6H\rho_{\text{kin}}$, which suggests $\rho_{\text{kin}} \equiv \frac{1}{2}\dot{\phi}^2 \propto a^{-6}$.* An exact solution to the above system is

$$\dot{\phi} = -\sqrt{2V_0} \sinh\left(\sqrt{\tfrac{3}{2}}\,\phi/m_P\right), \tag{8.44}$$

and

$$H = \frac{\sqrt{V_0}}{\sqrt{3}m_P} \cosh\left(\sqrt{\tfrac{3}{2}}\,\phi/m_P\right), \tag{8.45}$$

as can readily be checked. Thus, using Eq. (6.10), we find

$$\epsilon \equiv -\frac{\dot{H}}{H^2} = 3\tanh^2\left(\sqrt{\tfrac{3}{2}}\,\phi/m_P\right). \tag{8.46}$$

Because we need $\epsilon \ll 1$ for inflation, we require $\phi \simeq 0$. In this way we may have inflation, but it is not slow-roll inflation.

To understand this, we define (see also Sec. 7.4)

$$\mathcal{E} \equiv \frac{\dot{\epsilon}}{\epsilon H} = \frac{2\ddot{\phi}}{H\dot{\phi}} + 2\epsilon, \tag{8.47}$$

where we have employed Eqs. (6.2), (6.8) and (6.15) with $\rho = \rho_{\text{kin}} + V$. As we have shown in Sec. 6.2.4, in slow-roll (SR) we have $\epsilon \simeq \varepsilon$, where the latter is given by Eq. (6.21). Then, in view of Eq. (6.24) we see that in SR we have $\mathcal{E} \simeq 2(2\varepsilon - \eta)$ so that $|\mathcal{E}| \ll 1$. However, in the present case, Eq. (8.43) suggests that $\mathcal{E} = -6 + 2\epsilon$. Considering inflation with $\epsilon \ll 1$ we see that $|\mathcal{E}| \simeq 6$, which does not correspond to SR. In fact, using Eqs. (8.44) and (8.45), Eq. (8.47) suggests

$$\mathcal{E} = -\frac{6}{\cosh^2\left(\sqrt{\tfrac{3}{2}}\,\phi/m_P\right)}, \tag{8.48}$$

which approaches $\mathcal{E} = -6$ when $\phi \to 0$. From the definition of $\mathcal{E}$ in Eq. (8.47), we see that $\mathcal{E} = -d\ln\epsilon/dN$, which means that, during inflation, $\epsilon \propto a^{-6}$. This new type of inflation is called *ultra-slow-roll* (USR) inflation, a name coined by William H. Kinney in 2005 [Kin05].

To explore USR inflation, we look closely at the inflaton Klein-Gordon equation of motion (6.8). To help in our discussion, we name the three terms on the left-hand side of Eq. (6.8) as the acceleration, the friction and the slope terms, respectively. During SR inflation, the acceleration term is negligible, and the Klein-Gordon equation reduces to Eq. (6.19) $3H\dot{\phi} \simeq -V'$, while $|\mathcal{E}| \ll 1$. If the potential suddenly becomes extremely flat, then the slope term in the Klein-Gordon disappears so that the equation is now Eq. (8.43), while $|\mathcal{E}|$ jumps up to $|\mathcal{E}| \simeq 6$ and SR gives way to USR.

Intuitively, one can understand this as follows. If we are originally in SR but the slope $|V'|$ reduces drastically, it initially drags with it the friction term, by virtue of Eq. (6.19). This decreases the value of $|\dot{\phi}|$, i.e., the kinetic density $\rho_{\text{kin}} = \frac{1}{2}\dot{\phi}^2$, but this value cannot decrease arbitrarily quickly. The fastest it can decrease is $\rho_{\text{kin}} \propto a^{-6}$, which we call "freefall" because it corresponds to a field with no potential density $V = 0$, such that its equation of motion is Eq. (8.43). Therefore, if the inflaton kinetic density is forced (by the decreasing slope) to reduce faster than freefall then the system breaks away from SR. In SR the acceleration term is negligible, because it is very small compared to the friction and slope terms, which are locked

---

*A common misconception is that $\rho_{\text{kin}} \propto a^{-6}$ always. However, this is not so in general. For example, as shown in Sec. 8.2, in power-law inflation $\rho_{\text{kin}} \propto t^{-2} \propto a^{-2\epsilon}$ (cf. Eqs. (8.2) and (8.7)) with $\epsilon < 1$.

together as shown in Eq. (6.19). However, if the slope reduces drastically and drags the friction term with it, they both become small too and eventually comparable to the acceleration term. So all three terms in Eq. (6.8) are comparable. When this happens, the friction term changes allegiance and becomes locked with the acceleration term, resulting in Eq. (8.43) and USR.

Now, once in USR, the field becomes oblivious to the potential, as demonstrated by Eq. (8.43). This is similar to the kination period in quintessential inflation models, as we discuss in Sec. 9.7.2, but there is a crucial difference. In kination, the Universe is dominated by $\rho_{\rm kin}$, while in USR inflation we still have potential domination and $\rho_{\rm kin} < V \simeq V_0$. Being oblivious to the potential, the inflaton field can even climb up an ultra-shallow V. Indeed, when the system enters the USR regime, it "flies over" the flat patch of the potential, sliding on its decreasing kinetic density. In that sense, the term ultra-SR is actually a misnomer, because the field rolls faster than it would have done if SR were still applicable over the extremely flat region. Beause of this, during USR $\epsilon > \varepsilon$, where $\varepsilon \to 0$ in the flat patch of V (because $V' \to 0$). If the system were following SR it would stop when $V' = 0$ because SR would dictate $\dot\phi = -V'/3H = 0$. Instead, it slides over the flat patch with $\rho_{\rm kin} \propto a^{-6}$.

Because the number of elapsing e-folds is inversely proportional to ϵ (cf. Eq. (6.13))* and $\varepsilon \ll \epsilon < 1$ in USR, we see that if a period of USR inflation is ignored (so ε is assumed in place of ϵ) there is a danger of overestimating the number of elapsing e-folds, which could throw estimates of N_* completely off, seriously affecting inflationary observables, such as n_s and r. Thus, one needs to be careful when studying inflation models which feature extremely flat patches.

One prominent case of such models is *inflection-point* inflation [AEGBM06], where the inflationary plateau is due to the interplay of opposing contributions to the scalar potential which (almost) cancel each other out, generating a step (or ledge) on the otherwise steep potential wall (see Fig. 8.5). One example is *A-term* inflation [BSDL07] with potential

$$V(\phi) = \frac{1}{2}m^2\phi^2 - \frac{1}{3}\lambda A\phi^3 + \lambda^2\phi^4\,, \tag{8.49}$$

which features a flat inflection point when $A = 4m$ at the value $\phi_f = m/2\lambda$. The model exploits the soft quadratic and cubic terms due to supersymmetry breaking in supersymmetric theories. Another example is *loop inflection-point* inflation with potential [DOR18]

$$V = \beta\left[-\ln\left(\frac{\phi^2}{m_P^2}\right)\phi^4 + \alpha\frac{\phi^6}{m_P^2}\right], \tag{8.50}$$

which features a flat inflection point when $\alpha = 2/3\sqrt{e}$ at the value $\phi_f = e^{1/4}m_P$, such that $V'(\phi_f) = V''(\phi_f) = 0$. The potential is determined by the 1-loop expression of the Coleman-Weinberg result in non-supersymmetric theories, where $\beta = y^4/32\pi^2$, with y being a Yukawa coupling between the inflaton ϕ and a Weyl fermion.† Both these models allow no more than a tiny tilt at the inflection point. This really means that the cancellation which results in the flat inflection-point requires sizeable fine-tuning, because it is not implied by the theoretical background.

It is evident that USR depends on having substantial kinetic density, which cannot decrease faster than freefall. However, if one begins inflation at the extremely flat region with very small kinetic density, then SR may be attained quickly, even immediately. The bound on the initial kinetic density $\rho_{\rm kin}^0$ is

$$\rho_{\rm kin}^0 \leq \frac{(m_P V')^2}{6V} = \frac{1}{3}\varepsilon V\,, \tag{8.51}$$

---

*Using Eq. (4.24), $dN = -Hdt = -HdH/\dot H = dH/\epsilon H$.

†The theory features a quartic minimum at $\phi = 0$ because $-\beta\ln\left(\frac{\phi^2}{m_P^2}\right)\phi^4 \to \frac{1}{2}\beta\phi^4$ when $\phi \to 0$.

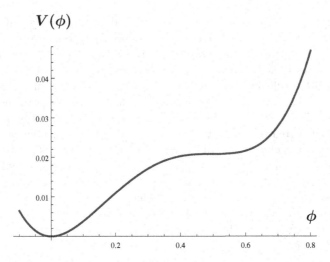

FIGURE 8.5   The scalar potential of inflection-point inflation. Fiducial units are used. The flat inflection-point is at $\phi_f = 1/2$.

where ε is defined in Eq. (6.21). This makes sense because in SR we have

$$\frac{(m_P V')^2}{6V} = \frac{m_P^2 (-3H\dot{\phi})^2}{18 m_P^2 H^2} = \frac{1}{2}\dot{\phi}^2 = \rho_{\text{kin}}^{\text{sr}}, \qquad (8.52)$$

where we used Eqs. (6.19) and (6.20). The above suggests that the initial kinetic density must not be bigger than the kinetic density in SR $\rho_{\text{kin}}^0 \leq \rho_{\text{kin}}^{\text{sr}}$. Of course, the SR kinetic density is zero at a flat inflection-point ϕ_f, so satisfying the bound in Eq. (8.51) would demand that $\rho_{\text{kin}}^0(\phi_f) = 0$. Because, $\rho_{\text{kin}} \propto a^{-6}$ during USR inflation the total number of e-folds is given by

$$N_{\text{usr}} = \frac{1}{6} \ln \left(\frac{\rho_{\text{kin}}^0}{\rho_{\text{kin}}^{\text{sr}}} \right). \qquad (8.53)$$

Let us see what the above imply for inflection-point inflation. While rolling from large to small values of ϕ, when the field reaches the flat segment there is an abrupt reduction in $|V'|$. Because the friction term in the Klein-Gordon was at least as large as the slope term before reaching the flat segment (that is we had SR or freefall), afterwards, the friction term cannot be balanced by the (substantially reduced) slope term. Thus, the acceleration term rushes to balance it and we have USR.

Now, during USR, we have $\rho_{\text{kin}} \propto a^{-6}$ so that

$$\dot{\phi}\ddot{\phi} = \dot{\rho}_{\text{kin}} = -6H\rho_{\text{kin}} \propto a^{-6}, \qquad (8.54)$$

where we took $H \simeq$ constant. Because $|\dot{\phi}| = \sqrt{2\rho_{\text{kin}}} \propto a^{-3}$, the above suggests that $|\ddot{\phi}| \propto a^{-3}$. After crossing the inflection point though, the (magnitude of the) slope of the potential begins to increase, while the (magnitude of the) inflaton's acceleration decreases, as we have seen. At some point, they meet each other and then the friction term changes allegiance and becomes locked with the slope term, so that SR is recovered. It must be stressed here that USR inflation is not an attractor, which means that the system would try to escape USR as fast as it can. But what if the evolution of the field begins at the flat patch? Then, provided the kinetic density is small enough and satisfies the bound in Eq. (8.51), one can immediately have SR inflation.

Before concluding this section we briefly refer to the curvature perturbation generated by USR inflation. Despite the fact that $|\mathcal{E}|$ is not small, the generated perturbations remain scale invariant. This is because the equation of motion of the quantum modes (see Appendix C) remains identical to the one in the de Sitter case, because the terms dependent on $\mathcal{E}$ cancel out. The spectral index is given by $n_s = 7 + \mathcal{E} = 1 + 2\epsilon$. The spectrum is scale invariant, if not marginally blue for $0 < \epsilon \ll 1$, which contradicts observations that suggest a red spectrum $n_s < 1$. The amplitude of the spectrum increases with time during USR inflation.* Indeed, Eq. (6.78) suggests that $\mathcal{P}_\zeta \propto 1/\epsilon \propto e^{6N_{\mathrm{usr}}}$, where we used that $\epsilon \propto a^{-6}$. Note that, while its overall amplitude increases, the spectrum remains scale-invariant.

So far, we have considered that the motion of the inflaton field in field space is only due to its classical equation of motion (6.8). However, if the slope of the scalar potential is very small the classical motion of the field is reduced to the point that it can become comparable to its quantum fluctuations. Next, we look into their effect on the evolution of the field.

8.7 Stochastic inflation

In Sec. 6.5.1, we discussed how a light scalar field is perturbed during inflation due to the particle production process. This process is inherently quantum mechanical, which means that the perturbations of the field are random. However, they are characterised by specific statistical properties, which for random perturbations are called *stochastic* [Lin82b]. If the scalar field in question is responsible for the generation of the curvature perturbation in the Universe (the inflaton field for example), then the resulting density perturbations are also stochastic. Thus, we have predictive power for statistical properties of structures in the Universe, e.g., the typical intergalactic distance, even though we are not able to calculate say the location of a particular galaxy. But we do not have to confine ourselves to the inflaton field only. In this section, we consider generically the evolution of a light scalar field during quasi-de Sitter inflation, when $H \simeq$ constant, under the influence of its quantum fluctuations.

First of all, in order for a scalar field to be subject to particle production it needs to be light, with mass $m < H$. This means that the scalar field (at most) slow-rolls down the potential. Its equation of motion is Eq. (6.19), but this is only when the perturbations of the scalar field are negligible. If not, then Eq. (6.19) assumes a source term and becomes the *Langevin* equation

$$\dot{\phi} = -\frac{V'}{3H} + \xi \,, \tag{8.55}$$

where $\xi = \xi(\mathbf{x}, t)$ is a noise term. The quantum noise is therefore superimposed on the classical roll, much like a mobile phone on vibration when rolling down the walls of a bathtub.

How important the noise term is depends on the slope of the potential. Indeed, we can compare the classical roll of the field and its perturbations due to its quantum fluctuations by comparing the magnitude of the classical "velocity" in field space $|\dot{\phi}|$ and the quantum displacement per unit time $\delta\phi/\delta t$. We find

$$\frac{1}{|\dot{\phi}|}\frac{\delta\phi}{\delta t} = \frac{H^2}{2\pi|\dot{\phi}|} = \frac{3H^3}{2\pi|V'|} \,, \tag{8.56}$$

where the quantum effect is determined by the fact that the field suffers a quantum "kick" $\delta\phi = H/2\pi$ in a Hubble time $\delta t = 1/H$, and we employed Eqs. (6.19) and (6.67). Thus, we see that that the condition for the quantum effects dominate the evolution of the scalar field is

$$|V'| \lesssim \frac{3}{2\pi} H^3 \,. \tag{8.57}$$

---

*The reason is that in USR inflation the curvature perturbation ζ continues to evolve when superhorizon, in contrast to SR inflation, where ζ freezes after horizon exit.

Now, the quantum kicks are random so they serve to spread the condensate over a region in field space, because each kick can send the field to a bigger or a smaller value. By spread we mean that the value of the classical field is perturbed such that it varies at different locations.* Therefore, the field performs a random walk with typical step $\delta\phi = H/2\pi$. After ΔN steps, the mean-squared field is

$$\langle\phi^2\rangle = \left(\frac{H}{2\pi}\right)^2 \Delta N = \frac{H^3}{4\pi^2}\Delta t\,, \tag{8.58}$$

where $\langle\phi^2\rangle = \langle\phi^2(\mathbf{x},t)\rangle$ and $\Delta N = H\Delta t$ is the number of elapsing e-folds after time Δt. The above assumes that $|V'|$ is negligible and the field is oblivious to the potential. However, if the scalar potential is not negligible in some regions in field space (that is for some values of ϕ), the spread of the field condensate cannot continue indefinitely. Indeed, the eventual probability distribution is [SY94]

$$P = P_0 \exp\left(-\frac{8\pi^2}{3}\frac{V(\phi)}{H^4}\right)\,, \tag{8.59}$$

where P_0 is a normalisation constant, such that the total probability is $\int_{-\infty}^{+\infty} P\,d\phi = 1 = 100\%$. For example, for a free, massive scalar field minimally coupled to gravity we have

$$\langle\phi^2\rangle = \frac{3H^4}{8\pi^2 m^2}\,. \tag{8.60}$$

The above can be understood as follows. Eq. (8.59) suggests that the field condensate diffuses such that

$$V \lesssim \frac{3}{8\pi^2}H^4 \sim H^4\,, \tag{8.61}$$

because otherwise the probability is exponentially suppressed. This implies that particle production creates a "puddle" with density $\sim H^4$, as suggested also in Eq. (6.62) (see Fig. 8.6). The "puddle" extends until the potential density V becomes comparable to H^4. Thus, the field is practically confined in the region $V \lesssim H^4$ with a practically flat distribution, meaning that every value of the field is equally probable as long as $V \lesssim H^4$. [†] When considering a free massive scalar field we have $V = \frac{1}{2}m^2\phi^2$, so that the spread of the diffused scalar field condensate is halted when $m^2\phi^2 \sim H^4$, i.e., $\phi^2 \sim H^4/m^2$, which is in agreement with Eq. (8.60). Following the same logic, when considering a quartic potential $V = \frac{1}{4}\lambda\phi^4$, we expect $\lambda\phi^4 \sim H^4$, which suggests $\phi^2 \sim H^2/\sqrt{\lambda}$. Indeed, the formal result is

$$\langle\phi^2\rangle = \sqrt{\frac{3}{2\pi^2}}\frac{\Gamma(3/4)}{\Gamma(1/4)}\frac{H^2}{\sqrt{\lambda}} = 0.1318\frac{H^2}{\sqrt{\lambda}}\,, \tag{8.62}$$

where $\Gamma(x)$ is the gamma function (see Appendix C).

However, Eq. (8.59) provides the late-time, eventual probability distribution. In real time, the probability distribution changes as the condensate spreads out. The change of P is determined by the *Fokker-Planck* equation

$$\frac{\partial P}{\partial t} = -\frac{\partial}{\partial\phi}\left(\dot{\phi}P\right) + \frac{\partial^2}{\partial\phi^2}\left(\frac{H^3 P}{8\pi^2}\right)\,. \tag{8.63}$$

---

*Note that the field perturbations, albeit being of quantum origin, become classical after horizon exit, because of the magic of quantum decoherence, as explained in Sec. 6.5.1.

[†]When we consider a potential barrier V_0 the field would diffuse over it if $H^4 > V_0$.

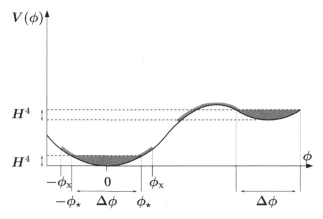

FIGURE 8.6 Schematic representation of the quantum diffusion of a spectator field condensate. The condensate assumes with equal probability, all the values for which $V \lesssim H^4$ around the origin (depicted by the left region $\Delta\phi$), meaning $\langle\phi^2\rangle \lesssim \phi_\star^2$, where $V(\phi_\star) \sim H^4$. This is like a "puddle" in the bottom of the potential. Because the value $|\phi| \ll \phi_\star$ is untypical, we expect $|\phi| \sim \phi_\star$. However, quantum diffusion is dominant when $|V'| \lesssim H^3$, which corresponds to a larger range $\langle\phi^2\rangle \lesssim \phi_x^2$, where $|V'(\phi_x)| \sim H^3$. This region is depicted by a thick grey line on top of the potential. Even though the classical roll is subdominant, it eventually prevails and sends the field into the "puddle." As shown, the quantum diffusion zone exists also on hilltops where $|V'| \sim 0$. Finally, we note that there may be other "puddles" at other local minima of the potential, as the region $\Delta\phi$ shown on the right. In the graph, we assumed that the contribution of the spectator field to the density budget is negligible, so that both the depicted "puddles" are of the same "depth" given by H^4, where $H = $ constant.

Using this equation, one can compute the mean-squared of the field at any given time. For example, in the case of a free, massive, minimally-coupled scalar field we have

$$\langle\phi^2\rangle = \frac{3H^4}{8\pi^2 m^2}\left[1 - \exp\left(-\frac{2m^2}{3H}\Delta t\right)\right]. \tag{8.64}$$

The above reduces to Eq. (8.60) when $\Delta t \to \infty$. Also, it reduces to Eq. (8.58) in the limit $m^2 \to 0$. Similarly, it is possible to calculate $\langle\phi^2\rangle$ for a scalar potential where the spread of the condensate does not become stationary. For example, considering the tachyonic potential $V = V_0 - \frac{1}{2}m^2\phi^2$ we have

$$\langle\phi^2\rangle = \frac{3H^4}{8\pi^2 m^2}e^{-\frac{2m^2}{3H^2}}\left[\exp\left(\frac{2m^2}{3H}\Delta t\right) - 1\right], \tag{8.65}$$

where we assumed that initially ϕ lies at the origin (the top of the potential hill).

Comparing the requirement $|V'| < H^3$ (as implied by Eq. (8.57)) for quantum to win over classical and the requirement $V \lesssim H^4$ (as implied by Eq. (8.61)) we see that they are not the same. Indeed, if we consider a massive scalar field with $V \sim m^2\phi^2$ we see that it is possible to be quantum dominated and still be outside the "puddle" with $V \lesssim H^4$ in the range $H^4/m^2 < \langle\phi^2\rangle < H^6/m^4$, where $H^6/m^4 > H^4/m^2$ because the field is light with $m < H$. Thus, we see that, in this range, even though the variation of the field is dominated by the quantum kicks, its classical motion eventually prevails and sends it into the "puddle" with $V \lesssim H^4$. The reason is that the direction of the quantum kicks is random, while the classical roll pushes the field in one direction only, into the "puddle."

In the next section, we see how the above can have dramatic effects for the inflationary scenario because it may lead to inflation that never ends.

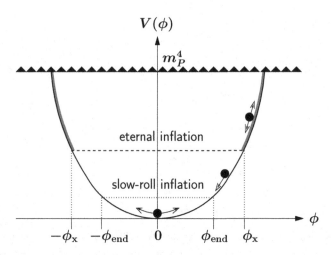

FIGURE 8.7 Schematic representation of the inflaton quantum diffusion in models of chaotic inflation $V \propto \phi^{2q}$. As $|\phi|$ grows, the Hubble scale increases accordingly $H^2 \propto V$. Consequently, in contrast to the case of spectator fields where H is determined by some other inflaton field, in chaotic inflation quantum diffusion dominates away from the minimum, at large values of V, above a critical value $|\phi| > \phi_x$, where $|V'(\phi_x)| = \frac{3}{2\pi} H^3(\phi_x)$. Once the inflaton exits the diffusion zone (depicted by a thick grey line on top of the potential) and $|\phi| < \phi_x$, the field engages in slow-roll until it decreases down to $|\phi| = \phi_{end}$, when inflation ends. Afterwards, the inflaton oscillates coherently around its VEV $\langle \phi \rangle = 0$ and may decay perturbatively, as explained in Sec. 6.3.1. While the inflaton field is in the diffusion zone $|\phi| > \phi_x$, inflation is eternal because there are always locations, where the quantum "kicks" conspire to keep the field at large values of V, i.e., above the dashed line. If the field escapes the diffusion zone it undergoes slow-roll inflation when $\phi_{end} < |\phi| < \phi_x$, which ends when $V(\phi)$ is reduced below the dotted horizontal line.

8.8   Eternal inflation

8.8.1   The basic idea

In the previous section we implicitly assumed that H remains (roughly) constant. This is not necessarily so when the scalar field in question is the inflaton, in which case $H = H(\phi)$, because in slow roll $H^2 = V(\phi)/3m_P^2$, cf. Eq. (6.20). For example, in chaotic inflation discussed in Sec. 7.2.1 we have $V \propto |\phi|^q$, with $q > 0$. Therefore, if ϕ is large enough, quantum diffusion might dominate even if the slope of the potential is substantial, because the condition in Eq. (8.57) is satisfied not by virtue of a small slope $|V'|$ but by having a huge H. Thus, for a chaotic inflation potential quantum diffusion dominates not near zero but up the slope. Therefore, if we consider an even chaotic potential, we expect the diffusion zone to be high. The inflaton at even-power chaotic inflation is not diffused in a "puddle" around the minimum. In fact, inflation ends if $|\phi| < \phi_{end}$, cf. Sec. 7.2.1. The diffusion zone lies up the slope and makes contact with the Planck scale, where $V \sim m_P^4$. What is the significance of this?

Well, if the inflaton field finds itself inside a quantum diffusion zone of its potential then there are always locations where inflation continues indefinitely, the reason being that there can always be places where the quantum kicks conspire to keep the inflaton field up the potential forever. This phenomenon was realised by Alexander Vilenkin [Vil83] in 1983, but it was Andrei Linde in 1986, who called it *eternal inflation* [Lin86]. So, eternal inflation is not a particular inflationary model but a distinct mode of inflation, much like slow-roll inflation. For chaotic inflation, the diffusion zone is high up the potential. Once the inflaton field finds itself there, inflation is eternal, see Fig. 8.7.

Eternal inflation proceeds roughly as follows. As we have seen in Sec. 6.5.1, the typical wavelength of inflaton quantum fluctuations generated per Hubble time $\delta t = H^{-1}$ is $\delta\phi^{-1} \sim H^{-1}$. The whole horizon volume domain H^{-3} in a Hubble time (an e-fold) grows to size $(e \times H^{-1})^3$, which means that it is divided into $e^3 \simeq 20$ new separate domains of radius H^{-1}. The variation of the inflaton field inside each of these domains is $\Delta\phi = \delta\phi + \dot{\phi}\delta t$, where the last term is due to the classical roll, while the first term is due to the quantum fluctuations. In contrast to the classical roll term, which always sends the field downwards in its scalar potential,[*] the quantum fluctuation term is completely random and may push the inflaton up or down the potential with equal probability. Inside the diffusion zone we have $\delta\phi > |\dot{\phi}|\delta t$, thus the quantum variation wins. Because the process is random, in almost half of the 20 separate domains ϕ jumps to a higher potential density. In the following step, these 10 domains expand even faster than before. Recall that processes separated by a distance greater than H^{-1} (superhorizon) are causally disconnected and proceed independently of one another because objects at a distance more than H^{-1} are pulled apart superluminally with respect to each other. Thus regions a superhorizon distance apart behave as if they are separate (mini) universes. The above process continues in the next step and the one after that, forever. Therefore, we see that eternal inflation is a quasi-stationary state of infinite self-reproduction.

So far, the way we have thought of cosmic inflation is that it takes a tiny domain of space and rapidly expands it to a size many orders of magnitude larger. This stretching removes all preexisting inhomogeneities and makes our Universe uniform. But this mechanism does not make the global universe uniform! This is the actual global universe, not the idealised one extrapolated from our observable Universe, as discussed in Sec. 2.2.2. Indeed, eternal inflation results in a highly inhomogeneous global universe with a fractal structure. Statistically, eternal inflation never ends, but locally the inflaton field always eventually exits the diffusion zone and after a period of classical (e.g., slow-roll) inflation, thermalises and gives rise to a local Big Bang universe, which is called a *pocket universe*. These thermalised domains (pocket universes) are separated by inflating regions and thus remain separated forever.[†] This is similar to the nonpercolating bubbles of old inflation, see Sec. 8.1. However, in contrast to the bubbles in old inflation, pocket universes are not empty and do not have to be round. The continuous spawning of pocket universes (our Universe is one of them) repeats forever resulting in infinite pocket universes, producing a fractal structure [LLM94]. We can say that the global universe is endlessly populated by pocket universes. However, inflating space never disappears, meaning that the stochastic evolution of the inflaton suggests that thermalisation never happens globally, while the total inflating volume grows exponentially in time ad infinitum. The eternally inflating global universe becomes a *multiverse* consisting of an infinite collection of pocket universes.

One way to understand the above better is through a biological analogy. Indeed, the dynamics of eternal inflation can be mimicked by a set of multiplying bacteria. Each bacterium represents a Hubble volume. Reproduction is analogous to Hubble expansion. In every step (an e-fold) a fixed number of independent offsprings (e^3) are produced by a parent bacterium. The quantum diffusion is the random hopping of the bacteria, while the thermalised volume corresponds to the number of dead bacteria (noninflating Hubble volumes) in the asymptotic future. Eternal inflation is never ending, which in the bacteria analogy corresponds to a nonzero probability that the population of bacteria never dies out. Thus, eternal inflation can be modelled with the theory of branching process used to study bacteria.

---

[*]We are not considering a phantom field here, see Sec. 9.3.

[†]They could be connected through wormholes though.

8.8.2   Eternal inflation in action

For chaotic inflation, the diffusion zone corresponds to $|\phi| > \phi_x$, where ϕ_x is defined as the value of the inflaton field where the equality in the condition in Eq. (8.57) is satisfied exactly. Using that $V = V_0(\phi/m_P)^q$ (cf. Eq. (7.8)) and that $V' = qV/\phi$ as well the slow-roll Friedmann equation (6.20), we find

$$\phi_x/m_P = \left(2\pi\sqrt{3}q\frac{m_P^2}{\sqrt{V_0}}\right)^{\frac{2}{2+q}}. \tag{8.66}$$

Further, employing Eq. (7.13) and after a little algebra, the above becomes

$$\phi_x/m_P = A_s^{-\frac{1}{q+2}}\sqrt{2qN_*}, \tag{8.67}$$

where $N_* \simeq 60$ and $A_s \simeq 2 \times 10^{-9}$ (cf. Eq (5.42)) correspond to the remaining e-folds of inflation and the curvature perturbation respectively, when the pivot scale exits the horizon during inflation, and we have taken $N_* > q/4$. For example, when $q = 4$ we readily find $\phi_x \simeq 620\,m_P$. This is significantly larger than $\phi_{\rm end} = 2.8\,m_P$, according to Eq. (7.10). In fact, using Eq. (7.10), one finds that to roll from ϕ_x down to $\phi_{\rm end}$, the inflaton needs about $N_x = A_s^{-\frac{2}{q+2}}N_* \simeq 4.8 \times 10^4$ e-folds. Thus, in quartic chaotic inflation, when the local value of the inflaton exits the diffusion zone and a pocket universe is generated, it undergoes slow-roll inflation for about 4.8×10^4 e-folds before inflation ends.

We can play the same game with plateau inflation, where the eternal inflation regime corresponds to the inflaton field being far along the plateau, such that $|V'|$ is very small. Considering Starobinsky-type inflation in Eq. (7.49) and repeating the above steps we find that

$$\phi_x/m_P = \frac{1}{q}\ln\left(4\pi\sqrt{3}q\frac{m_P^2}{\sqrt{V_0}}\right), \tag{8.68}$$

where we have taken $q\phi > m_P$ because we are far along the plateau. Using Eq. (7.55) and after a little algebra, the above becomes

$$\phi_x/m_P = \frac{2}{q}\ln\left(A_s^{-1/4}\sqrt{2q^2N_*}\right). \tag{8.69}$$

Taking the Starobinsky value $q = \sqrt{2/3}$ (cf. Sec. 7.3.4) we find $\phi_x \simeq 18\,m_P$, which is significantly larger than $\phi_{\rm end} \simeq 0.94\,m_P$, where we used Eq. (7.51). Thus, using Eq. (7.52) it can be shown that, after exiting the diffusion zone, the inflaton field undergoes normal slow-roll for about $N_x = A_s^{-1/2}N_* \simeq 1.3 \times 10^6$ e-folds until inflation ends. The situation is similar with other types of inflation models, e.g., hybrid inflation. As we see in the above examples, there is ample slow-roll inflation after the end of eternal inflation. Recall that our observable Universe corresponds only to the last $N_* \simeq 60$ e-folds of slow-roll inflation. Therefore, by the time the pivot scale exits the horizon, there are no traces left of the eternal inflation regime.

In both chaotic and in plateau inflation models the diffusion zone is infinite or touches the Planck scale. In contrast, hilltop inflation has a finite diffusion zone at the top of the potential hill. Unlike chaotic inflation, hilltop inflation can be eternal regardless of the inflation energy scale because the condition in Eq. (8.57) is satisfied not because H is large, but because $|V'| \to 0$ near the maximum. The extent of the region of quantum diffusion can be estimated as follows. First we Taylor expand the potential near the top, where $V' = 0$, to get $V'(\phi) = \phi V''$. From Eq. (6.23) we have $V'' = \eta V/m_P^2$. Thus, we readily find $V' = 3\eta H^2\phi$, where we used the slow-roll Friedmann equation (6.20). Saturating the eternal inflation condition in Eq. (8.57) we obtain

$$|\phi_x| = \frac{1}{|\eta|}\frac{H}{2\pi}. \tag{8.70}$$

Because $|\eta| \ll 1$ in hilltop inflation, we see that $|\phi_x| \gg H/2\pi$ and there is a diffusion zone on top of the potential hill, which the inflaton may exit only after several quantum jumps. In contrast, when $|\eta| \sim 1$ we have fast-roll inflation (cf. Sec. 8.3) and the diffusion zone does not extend beyond a single quantum jump of the inflaton. Eternal inflation is not possible in this case.

As we have seen, hilltop inflation is eternal because $\dot{\phi} = -V'/3H = 0$ on top of the hill, where we used the slow-roll equation Eq. (6.19). Indeed, it seems that whenever $\dot{\phi} \to 0$ the quantum variation $\delta\phi/\delta t \sim H^2$ would always dominate and eternal inflation would occur. Could we have eternal inflation even if $|V'| > H^3$ if the initial conditions for the inflaton are such to send it upwards at first, so that it inevitably stops and rolls back down? In this case, momentarily $\dot{\phi} \to 0$ so the quantum variation would dominate. Thus, it seems that the condition in Eq. (8.57) can be evaded. Looking more closely into this possibility however, reveals that $|V'| > H^3$ would not lead to eternal inflation in this case. The reason is that the time period when the quantum variation of the field becomes dominant compared to its classical roll is more than a Hubble time only when Eq. (8.57) is satisfied and the slope is small enough for the turnaround point to lie in a diffusion zone.

Yet, it is possible to avoid eternal inflation even if $|V'| < H^3$ when slow-roll is inapplicable. A prominent example is inflection-point inflation where, as we have explained in Sec. 8.6, slow-roll is terminated and a phase of ultra-slow-roll takes place. Then, the value of $|\dot{\phi}|$ may actually be larger than what slow-roll would suggest, because $\epsilon > \varepsilon$ and the inflaton field "flies over" the inflection point riding on its decreasing kinetic density oblivious of the potential. Thus, we may have $|V'| < H^3$ so that the eternal inflation condition in Eq. (8.57) is satisfied but still have $|\dot{\phi}| > \delta\phi/\delta t \sim H^2$ because $|\dot{\phi}| > |V'|/3H$ in ultra-slow roll. This means that eternal inflation may not happen because the classical roll remains dominant against quantum diffusion, despite $|V'| < H^3$.

So far we have been using the condition in Eq. (8.57) to determine when inflation becomes eternal. This condition presupposes slow-roll, but this is not always the case, as we just saw with inflection-point inflation. A more generic condition is obtained as follows. The condition for eternal inflation to occur is linked with the generated curvature perturbation. Indeed the ratio of the quantum variation of the inflaton field over its classical motion of in field space is

$$\frac{\delta\phi}{\dot{\phi}\delta t} = \frac{H^2}{2\pi\dot{\phi}} = \frac{5}{2}\frac{\delta\rho}{\rho} = \sqrt{\mathcal{P}_\zeta}\,, \tag{8.71}$$

where we used Eqs. (5.32) and (6.76). Thus, we see that when classical roll is comparable to quantum variation, we have $\mathcal{P}_\zeta \sim 1$. Therefore, eternal inflation occurs precisely when the curvature perturbation is of order unity. This leads to sizeable deformations of the geometry, which are of fractal form. However, even though the curvature perturbation is large, non-linearities remain small in slow-roll inflation, which means that perturbative treatment is still allowed.* Thus, we see that the stochastic quantum process can have a large influence in the cosmological spacetime, even though all the energy scales in the Lagrangian density of the system are far below the scale of quantum gravity.

Using Eq. (6.78), we can write

$$\epsilon = \frac{H^2}{8\pi^2 m_P^2 \mathcal{P}_\zeta}\,. \tag{8.72}$$

Because $\mathcal{P}_\zeta \sim 1$ in eternal inflation, we find $\epsilon \sim (H/m_P)^2 \ll 1$, where $H < m_P$ to avoid quantum gravity corrections. Hence, we see that eternal inflation is quasi-de Sitter.

---

*This is not necessarily so in other kinds of inflation, e.g., k-inflation, where eternal inflation may lie outside the regime of validity of effective field theory.

We can use the above to study eternal inflation in models which do not necessarily lead to slow-roll. For example, in power-law inflation, which was introduced in Sec. 8.2, the potential is $V \propto e^{-\lambda\phi/m_P}$ (cf. Eq. (8.1)) and we have $\sqrt{2\epsilon} = \lambda$ (cf. Eq. (8.6)). Using this in Eq. (8.72) and considering that eternal inflation happens when $\mathcal{P}_\zeta = 1$, we find that the Hubble parameter at the border of the diffusion zone is

$$H_{\mathrm{x}} = 2\pi\lambda m_P \, , \tag{8.73}$$

where $\lambda < \sqrt{2}$ for inflation. We see that, if λ is not exponentially small, $H_{\mathrm{x}} \sim m_P$ and quantum gravity corrections cannot be ignored because the Friedmann equation (6.2) suggests that $\rho \sim m_P^4$. This really means that eternal inflation does not happen when λ is not tiny.

The above suggest that, in general, inflation models allow the possibility of eternal inflation, although there may be exceptions such as inflection point inflation or power-law inflation. Eternal inflation is realised if the inflaton field finds itself inside the quantum diffusion zone. Once eternal inflation is triggered it cannot be stopped. Instead it inevitably produces a multiverse of infinite pocket universes. The eternally inflating universe is indefinitely large. So, it is likely that we find ourselves inside the eternally inflating volume, even if the onset of eternal inflation is unlikely. In that sense, the occurance of eternal inflation does not need to be probable.

8.8.3   The multiverse hypothesis

In general, inflation suffers from a *measure problem* [LN10] because, even if finding oneself in the spacetime patch which is about to inflate might be unlikely, once inflation occurs and the volume of the inflated space is enlarged enormously, one is more likely to be in a location which has undergone inflation than not. So, when are we supposed to estimate the probability, before or after inflation takes place? With eternal inflation the measure problem becomes even worse because the above volume weighing does not work; it is more probable to find oneself in eternally inflating space than inside a pocket universe. Furthermore, the underlying theory may allow different types of pocket universes to form, corresponding to different vacuum states of the same theory, with possibly different laws of low-energy physics. Given that pocket universes of any kind are infinite, comparison between different types of pocket universes amounts to a comparison of infinities, which is inherently ambiguous. Thus, the fraction of pocket universes with a particular property corresponds to infinity divided by infinity, which is an indefinite ratio.

To better understand what we mean with pocket universes having different laws of physics, let us first consider even-powered chaotic inflation, which has a single vacuum, the minimum of the potential at the origin. Suppose that we start high up the potential walls such that we trigger eternal inflation. Eternal inflation spawns infinite pocket universes but each and every one of them features the same laws of physics, because there is only one vacuum, corresponding to $\phi_{\mathrm{vev}} = 0$. However, if we did the same exercise considering hilltop inflation instead, we would end up with infinite pocket universes of two kinds, corresponding to the two possible vacua with VEV $\langle\phi\rangle = \pm\phi_{\mathrm{vev}} \neq 0$. The laws of physics are not necessarily the same in both these kinds of pocket universes, because it might be important what the sign of the inflaton VEV is. The same would be true in the case of hybrid inflation, natural inflation and so on. Now consider a theory that is much more complicated than these toy-models, a more "realistic" theory. This one might have a large number of different vacua, i.e., minima of the (possibly multi-field) scalar potential. In these different vacua the theory may feature different low-energy physics. For example, in some vacua the symmetry of electromagnetism might be broken and the photon might be massive, while in others the electroweak symmetry might be restored and the Higgs field set to zero (e.g., because it is coupled to a large nonzero VEV of the inflaton field). It might also be possible that low-energy physics features symmetries completely different from the standard model of particle physics which applies to our world. Indeed, string theory might suggest that the number of these different vacua is enormous; of the order of 10^{500} or so* corresponding to a complicated multifield scalar potential, which forms a landscape featuring many sinkholes and valleys. In my opinion, the string landscape should not be taken too seriously though, because string theory amounts to merely the weak limits of some fundamental theory that is as yet unknown. This means that the string landscape is probably not the final answer, and certainly not a "prediction" of string theory. Still, one does not have to endorse the string landscape in order to consider a multiverse of different kinds of infinite pocket universes. This seems to be the inevitable result of a fundamental theory which has many different vacua and gives rise to eternal inflation.

Thus, if eternal inflation occurs (meaning if the inflaton enters a diffusion zone), then all the possible vacua of the fundamental theory are realised by infinite pocket universes, because infinite time implies that anything that can happen does happen an infinite number of times. In other words, anything possible becomes inevitable[†] and one cannot distinguish the probable from the improbable. Realisation of all the vacua might happen by quantum tunnelling from one metastable minimum of the scalar potential to another, or by quantum diffusion at local maxima or saddle points in the scalar potential, where the classical inflaton trajectory bifurcates as in hybrid inflation for example. Thus, we can say that eternal inflation populates all the possible vacua and therefore all possible low-energy universes consistent with fundamental theory actually exist. If the fundamental theory allows locally closed universes or universes with a negative cosmological constant (string theory, for example, vastly prefers anti-de Sitter vacua), then many pocket universes collapse into black holes embedded in the eternally inflating space. Thus, apart from infinite different pocket universes eternal inflation also generates swarms of black holes inside the eternally inflating space. Therefore, eternal inflation realises a diverse multiverse of endless worlds within worlds.[‡]

The multiverse hypothesis allows an interesting philosophical proposal, that of anthropic reasoning. In a nutshell, this goes as follows. Amid the myriads of pocket universes, life evolves only in those where the local laws of physics have the right properties for life. This idea gives a plausible "explanation" for why our Universe appears to have just the right properties for our version of life. As such, it provides the illusion of intelligent design, without the need for the actual existence of an intelligent designer (aka God). However, such anthropic reasoning means that logical deduction cannot lead to precise and unique predictions, so it seems to lie beyond science (see also the discussion in Sec. 9.1.1 about the anthropic "explanation" of the cosmological constant). Moreover, the multiverse hypothesis gives rise to a number of fascinating but also disturbing paradoxes.

One of them is called the *youngness paradox*. This considers that, since the rate of formation of pocket universes is proportional to the volume of the eternally inflating space, which increases exponentially, this rate would increase exponentially too. This means that, at any given time almost all existing pocket universes have formed very recently, while more mature pocket universes are vastly outnumbered by these youngsters. This is despite the fact that a mature pocket universe has larger volume (it has expanded since its formation) compared to a young pocket universe, because the number of the young ones is so much larger than the number of older ones. By this reasoning, it is much more probable to find yourself in a pocket universe

---

*Even more ridiculously large numbers have been suggested, e.g., 10^{10^6}.

[†]However, this does not mean anything is possible, only what the fundamental theory allows.

[‡]A criticism of inflation is that, since most inflation models can result in eternal inflation, which leads to a multiverse of infinite pocket universes, inflationary theory does not seem to make any predictions because almost anything is possible in say 10^{500} types of universes. However, this is not a problem of inflationary theory but of the underlying fundamental theory. If the fundamental theory featured a single potential minimum then all the pocket universes would be identical and the predictivity of eternal inflation would be unquestionable.

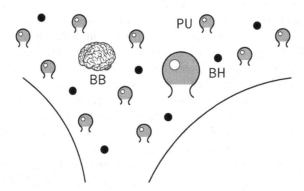

FIGURE 8.8 Schematic representation of the multiverse according to eternal inflation. The eternally inflating space continuously spawns pocket universes (denoted by PU), which are engulfed in a swarm of black holes (denoted by BH) and the occasional Boltzmann brain (denoted by BB). The youngness paradox suggests that the larger mature pocket universes are less populous than youngsters, with only one mature such pocket universe depicted, amid numerous younger and therefore smaller pocket universes.

that has just been born. Coupling this to anthropic reasoning, our civilisation is the first one to evolve in our Universe, because this is more probable than being second or even later. This might "explain" the famous *Fermi paradox*,* which states that there must be no other intelligent life in the cosmos because otherwise "they" (the intelligent E.T.s) would be here already.

A more disturbing paradox is the proliferation of *Boltzmann brains*, named after famous physicist Ludwig Boltzmann (1844–1906), not because he put the idea forward, but as a response to his suggestion that the low entropy of the Universe is due to a quantum fluctuation. Let us briefly explore the idea of Boltzmann brains here. Consider an empty pocket universe. What is the probability of quantum nucleating an entire Universe filled with galaxies, stars and planets? Tiny, you might think. Couple this with the probability that in this Universe, which has just appeared, there exists an intelligent species with cosmologists who happen to come up with the multiverse hypothesis and with you, dear reader, who also just popped into existence while reading this sentence, with your life memories and your loved ones, who also just appeared. The probability is infinitesimally small, but if you consider it in the context of infinite pocket universes, then it cannot be dismissed. The Boltzmann brain paradox considers that typical observers might not be the result of the usual cosmological evolution but may arise as a quantum fluctuation in an empty post-inflation universe. To avoid the proliferation of the multiverse with Boltzmann brains, the thermalisation rate of the various vacua must be greater than the rate of production of Boltzmann brains. A related paradox, linked to the *fake Universe paradox*, states that your brain is the only one that has just quantum nucleated with the surrounding world being a simulation run by a sophisticated computer, which also just appeared. It may be argued that this is more probable that the traditional Boltzmann brains paradox (which involves the actual nucleation of trillions of galaxies), although the word "probable" is probably (!) meaningless in this context. A schematic representation of the eternal multiverse is shown in Fig. 8.8.

I think we should stop here. There are other paradoxes with cool names like the *Q-catastrophe* for example, but we have probably ventured too far into speculation, away from science. I promised you exotica in this chapter, but maybe this is too far out there.

---

*Named as such by physicist Enrico Fermi (1901–1954), who famously exclaimed "But where is everybody?" in one discussion about extraterrestrials. Among many other things, he is the reason that particles with half-integer spin are called fermions.

Wrapping up, it seems that cosmic inflation generally allows the realisation of eternal inflation if the inflaton field lies initially inside a quantum diffusion zone. There might be some exceptions to this rule like inflection-point inflation, which may be tilted too much or mask behind a phase of ultra-slow-roll, or steep power-law inflation, although the latter is not compatible with observations. Alternatively, eternal inflation might accidentally not be realised in certain domains, even though the inflaton field does find itself inside the diffusion zone initially. If eternal inflation is indeed realised, this is either because H is large enough, or $|\dot\phi|$ is small enough or both. As a last comment, it should be emphasised that the above scenario considers a sub-Planckian energy density, which means quantum gravitational corrections are ignored, even though eternal inflation generates curvature perturbation of order unity. If the quantum effects of gravity become important we are back to spacetime foam, where causality and all our theories so far break down.

8.9   The beginning of time

In classical Greek cosmogony, before the appearance of the Universe, there was Chaos (ΧΑΟΣ). Hesiod wrote that Chaos was the first thing to exist. Today, the Hot Big Bang theory suggests that the Universe emerges from spacetime foam (see Sec. 3.4.5), which we do not really understand and does not seem too different from the ancient Chaos. However, as we have seen in Chapter 6, the Hot Big Bang suffers from a number of fine-tuning problems (the horizon and flatness problems), which is one of the main motivations behind cosmic inflation (the other one being the origin of the cosmological perturbations, necessary for structure formation). Inflation precedes the Hot Big Bang, but it is still classical, meaning that inflation juts in between spacetime foam and the Hot Big Bang.

Since inflation is brought in to explain the fine-tuning of the Hot Big Bang, it is legitimate to enquire whether the initial conditions of inflation itself are fine-tuned. This is despite the fact that the no-hair theorem suggests that information on inflationary initial conditions is beyond reach (it lies behind the event horizon of the "inverted black hole," cf. Sec 4.3.4) and also despite the fact that the slow-roll attractor erases memory of initial conditions.[*]

8.9.1   Chaotic initial conditions

The good news is that chaotic inflation is not in need of any fine-tuning, as it naturally delivers the Universe from spacetime foam (akin to Chaos, hence the name) [Lin85]. In spacetime foam all forms of energy density of the inflaton scalar field are expected to be of the same order

$$\rho_{\rm kin} \sim \rho_\nabla \sim V \sim m_P^4\,, \tag{8.74}$$

where $\rho_{\rm kin} = \frac{1}{2}\dot\phi^2$ and $\rho_\nabla \equiv \frac{1}{2}(a^{-1}\nabla\phi)^2$ is the gradient density, cf. Eq. (D.11). This is because all the contributions to the energy density range between 0 and m_P^4, so their typical value is $\sim m_P^4$ unless fine-tuned.[†] But, of course, not all density contributions are expected to be identical. It is reasonable to consider that some density contributions are larger than others at different regions, fluctuating at random. All that is needed is a single domain where, by chance, $V \gtrsim \rho_{\rm kin}, \rho_\nabla$. Then, if the potential is sufficiently flat, chaotic inflation begins and exponentially dilutes both $\rho_{\rm kin}$ and ρ_∇, before they have a chance to prevent it. This is when the classical description of the Universe becomes possible and the cosmic clock starts ticking. Thus, with chaotic inflation, the Universe is but an expanded (inflated) domain which escapes spacetime foam.

---

[*]By definition an attractor is reached for a range of initial conditions.

[†]If a variable ranges between zero and one it is of order unity, unless fine-tuned near zero (e.g., $\sim 10^{-3}$).

For the above scenario, it is not necessary to consider only a monomial potential as in Sec. 7.2.1. In fact, any inflationary potential which makes contact with the Planck scale, i.e., $V \sim m_P^4$, for some inflaton field values can have chaotic initial conditions. So the term chaotic inflation can be extended beyond monomial inflation. Now, as we explain in the previous section, for monomial chaotic inflation, when the inflaton field exits spacetime foam and the potential density is close to the Planck scale $V(\phi_P) \equiv m_P^4$, the field lies inside the diffusion zone $\phi_x < |\phi| < |\phi_P|$ and eternal inflation is unavoidable. However, this is not necessarily so for other kinds of chaotic inflation. For example, power-law inflation (see Sec. 8.2) can reach the Planck scale and have chaotic initial conditions, while maintaining $|V'| > H^3$ throughout. Indeed, considering $V \propto \exp(-\lambda\phi/m_P)$ we find that the requirement is $\lambda > \sqrt{V}/m_P^2$ (cf. Eq. (8.1)), where $|V'| = \lambda V/m_P$ and we used the flat Friedmann equation $V \sim (Hm_P)^2$. Thus, we see that, with $\lambda \sim 1$ eternal inflation can be avoided for any $V < m_P^4$. Therefore, chaotic inflation (but not monomial chaotic inflation) without eternal inflation is possible.[*]

8.9.2   A Universe from nothing

The Gospel of John begins as: "Ἐν ἀρχῇ ἦν ὁ λόγος" (John 1:1). This is usually translated as "In the beginning was the Word," but the actual meaning is richer because the word λόγος (logos) in Greek means both Word and Reason (logic). In the same spirit, is it possible that the Laws of Nature exist beyond spacetime? If so, the Universe can be thought not to emerge from spacetime foam or from a singularity but from "nothing," meaning a state without spacetime.

Consider a closed de Sitter universe (meaning dominated by positive vacuum density). Then the radius of this universe is given by $r(t) = H^{-1}\cosh(Ht)$, where $r = aX_c$ with X_c being the constant comoving radius of curvature (cf. Sec. 2.5.2). This implies that this universe contracts when $t < 0$, bounces at $t = 0$, where it reaches a minimum radius $r_{\min} = H^{-1}$ and then expands exponentially when $t > 0$. This is similar to the evolution of a bubble of true vacuum, surrounded by false vacuum, with radius $r(t) = \sqrt{r_{\min}^2 + t^2}$, which also "bounces" at $t = 0$, when $r = r_{\min}$. However, in the case of a bubble, the solution is valid only when $t \geq 0$, because the bubble tunnels quantum mechanically from $r = 0$ to $r = r_{\min}$ at $t = 0$. This is the picture at the end of old inflation, cf. Sec. 8.1. By analogy, the closed de Sitter universe can be thought to appear at the bounce, with finite size H^{-1} and also with $\dot{r} = 0$ meaning $\dot{a} = 0$. Subsequent expansion requires $\ddot{a} > 0$ for $t > 0$. This transition is called de Sitter *instanton* and amounts to the Universe quantum mechanically tunnelling from "nothing" into existence, provided the rules of quantum tunnelling are already in place [Vil82].

The probability of this event has been calculated as

$$P \sim \exp\left(-24\pi^2 \frac{m_P^4}{V}\right). \tag{8.75}$$

The above expression suggests that, if fundamental theory has many de Sitter vacua then it is more probable that the Universe tunnels at the vacuum with the largest potential density. If $V \sim m_P^4$, then we find $P \sim 1$, which suggests that nucleating the Universe at the Planck scale is quite likely. The reason is that the positive energy of matter filling the Universe is compensated by the negative energy of the gravitational self-interaction of this matter. This is why Alan H. Guth is quoted as saying "The Universe is the ultimate free lunch."

We may understand the above as follows. From the Friedmann equation we estimate roughly that $V \sim (Hm_P)^2 \Rightarrow H^{-1} \sim m_P/\sqrt{V}$. The volume of the closed universe (of a hypersphere) is

---

[*]Of course, monomial chaotic inflation $V \propto |\phi|^q$ without eternal inflation is also possible, but only if initially the inflaton is $|\phi| < \phi_x$ so that the initial potential density is $V \ll m_P^4$. In this case, we are never in contact with the Planck scale, and we cannot assume chaotic initial conditions.

$\Delta \mathcal{V} = 2\pi^2 r^3 \sim H^{-3}$ at nucleation. Thus, the energy required is $\Delta E = V \Delta \mathcal{V} \sim m_P^3 / \sqrt{V}$. Considering that the timescale of nucleation is $\Delta t \sim H^{-1} \sim m_P / \sqrt{V}$, we find $\Delta E \cdot \Delta t \sim m_P^4 / V$. Quantum field theory suggests (cf. Sec. 4.3.4) $\Delta E \cdot \Delta t \sim 1$, which means $V \sim m_P^4$. If $V \ll m_P^4$ then, because $\Delta E \propto 1 / \sqrt{V}$, the energy ΔE would be large and the Universe heavy and difficult to create. This would make the nucleation of a sub-Planckian Universe improbable, which agrees with Eq. (8.75), that suggests $P \ll 1$ if $V \ll m_P^4$. For example, in hilltop inflation $V \leq V_0 \sim 10^{-10} m_P^4$, which implies $P \sim \exp(-10^{12}) \ll 1$. Does this mean that nonchaotic inflation models, which do not come in contact with the Planck scale but have $V \ll m_P^4$, are improbable? We discuss this next.

8.9.3   Sub-Planckian inflation

Observations support inflation models which feature an inflationary plateau with potential density $V \sim 10^{-10} m_P^4$. As we discussed above, quantum nucleation of such an inflationary Universe is exponentially suppressed. Alternatively, if the Universe exited the spacetime foam with $\rho_{\rm kin}, \rho_\nabla \sim m_P^4 \gg V \sim 10^{-10} m_P^4$, this would mean that the kinetic and gradient densities would overwhelm the potential density and block inflation from ever happening. In fact, it has been known for some time that, when $V < m_P^4$, inflation can only start in a homogeneous patch roughly a few times larger that the Hubble volume with radius H^{-1}. This means that, in order to begin, inflation must assume homogeneity over superhorizon scales [GP92].*

This is rather serious, for if we are to assume superhorizon homogeneity for the onset of inflation, why can't we assume the same today? Does this imply that inflation does not really solve the horizon problem after all? Of course, the observational confirmation of the Hot Big Bang as far back as BBN implies that the level of fine-tuning required is much higher, since at the time of BBN the Universe must be homogenised over about 10^{23} uncorrelated horizon volumes so as to remain homogenised today. This is estimated as follows. Considering that the cosmological horizon grows as $D_H \sim t$, while lengthscales grow only as $\ell \propto a(t)$ we find that the ratio of the range of causal correlations over the enlarged (by the expansion) lengthscales is

$$F = \frac{t_0 / t_{\rm BBN}}{a_0 / a_{\rm BBN}} = \frac{t_0 / t_{\rm BBN}}{(t_{\rm eq} / t_{\rm BBN})^{1/2} (t_0 / t_{\rm eq})^{2/3}} \sim 10^{7.5} , \tag{8.76}$$

where we considered that the present time is $t_0 \sim 10^{10} \, {\rm y} \sim 10^{17} \, {\rm sec}$, the time of equality is $t_{\rm eq} \sim 10^4 \, {\rm y} \sim 10^{11} \, {\rm sec}$, the time of BBN is $t_{\rm BBN} \sim 1 \, {\rm sec}$ and $a \propto \sqrt{t}$ in the radiation era and $a \propto t^{2/3}$ in the matter era (we ignored recent dark energy). So the horizon volume today includes $F^3 \sim 10^{23}$ horizon volumes at the time of BBN. Obviously, if we consider that the Hot Big Bang extends well before BBN then this number can further grow considerably. Thus, it can be said that inflation substantially ameliorates the horizon problem.

However, we can apply the same reasoning for the onset of inflation. If we assume that the Universe begins at the Planck scale correlated over a Planck length $\ell_P = m_P^{-1}$ we can estimate the number of correlated volumes which correspond to the horizon size at the energy scale of inflation. We thus assume a subluminal expansion, parameterised as $a \propto t^n$ with $0 < n < 1$. Then the ratio of the range of causal correlations over the enlarged lengthscales is

$$F = \frac{t_{\rm inf} / t_P}{a_{\rm inf} / a_P} = \left(\frac{t_{\rm inf}}{t_P} \right)^{1-n} = \left(\frac{\rho_P}{\rho_{\rm inf}} \right)^{\frac{1}{2}(1-n)} = \left(\frac{m_P^4}{V} \right)^{\frac{1}{2}(1-n)} \sim 10^{5(1-n)} , \tag{8.77}$$

where the subscripts "P" and "inf" denote the Planck time $t_P \sim m_P^{-1}$ and the moment when the density of the Universe becomes comparable to $V \sim 10^{-10} m_P^4$, respectively, and we considered

---

*Hence, local (subhorizon) inflation is impossible.

that the density scales as $\rho \propto t^{-2}$ (cf. Sec. 2.8). In the above, for simplicity, we ignored the curvature of the Universe, noting that at early times the curvature may be negligible (see Sec. 2.4.2). Thus, the number of uncorrelated volumes inside the horizon when $\rho \sim V \sim 10^{-10}m_P^4$ is $F^3 \sim 10^{15(1-n)}$. If the Universe is dominated by relativistic particles then $a \propto \sqrt{t}$ and $n = \frac{1}{2}$. This suggests that $F^3 \sim 10^8$. Note that the perturbations of the inflaton when subhorizon behave as relativistic particles, so $\rho_\nabla \propto a^{-4}$. This overwhelms the kinetic density, which scales as $\rho_{\rm kin} \propto a^{-6}$ (see Sec. 9.7.2). However, if it were $\rho_{\rm kin}$ that dominated, then $a \propto t^{1/3}$ and $F^3 \sim 10^{10}$. Thus, we conclude that the horizon volume when the density decreases down to $10^{-10}m_P^4$ is comprised by roughly a billion uncorrelated volumes. This makes it rather difficult to assume homogeneity over the scale H^{-1}, let alone beyond it.

8.9.4 Homogenising the horizon before inflation

8.9.4.1 Starting with an open Universe

Despite our simplifying assumption of a flat Universe above, in principle we should expect a non-flat geometry before inflation. If the Universe is closed ($k > 0$) before inflation then the chances that it recollapses before $t_{\rm inf}$ are very high. On the other hand, if the Universe is open ($k < 0$) then something curious happens. Suppose that the Universe is nucleated at the Planck scale, empty but with $k < 0$. Then the Friedmann equation (2.10) can be written as

$$H^2 = -\frac{k}{a^2} \equiv \frac{8\pi G}{3}\rho_k \;\Rightarrow\; \rho_k = \frac{3|k|m_P^2}{a^2}, \tag{8.78}$$

which implies that the Universe can be modelled as a flat Universe dominated by a substance with density $\rho_k \propto a^{-2}$. In view of Eq. (2.17), we see that the barotropic parameter for such a Universe is $w = -\frac{1}{3}$. As a result, Eq. (2.39) gives that $a \propto t$ (Milne universe, see Sec. 2.5.1). Thus, both the size of the homogeneous region and the horizon grow proportional to time. This means that we have $n = 1$ and Eq. (8.77) suggests $F \sim 1$. Consequently, the horizon volume remains homogeneous and no more tuning is needed than the case when inflation started at the Planck scale. Even if the Universe started without being empty, with $\rho_k \sim \rho_\nabla \sim \rho_{\rm kin}$, the curvature term $\rho_k \propto a^{-2}$ would immediately dominate $\rho_\nabla \propto a^{-4}$ and $\rho_{\rm kin} \propto a^{-6}$, and the above conclusion stands.

8.9.4.2 Considering a compact Universe

Another way to achieve a homogeneous horizon at sub-Planckian density is by considering a compact Universe, with a nontrivial topology. What do we mean by this? Consider a square sheet of paper. You can glue the two opposite edges of the paper to make a cylinder. The surface of the cylinder is still flat in that parallel lines stay a constant distance apart and the sum of the angles of a triangle equals 180° (see Sec. 2.4.1).* However, the topology of the surface has changed. Indeed, two straight lines crossing at an angle will cross each other again and again as they wrap around the cylinder. This would never happen on a "truly" flat sheet of paper. Now, consider gluing the remaining opposite edges. This is hard to imagine, but the result is a kind of doughnut, which the mathematicians call a torus. Now, upgrade to three dimensions and consider a cube of space. If one identifies all the edges of the cube with their opposite edges then the topology is that of a 3-D torus, which can still remain flat but it is not infinite; it is compact, in that if one crosses one edge they would appear at the opposite edge as if nothing had happened. People playing video games are quite familiar with this behaviour.

---

*And you did not need to crush the paper, as you would have to do if you wanted to make the surface of a sphere, i.e., wrap a ball with it.

Can the Universe be like this? Not only it can, but it is easier to nucleate, because it would not need to tunnel through a forbidden region any more. Instead, the quantum mechanic description of a compact Universe is that it appears from "nothing" without exponential suppression, even if its initial density is sub-Planckian. The reason is that the volume of the compact Universe is not necessarily larger when the density is sub-Planckian, so the energy needed to nucleate such a universe is not enlarged when $V \ll m_P^4$. Considering a compact Universe agrees well with the thinking in string theory, which considers (at least) six extra spatial dimensions, which are compactified. The idea of compactification when applied to cosmology might imply that the remaining three spatial dimensions are compact too, but they have been enlarged to superhorizon size by inflation.[*]

Why would a compact Universe make it easier for inflation to begin? Well, suppose that the Universe was nucleated as a Planck-sized domain with compactified edges such that we can write $x_i = x_i + \ell_P$ for the spatial coordinates, where $\ell_P = m_P^{-1}$ is the Planck length. If originally the compact Universe expanded subluminally, the scale factor would grow less quickly than the horizon. This would mean that soon particles would be able to run all over the torus multiple times, which would rapidly become (or even remain from the beginning) homogeneous. Equivalently, we could say that the compact domain (which is the entire Universe) rapidly becomes causally connected and homogeneous. The process is called *chaotic mixing* [Ell71, CSS96]. Homogenisation due to chaotic mixing applies to all light fields including the inflaton field for which $|V''| < V/m_P^2$. As the Universe expands, its density decreases until it reaches the density scale of inflation. Within a homogeneous Universe inflation can begin without problems.

In fact, the idea of chaotic mixing works too well as it can eliminate the horizon problem without the need for inflation. However, inflation would still be needed to overcome the flatness problem as well as generate the primordial density perturbations. Also, without inflation the compact nature of the Universe would have been apparent, as we would be able to observe multiple copies of the Milky Way further and further away (so earlier and earlier versions of it). Similarly, there would be characteristic "rings" in the CMB, where the CMB signal would repeat itself. Many researchers have indeed looked for such patterns and could not find them, which suggests that either inflation enlarged the Universe to the extent that its topological structure is unobservable (the size of the compact domain is superhorizon today) or that the topology of the Universe is trivial after all.

8.9.4.3 Introducing additional early inflation

One other way to make sure inflation does not suffer from an initial conditions problem (because space is homogenised over superhorizon scales) is by considering an early stage of inflation, preceding the inflationary stage we can observe. Indeed, as we have discussed, chaotic initial conditions are quite natural. But the problem is that the observations favour plateau and not chaotic inflation. Well, we may have inflation at two stages; first chaotic inflation, which begins at the Planck scale with $V \sim m_P^4$, followed by slow-roll plateau inflation with $V \sim 10^{-10} m_P^4$, i.e., close to the GUT scale.[†]

Fundamental theory seems to suggest that the scalar potential is probably complicated, with multiple fields and many local minima. In this context, the assumption that there was a single phase of inflation from the Planck-scale down to the GUT scale may be "unrealistic" because it is like suggesting that V is featureless. Therefore, early inflation may have been quite different

---

[*]The inflaton field can be thought to be confined ("live") only in our three familiar spatial dimensions, similarly to the photon field of electromagnetism.

[†]It has been argued that inflation is necessary in order to take the Universe out of spacetime foam. If this is so, then preinflationary subluminal expansion may not be justified.

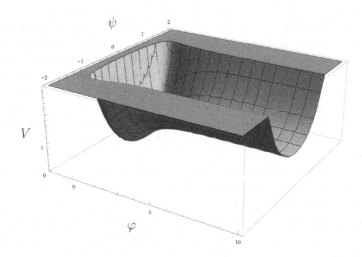

FIGURE 8.9 The scalar potential in a model of plateau inflation driven by the inflaton field ϕ supplemented by early chaotic inflation driven by another inflaton field ψ. The inflationary plateau lies at the bottom of a valley whose slopes support chaotic inflation as the field ψ rolls down to its minimum and brings the system from the Planck-scale to the plateau GUT-scale.

to plateau inflation. Such early inflation would ensure spatial homogeneity prior to the plateau inflation stage. More importantly though, it would overcome the danger that inhomogeneities propagate inwards as the horizon grows, because early inflation features an event horizon which protects the initially smooth patch against outside inhomogeneous regions.

There are two ways to realise early inflation. One is to consider it due to another scalar inflaton field ψ. In this case, the inflationary plateau becomes the bottom of a valley, whose side slopes connect with the Planck scale and support chaotic inflation (but not necessarily monomial chaotic inflation) as the system rolls from the Planck-scale down to the GUT-scale valley which is the inflationary plateau, as shown in Fig. 8.9. The valley slopes can be exponential such that early inflation is power-law and eternal inflation is avoided. However, the plateau can be very flat far from the minimum, which makes eternal inflation very likely once the system reaches the GUT-scale valley. In fact, except from the shaft around the potential minimum, the inflationary plateau is similar to an effective cosmological constant. This implies that the theory is almost shift-symmetric with $\phi = \phi + C$, where C is some constant. Consequently, almost all values of ϕ on the plateau are equally probable. However, while the global Universe features a variety of ϕ-values, locally the inflaton field gradients are rapidly becoming exponentially suppressed and the Universe becomes uniform.

The above picture is less successful if the GUT-scale inflation is hilltop, where the field, while rolling down the valley slopes must land near the top of the potential hill at $\phi = 0$. This might be facilitated if there is an interaction between the two inflatons (ϕ and ψ in Fig. 8.9) and the origin is an enhanced symmetry point (ESP).

The other way to realise early inflation avoids the introduction of an additional degree of freedom (another inflaton field, that is) but considers the possibility of the inflationary plateau turning upwards at high values to make contact with the Planck scale. The form of the potential is similar to inflection-point inflation, shown in Fig. 8.5, where the ledge on the potential slope is lengthened to become the inflationary plateau, which however, for very large values eventually rises upwards towards the Planck scale.

8.9.5 Inflation after the bounce

In an effort to evade the Big Bang singularity and the embarrassing ignorance we have of quantum gravity and spacetime foam, many cosmologists have investigated the possibility that our expanding Universe is the outcome of a "bounce", that is of a collapsing universe which, just before hitting the Big Crunch, reverses its collapse into expansion. How can this come about? Well, consider a closed universe with $k > 0$. The Fridemann equation (2.10) suggests that there is a special value of the density $8\pi G\rho = 3kc^2/a^2$, for which the variation of the scale factor momentarily stops and $\dot{a} = 0$. What happens afterwards depends on what kind of density dominates this universe. If this universe is dominated by matter, then $\rho + 3p/c^2 = \rho > 0$ and the acceleration equation (2.21) suggests that $\ddot{a} < 0$ at the moment when $\dot{a} = 0$. This means that a previously expanding universe stops its expansion and begins to collapse. The density reaches a minimum value $\rho_{\min}$ and the scale factor reaches a maximum value $a_{\max}$, as we have discussed in Sec. 2.5.2. However, if this universe is dominated by dark energy, by definition $\rho + 3p/c^2 < 0$ and Eq. (2.21) suggests that $\ddot{a} > 0$ at the moment when $\dot{a} = 0$. This means that a previously contracting universe stops its contraction and begins to expand. Thus, the collapsing universe bounces back into expansion. The density reaches a maximum value $\rho_{\max}$ and the scale factor reaches a minimum value $a_{\min}$. If $\rho_{\max}$ is smaller than the Planck density ρ_P, then the quantum effects of gravity are kept negligible, spacetime foam is avoided and the Big Bang singularity has been pushed at least to the beginning of the previous universe, if not the one before. This nice "kicking the can" scenario has a fortunate by-product. At the bounce (when $\dot{a} = 0$) and presumably right after it, the Universe is dominated by dark energy and $\ddot{a} > 0$. Thus, the expanding Universe is initially undergoing inflation and we can say that inflation is the natural outcome of a bouncing Universe. This inflation may no more be necessary to explain the horizon problem, because the previously collapsing phase implies that the actual age of the Universe is more than 14 Gy, so there might be time for the observable Universe to be causally connected after all. However, inflation is still needed to account for the flatness problem (recall that we started with a closed universe) and for the generation of the observed primordial density perturbations.

As we have seen there are ways to account for the initial conditions of inflation. These proposals, however, have been criticised over the degree that they are "realistic" in the framework of fundamental theories. On the other hand, I believe that if there are contradictions between cosmic inflation and speculative ideas on Planck-scale physics, inflation being observationally supported, has an edge. This is an example of how cosmological observations can constrain fundamental theory. Yet, even if inflation did suffer from an initial conditions problem, it is possible that this is not for inflation to solve. After all, as well said by Alan H. Guth, David I. Kaiser, and Yasunori Nomura in 2014 [GKN14], "we do not reject Darwinian evolution because it does not explain the actual origin of life."

EXERCISES

1. The scalar spectral index in minimal hybrid inflation in supergravity is

$$n_s = 1 - \frac{1}{N_*}.$$

When $N_* = 50 - 60$ the above is not compatible with observations, which suggest that $n_s = 0.965 \pm 0.004$ for negligible tensors. One way to save the model is considering that, after reheating, there is a period of modular thermal inflation followed by a period of fast-roll inflation.

   (a) Modular thermal inflation corresponds to VEV: $\langle \phi \rangle \sim m_P$ for the thermal inflaton field.

   Calculate the total number of e-folds of thermal inflation N_T assuming that the coupling of the thermal inflaton to the thermal bath prior to thermal inflation is of order unity $g \sim 1$ and the tachyonic mass of the thermal inflaton field is $m \sim 10\,\text{TeV}$.

   (b) After the end of thermal inflation, the thermal inflaton field is released from the top of the potential hill (where it was hitherto held by thermal corrections) and rolls downhill driving a brief period of fast-roll inflation.

   Assuming that the original hybrid inflation took place at energies close to the grand unification scale and that primordial reheating is prompt (because of parametric resonance effects between the primordial inflaton field and the waterfall field), the total inflationary e-folds corresponding to the cosmological scales is

$$N_{\text{tot}} = N_* + N_T + N_{\text{FR}} \simeq 60.$$

   Using this and the observational requirements for the value of the spectral index, find the available range of values of the total number of the fast-roll e-folds N_{FR}.
   Using this, find the allowed range for the ratio m/H, where H is the Hubble scale during thermal and fast-roll inflation.

2. The initial conditions problem of plateau inflation may be overcome if, before the system finds itself on the inflationary plateau, there is a phase of proto-inflation, connecting with the Planck scale. An example of this can be studied in the context of α-attractors, when considering two non-interacting, scalar fields driving proto-inflation and primordial inflation, respectively.
   Consider the Lagrangian density

$$\mathcal{L} = \frac{1}{2}m_P^2 R + \frac{\frac{1}{2}(\partial\varphi)^2}{\left(1 - \frac{\varphi^2}{6\alpha\,m_P^2}\right)^2} + \frac{\frac{1}{2}(\partial s)^2}{\left(1 - \frac{s^2}{6\beta\,m_P^2}\right)^2} - \frac{\frac{1}{2}m_\phi^2\varphi^2}{\left(1 + \frac{\varphi}{\sqrt{6\alpha}\,m_P}\right)^2} - \frac{\frac{1}{2}m^2 s^2}{1 - \frac{s^2}{6\beta\,m_P^2}},$$

   (a) Demonstrate that the corresponding canonically normalised scalar fields are obtained through the following field redefinitions

$$\frac{\varphi}{m_P} = \sqrt{6\alpha}\,\tanh\left(\frac{\phi/m_P}{\sqrt{6\alpha}}\right) \quad \text{and} \quad \frac{s}{m_P} = \sqrt{6\beta}\,\tanh\left(\frac{\psi/m_P}{\sqrt{6\beta}}\right).$$

   (b) Show that in terms of the canonical fields the potentials become

$$V(\phi) = V_C\left(1 - e^{-\sqrt{\frac{2}{3\alpha}}\,\phi/m_P}\right)^2 \quad \text{and} \quad V(\psi) = V_0 \sinh^2\left(\frac{\psi/m_P}{\sqrt{6\beta}}\right),$$

where

$$V_C = \frac{3}{4}\alpha m_\phi^2 m_P^2. \quad \text{and} \quad V_0 = 3\beta\, m^2 m_P^2$$

Thus, ϕ is the inflaton field of primordial inflation, which rolls down a Starobinsky-type inflationary plateau, while ψ is the proto-inflaton field, which rolls down the sides of a potential valley from the Planck scale down to the inflationary plateau scale, where the latter is the bottom of the valley. The form of the potential is shown in the last figure of Chapter 8. Note that, despite a kinetic pole, the canonical potential of ψ does not feature a plateau because the noncanonical potential of s has the same pole as the corresponding kinetic term.

Show that proto-inflation interpolates between power-law inflation when $\psi_c < \psi \lesssim m_P$ and quadratic chaotic inflation when $\psi_{\text{end}} \leq \psi \lesssim \psi_c$, with $\psi_c \equiv \sqrt{6\beta}\, m_P$ and $\psi_{\text{end}} = \sqrt{2}\, m_P$.

(c) Calculate the total number of e-folds of proto-inflation.

(d) Find a lower bound on the value of m such that proto-inflation has ended by the time primordial inflation can begin.

[Hint: The observations suggest that for the inflationary plateau we have $V_C \sim 10^{-10} m_P^4$.]

9

Dynamic Dark Energy

9.1   Dark energy today 205
Dark energy and the cosmological constant • Dark
energy beyond the cosmological constant

9.2   Quintessence 209
Dynamics of a spectator scalar field • Scaling and
tracking quintessence • Exponential quintessence •
Inverse-power-law quintessence • Freezing versus
thawing quintessence • PNGB quintessence • The fifth
force problem and beyond

9.3   Phantom and quintom.............................. 222

9.4   k-essence.. 224

9.5   Coupled dark energy................................ 226
Coupled exponential quintessence • Coupled
nonminimal quintessence • Mass-varying neutrinos

9.6   Other proposals...................................... 229
Chaplygin gas • Holographic dark energy

9.7   Quintessential inflation 232
The scenario in general • Kination • Spike of
gravitational waves • Quintessential inflation with
α-attractors

9.8   The end of time 243
Open-ended fate • Cataclysmic end

Out of (or inside) spacetime foam a blob of classical spacetime appeared and exploded exponentially in a Big Bang. This lasted a tiny fraction of a second (unless the explosion is eternal).
At some point inflation ended,* and the expansion of the blob, our Universe, which by now
was enormous in size (at least compared to its original size), continued in a milder, subluminal
manner. Reheating happened, and our Universe was filled with a bath comprised of a zoo of
massless particles. Then followed a series of phase transitions, which rendered most of these
particles massive, slow and heavy. Just after a millionth of a second, the hitherto free quarks
condensed into protons and neutrons and the Universe was filled with the primordial soup. A
few seconds into the story, the temperature of the bath decreased enough for conditions to be
comparable to the centre of stars today. From then on, we are on familiar territory. The world
is not too different from the one we live in. Hundreds of thousands of years passed, and the
Universe expanded and cooled down until, suddenly, the plasma filling the Universe morphed
into atoms, a new kind of structure, which would eventually lead to molecules and chemistry.
The Universe became transparent, filled with the light of the original explosion. Further on,

---

*At least in some part of the blob.

millions of years passed. Gradually, the gas and the dark matter filling the Universe collapsed into lumps, while the cooling CMB became darker and darker. The Universe entered the dark ages. A few hundred million years later the first stars were born, and new light appeared in the heavens. As time passed, star clusters became galaxies comprised of hundreds of billions of stars, most of which were circled by planets, which are by-products of star formation. In one such planet several billion years ago, life appeared. Probably, this has also happened countless times elsewhere in the vast expanses of the Universe. It took roughly two billion years for multicellular organisms to evolve. Less than one billion years ago, the first animals and plants formed. And then, things changed in the Universe. As if there was a gear shift, the expansion of space started to speed up as the Universe was becoming filled with a new and exotic substance. Today, almost 14 billion years after the beginning of time, a species of intelligent ape on planet Earth measured that this mysterious substance has grown up to about 70% of the total content of the Universe. The ape called the substance *dark energy*, in the hope that a name would dispel some of its mystery. It didn't.

9.1 Dark energy today

Some claim that the surprise discovery of dark energy in 1998 by Riess et al. and independently by Perlmutter et al. is comparable to the discovery of the Universe expansion by Hubble in 1929. It is certainly a shift in paradigm, although I think going from a static and eternal to an expanding and (possibly) temporary Universe is more dramatic. Still, it forces physicists to confront the unsettling fact that the vast majority of stuff in the Universe today is unknown (dark matter and dark energy) and most of it (dark energy) seems rather strange and, well, otherworldly. This is why, at first, many cosmologists thought that the discovery was a mistake and it would go away. Soon, however, the existence of dark energy was independently confirmed, beyond the SNe-Ia Hubble diagram, by CMB and BAO observations. It also resolved the age problem (cf. Sec. 2.11), which, when turned around, is another proof of dark energy. Saul Perlmutter, Brian P. Schmidt, and Adam G. Riess, won the Nobel prize in 2011 for the discovery of late dark energy.

9.1.1 Dark energy and the cosmological constant

As we have discussed, by far the easiest way to account for dark energy is to assume a nonzero, suitable value for the cosmological constant Λ. Thus, the ΛCDM paradigm was born, which is also called the *concordance model*. One would naïvely expect that, with the dark energy observations, physicists would triumphantly exclaim that the value of the cosmological constant has been observed. However, this is far from what happened. The reason is that for ΛCDM to work Λ needs to be incredibly fine-tuned. This is the infamous cosmological constant problem, which has been known well before the observation of dark energy [Wei89] (see also Sec. 2.6.2).

A nonzero value of Λ can have two origins. First, Λ is classically introduced in general relativity, which is our theory of gravity. The only mass-scale in general relativity is due to Newton's gravitational constant and is the Planck mass $G = 1/M_P^2$. But this cannot be the mass-scale of the cosmological constant, so a new scale must be included, which is at most $10^{-30} M_P$, for otherwise structure formation could not have occurred. Thirty orders of magnitude! Why do the two scales differ so much? The second origin of Λ is from quantum physics. Quantum fields introduce a contribution to vacuum energy which diverges and is presumed capped at the cutoff energy scale of the theory, resulting in a nonzero cosmological constant. This is again the Planck scale, where quantum field theory in curved spacetime breaks down, due to quantum gravity effects. Thus, we find the same excessive fine-tuning. These two contributions to Λ can coexist, and it is reasonable to assume they counter each other. However, because the vacuum density $\rho_\Lambda = \Lambda m_P^2$ is at most as large as the current density of the Universe $\rho_\Lambda \lesssim \rho_0 \sim (10^{-3}\,\text{eV})^4 \sim 10^{-120}\,m_P^4$, it seems they must cancel out down to 120 (or at least 60, if one believes in supersymmetry) orders of magnitude (see Appendix A) because the dimensions of Λ are $[\Lambda] = E^2$.

In the past, the way the cosmological constant problem was dealt with was by assuming that an *unknown* symmetry made the above cancellation exact and resulted in zero vacuum energy. With this postulate, physicists simply ignored the elephant in the room and just assumed that $\Lambda = 0$. However, after the existence of dark energy became undeniable, theorists faced a dilemma. Either assume that $\Lambda \sim 10^{-120} m_P^2 \neq 0$ and try to explain the extreme 120-orders of magnitude fine-tuning, or continue to take $\Lambda = 0$ and explain dark energy observations by other means.

The first method was seriously considered in the framework of string theory, which may result in an enormous number of minima for the scalar potential, with the cosmological constant assuming different values in each of them. Then, the argument went, the system tunnels through from one minimum to another, progressively leading to a smaller and smaller value of Λ. Different regions in space would correspond to different values of Λ (see also Sec. 8.8). If $|\Lambda|$ were too large in a region then this would result either in the early collapse of that region into a black hole if $\Lambda < 0$ or in quasi de Sitter expansion taking over too early for the formation of structure, leading to never-ending inflation if $\Lambda > 0$. We could not find ourselves in such a region. Ergo, we could only find ourselves in regions where $|\Lambda|$ is small enough. Thus, the trick was to make the cosmological constant an environmental quantity (like the positions of the planets in the Solar System), which may have a vast range of values at different locations (akin to other Solar Systems) and then select the one which works for us. This "explanation" is called *anthropic* and boils down to saying that "Things are as they are because if they were different we would not be here." It is the answer to the squirrel asking God "Why are humans more intelligent than squirrels?" with God replying "Were it the other way around, the humans would ask the same question" as opposed to answers like "They have binocular vision and opposable thumbs" or even "They have big brains." However, many physicists question the scientific merit of anthropic reasoning, because it seems to imply that there is no point in looking for deeper explanations of the observed phenomena.* They are what they are! Moreover, recent developments in string theory seem to suggest that $\Lambda > 0$ is not possible in a minimum of the potential and the whole idea of dismissing the Λ-problem in an anthropic way unravels.

9.1.2 Dark energy beyond the cosmological constant

This brings us to the second possibility. This one suggests that an unknown symmetry keeps $\Lambda = 0$, while dark energy has a different origin. Thus, this possibility *does not* solve the cosmological constant problem. It deals with it in the same double-think way that physicists managed to ignore it before the observations of dark energy. Instead of Λ, a new substance is assumed, more like the one which was driving inflation, which cannot itself be Λ because inflation needs to end. Dark energy beyond Λ must be dynamic (its density cannot always be constant) for otherwise it would not be different from vacuum density.

An enormous effort has been put into finding out the characteristics of dark energy in the hope of discerning it from a pure Λ. For example, if $w_{DE} \neq -1$, or if $\dot{w}_{DE} \neq 0$ (which is not the same thing) is observed then ΛCDM will be ruled out. Unfortunately, at the time of writing, this has not yet happened. The main focus is into refining and expanding the observational evidence coming from SNe-Ia, CMB and BAO. We briefly regard these:

---

*Having said that, this does not mean that the position of the planets in our Solar System must have a deeper explanation. Hundreds of exoplanets in extra-Solar Systems have been observed, and they confirm that the location of planets on the planetary disk (the ecliptic) is random. However, there is no shred of evidence substantiating the existence of the gazillions of potential minima claimed by string theory. And there lies the difference.

- **SNe-Ia:** (see Secs. 2.1.2 and 2.6.1) These are standard candles that enable observers to construct the *Hubble diagram*, which plots the Hubble parameter in terms of redshift $H(z)$ (see Sec. 2.2.4). For large values of z, H departs from the Hubble constant H_0. This departure encodes information about the rate of expansion of the Universe in the past, which in turn reveals the characteristics of dark energy. Several thousands of SNe-Ia have been observed to date, with the furthest out at redshift $z \simeq 2$.

- **CMB:** (see Sec. 5.1.3) The existence and recent dominance of dark energy affects the distance to the last scattering surface, because the Universe with dark energy is older than the Universe without dark energy. In turn, this shifts the location of the CMB acoustic peaks because the angle subtended by CMB anisotropies is affected. Thus, the CMB peak structure can be used to constrain dark energy. Two distance ratios of the acoustic peaks are used: the *acoustic scale* and the *shift parameter*. Additionally, dark energy affects the large-scale anisotropy in the CMB through the integrated Sachs-Wolfe effect (ISW). This is due to the variation of the gravitational potential at the epoch when acceleration begins. ISW helps distinguish dynamical dark energy from pure Λ because it provides information on the evolution of the dark energy density parameter Ω_{DE}.

- **BAO:** (see Sec. 5.2) BAO generate a standard ruler (yardstick), which is used to measure cosmological distances and thus construct the Hubble diagram $H(z)$. This yardstick is given by the comoving sound horizon (cf. Eq. (5.22))

$$r_s = \int_0^{t_{\mathrm{dec}}} c_s \frac{a_0}{a(t')} dt' = \int_{z_{\mathrm{dec}}}^{\infty} \frac{c_s}{H(z)} dz = 144.6 \pm 0.5 \,, \qquad (9.1)$$

where $a_0 = a(t_0)$ is the scale factor at present and we used that $\dot{z} = -(1+z)H(z)$, with 'dec' denoting decoupling and recombination and $c_s \simeq c/\sqrt{3}$ being the speed of sound of the baryon-photon fluid before decoupling (c is reinstated here). Observations along the line of sight provide information for the ratio $r_s/H^{-1}(z)$, while observations transverse to the line of sight provide information for the ratio $r(z)/r_s$, where $r(z)$ is the comoving distance.* BAO provide an absolute distance measurement and can be used for $z \lesssim 1$.

Apart from the above, other probes for the nature of dark energy include:

- **Weak lensing:** This refers to slight distortions of the images of distant galaxies due to the gravitational bending of light by the potential wells of structures in the Universe. Weak lensing amounts to distortions in shape, size and brightness, with the distortions in shape also called *cosmic shear*. Weak lensing probes dark energy by providing information on the expansion history and the growth of structure, meaning the evolution of the density perturbations. However, the method is not as powerful as the use of SNe-Ia or BAO. Also, it suffers from many uncertainties, for example the possible intrinsic alignment of galaxy shapes due to tidal gravitational effects.

- **Galaxy clusters:** The comoving volume of space is affected by spacetime geometry. The abundance of galaxy clusters reveals spacetime geometry, which in turn encodes information for the expansion history of the Universe. This distribution is inferred by optical, infrared, X-ray, and gravitational lensing observations, as well as through

---

*The distance which also takes into account the expansion of the Universe while the light travels until observed today.

the Sunyaev-Zel'dovich effect.[*] The observed cluster distribution is compared with accurate N-body simulations of structure formation. The influence of dark energy is similar to the case of weak lensing in that it affects both the history/geometry and the growth of structure. The sensitivity of the two methods is also comparable.

- **X-rays:** Observations of X-ray emitting gas in between galactic clusters may probe the redshift-distance relation. This intercluster gas corresponds to most of the baryonic matter in the Universe. Because of this, the baryonic mass function defined as $f_B \equiv \Omega_B/\Omega_m$ is expected to be approximately the same as the gas mass function $f_{\text{gas}} \approx f_B$, where $f_{\text{gas}} = M_{\text{gas}}/M_{\text{tot}}$, where M_{gas} is the mass of the intercluster gas and M_{tot} is the total mass. Observations of f_{gas} may constrain the geometry of the Universe and hence dark energy.

- **GRBs:** Gamma-ray bursts (GRBs) are flashes of γ-rays due to powerful explosions in distant galaxies, associated with compact objects, e.g., the formation of black holes from stellar collapse. GRBs are treated as standard candles, used to form a Hubble diagram complementary to SNe-Ia. Advantages of the GRBs are that γ-rays are almost immune from dust extinction and probe redshifts in the range $1 \lesssim z \lesssim 8$, well beyond most other methods. However, they are not fully understood and their standard candle status is questionable.

- **Growth factor:** This investigates how the growth of structure is affected by the presence of dark energy. The growth equation is Eq. (5.46) and the influence of dominating dark energy is encoded in the $2H\dot{\delta}$ term, through the Hubble parameter. The growth factor is defined as

$$f \equiv \frac{\mathrm{d}\ln\delta}{\mathrm{d}\ln a} = \Omega_m^\gamma(z)\,, \qquad (9.2)$$

where γ is called the growth index and it is measured from redshift distortions[†] and Lyman-α forest observations.[‡]

- **GWs:** Gravitational waves (GWs) have become recently observable. The characteristic "chirp" (GW pulse) from black hole or neutron star mergers can be used to obtain the luminosity distance in the same way as standard candles. In fact, for this reason they are called "standard sirens". The characteristics of dark energy are obtained through the construction of a Hubble diagram, like in the case of SNe-Ia.

The above methods can be divided in two broad categories: a) The ones which aim to produce the Hubble diagram and find $H(z)$. This includes SNe-Ia and GRBs for example. b) The ones which intend to probe the Universe geometry and the growth of structure. Examples are weak lensing and galaxy clusters. The hope is that, as the precision of observations increases, it will become possible to discriminate between models of dynamic dark energy and exclude various possibilities. Ideally, vacuum energy might also be excluded by future observations.

In the following, we explore the main ideas regarding dynamic dark energy. In many cases we consider that the mysterious substance is a scalar field, as in inflation. As we have seen, modified gravity can be reduced to Einstein gravity with a suitable scalar field, through a conformal transformation (see Secs. 7.3.2 – 7.3.4). This is why we will not consider explicit modified gravity

---

[*]The Sunyaev-Zel'dovich effect is due to hot electrons in the interstellar medium, which scatter CMB radiation through inverse Compton scattering, This creates a "shadow" on the observed CMB in some frequencies.

[†]Galactic redshifts are modulated due to peculiar velocities, which are affected by the gravitational potential of large scale structure in a coherent and quantifiable way.

[‡]The Lyman-α forest corresponds to absorption lines seen in spectra of distant galaxies or quasars due to hydrogen gas clouds along the line of sight. They show up as multiple lines superimposed to each other, like trees in a "forest," hence the name.

models for dark energy.* There is an additional reason for this. A recent observation of a neutron star merger demonstrated that many hitherto popular modified gravity models are excluded. Finally, dynamic dark energy may not be uniform in space. However, dark energy perturbations would be apparent only on very large scales. They might influence the formation of large scale structure and the CMB through ISW, but this influence is mild and we do not consider it further. For all practical purposes, we assume that dark energy is homogeneous, whatever it is.

9.2   Quintessence

Aristotle famously believed that there were more than the four classical elements of earth, air, fire and water. He considered also a fifth element, which he called quintessence; the fifth element.[†] Well, it seems that the Universe at present is becoming filled with a strange substance, which forces the expansion to accelerate. It is also the fifth substance, after baryonic matter, cold dark matter, photons, and neutrinos, which comprises the rest of the late Universe content. Hence, mimicking Aristotle, Robert R. Caldwell, Rahul Dave, and Paul J. Steinhardt in 1998 [CDS98] called this substance *quintessence*, assuming it is a scalar field. Early attempts to consider a scalar field as dark energy date back to the 1980s, well before late accelerated expansion was observed (e.g., Y. Fujii (1982) [Fuj82], P.D. Peccei, J. Sola, and C. Wetterich (1987) [PSW87], L.H. Ford (1987) [For87a], C. Wetterich (1988) [Wet88], B. Ratra, and P.J.E. Peebles (1988) [RP88]).

Much like in the inflationary paradigm, the barotropic parameter of quintessence is (cf. Eq. (6.7))

$$w_\phi = \frac{\frac{1}{2}\dot{\phi}^2 - V(\phi)}{\frac{1}{2}\dot{\phi}^2 + V(\phi)} \geq -1 \,, \tag{9.3}$$

while its equation of motion is given by the Klein-Gordon equation (6.8) repeated here

$$\ddot{\phi} + 3H\dot{\phi} + V' = 0 \,. \tag{9.4}$$

However, in contrast to the inflationary paradigm, where the inflaton scalar field dominates the Universe and determines its evolution during inflation, quintessence has been subdominant throughout most of the history of the Universe and is becoming significant only recently. This is why we first need to consider how a scalar field behaves when subdominant.

9.2.1   Dynamics of a spectator scalar field

There are three possibilities for a subdominant scalar field. First, it depends on whether the field's evolution is dominated by its kinetic or its potential density. If the effect of the potential is negligible, the equation of motion of the scalar field (9.4) reduces to

$$\ddot{\phi} + 3H\dot{\phi} \simeq 0 \,. \tag{9.5}$$

To solve the above, we use that, in the Hot Big Bang, the Hubble parameter scales as $H(t) = 2/3(1 + w_B)t$ (cf. Eq. (2.40)), where w_B is the barotropic parameter of the Universe, with $w_B = \frac{1}{3}$ in the radiation era and $w_B = 0$ in the matter era. Then the solution to Eq. (9.5) suggests

$$\dot{\phi} \propto t^{-2/(1+w_B)} \;\Rightarrow\; \rho_\phi \simeq \rho_{\rm kin} = \frac{1}{2}\dot{\phi}^2 \propto t^{-4/(1+w_B)} \,, \tag{9.6}$$

---

*Having said that, modified gravity and scalar-field dark energy are not identical. One can discriminate between them by considering, on one hand, the Universe expansion history and, on the other hand, the details of structure growth.

[†]He actually called it *pemptousia* (ΠΕΜΠΤΟΥΣΙΑ), which literally means "the fifth substance" in Greek.

where ρ_ϕ is the density of the field. Using $t \propto a^{\frac{3}{2}(1+w_B)}$ (cf. Eq. (2.39)), we readily find $\rho_\phi \propto a^{-6}$. Because $\rho_\phi \propto a^{-3(1+w_\phi)}$ (cf. Eq. (2.17)), this means that the kinetically dominated, subdominant scalar field has barotropic parameter $w_\phi = 1$.

The above also hold true during power law inflation, when $-1 < w_B < -\frac{1}{3}$ (cf. Sec. 8.2). What if we are during quasi-de Sitter inflation instead? Then we can approximate $H \simeq$ constant and $a \propto a^{Ht}$ (cf. Eq (2.33)). In this case, the solution to Eq. (9.5) is $\dot\phi \propto e^{-3Ht} \propto a^{-3}$. Thus, $\rho_\phi \simeq \frac{1}{2}\dot\phi^2 \propto a^{-6}$, which means $w_\phi = 1$ again.

Following the above scaling, we say that the scalar field is in "freefall" down its potential. But, typically, this cannot continue forever.* At some point, the potential becomes important and freefall ends. Then, there are two remaining possibilities. If the scalar field is massive, meaning $|V''| \gg H^2(t)$, then the friction term in the Klein-Gordon equation (9.4) becomes negligible and it does not make a difference whether the field is dominant or not. Therefore, in this case, the scalar field rapidly oscillates around the minimum of the scalar potential. The oscillations are eventually quasi-harmonic because, near the minimum, the potential is typically approximately quadratic.[†] Thus, we have $\rho_\phi \propto a^{-3}$ as we have shown in Sec. 6.3.1. The coherently oscillating field corresponds to massive particles which can decay, confining the field at the minimum.[‡] This can be its VEV or a metastable local minimum of the potential. In the latter case, the field eventually tunnels through the potential barrier and begins rolling again. If, on the other hand, the field is light, meaning $|V''| \ll H^2(t)$, then naïvely we might expect it to undergo slow-roll like the inflaton does in this case. But it does not. Instead it freezes at a constant value.

To see this, we consider the slow-roll approximation, under which Eq. (9.4) during the Hot Big Bang (of more generally, when $-1 < w_B < 1$) is written as

$$\frac{2}{1+w_B}\frac{\dot\phi}{t} \simeq -m^2\phi \,, \tag{9.7}$$

where we used Eq. (6.19) and have Taylor approximated $V' \simeq V''\phi$ with $V'' = m^2$, because we expect the field to slow-roll (at-most) so that $V'' \simeq$ constant. Solving the above, we find

$$\phi = \phi_0 \exp\left\{-\frac{1+w_B}{4}(mt)^2\left[1 - \left(\frac{t_0}{t}\right)^2\right]\right\} \simeq \phi_0 e^{-\frac{1}{4}(1+w_B)(mt)^2}, \tag{9.8}$$

where the subscript '0' denotes some initial value and in the last equation we considered $t \gg t_0$. Now, the condition $m^2 \ll H^2(t)$ suggests that $\frac{1}{4}(1+w_B)(mt)^2 \ll 1/9(1+w_B) < 1$ (recall that $H(t) = 2/3(1+w_B)t$, cf. Eq. (2.40)), which means that $\phi \simeq \phi_0$ and the field is frozen.

Similarly, during quasi-de Sitter inflation with $H \simeq$ constant and $a \propto e^{Ht}$, the Klein-Gordon equation becomes $3H\dot\phi \simeq -m^2\phi$, whose solution is $\phi \propto e^{-m^2t/3H} \propto a^{-\frac{1}{3}(m/H)^2}$. With $m^2 \ll H^2$ we see that $\phi \simeq$ constant and again the scalar field is frozen. Therefore, as we have shown, when a spectator scalar field is light it does *not* slow-roll down its potential but it is frozen at a constant value.[§]

Because quintessence should cause accelerated expansion when it comes to dominate the Universe (that is today), it cannot be kinetically dominated when it does so, and neither should

---

*Of course, if $V(\phi) = 0$ freefall can continue forever.

[†]Unless $V'' = 0$ at the minimum.

[‡]If the field decays late, then it can act like a curvaton field, whose evolution we discussed in Sec. 6.5.5. If the field does not decay at all, then if such a field dominated the Universe it would behave as cold dark matter, similar to axions. Therefore, it would not be useful as quintessence, because it would not cause accelerated expansion.

[§]Here we consider only the classical variation of the field. In inflation, the field is also perurbed by its quantum fluctuations (see Sec. 8.7).

it be oscillating.* Therefore, it has to be light so as to lead to slow-roll when it becomes dominant. This means that $|V''| \lesssim H_0^2$ so that $|V''| \ll H^2$ in the past because $H_0 < H(t)$, where $H_0 = H(t_0)$ is the Hubble constant and $t_0 \simeq 13.8\,\mathrm{Gy}$ is the present time, with $t < t_0$. Thus, we expect quintessence to be frozen in the near past and unfreezing at present, when it is becoming dominant and is beginning to slow-roll. Such quintessence is called *thawing* but it is not the only possibility.

Indeed, it is also possible that quintessence has been steadily rolling down its relatively steep potential, approaching a freezing point as its potential becomes shallow near the present. Such quintessence is called *freezing*. It sounds unlikely that the potential is steep enough for quintessence to be rolling and not frozen, but not so steep as for the field to be oscillating or in freefall. Our discussion in this section seems to imply that we would need to tune the potential to get it so, ensuring somehow that $|V''| \sim H^2$ is maintained, so as to avoid freefall, oscillations and freezing. However, there is another possibility which we have not considered yet, when all the three terms in the Klein-Gordon equation (9.4) are important and stay important during the field evolution. This possibility corresponds to a possible attractor of the system and we investigate it next.

9.2.2 Scaling and tracking quintessence

Rendering late dark energy into a dynamical degree of freedom instead of a simple (albeit incredibly fine-tuned) cosmological constant introduces an additional complication. That of its initial conditions. One way to alleviate this problem is by considering dynamical dark energy with attractor properties. As with the case of the inflationary attractor, a wide range of initial conditions can be funnelled into a common final evolutionary path.[†] This path may lead the originally spectator quintessence field to eventual domination. If this is the case, then the attractor is called a *tracker*. Let's look into the conditions for such tracking behaviour to appear.

We define the following dimensionless parameters

$$\lambda \equiv -\frac{m_P V'}{V} \tag{9.9}$$

and

$$\Gamma \equiv \frac{V V''}{(V')^2}, \tag{9.10}$$

which are, in general, functions of the field $\lambda = \lambda(\phi)$ and $\Gamma = \Gamma(\phi)$. Locally in field space, $\lambda(\phi)$ is roughly constant so we can write $V \propto \exp(-\lambda\phi/m_P)$. This means that λ parameterises the slope of the potential. A small value of λ implies that the potential is shallow and the field could become quintessence if it came to dominate the Universe. We show below how a tracker solution leads to small λ. Γ is sometimes called the *tracking parameter*. If $\Gamma \geq 1$, there is an attractor solution to the evolution of the field. If $\Gamma > 1$, this attractor is a tracker. Let us demonstrate all this.

Taking the time derivative of Eq. (9.9), we obtain

$$\frac{\dot{\lambda}}{\lambda^2} = -\frac{\dot{\phi}}{m_P}(\Gamma - 1), \tag{9.11}$$

where we employed both Eqs. (9.9) and (9.10). If $V' < 0$ then $\dot{\phi} > 0$ as the field moves to larger values to minimise V, while Eq. (9.9) suggests that $\lambda > 0$. In this case, Eq. (9.11) suggests that

---

*Except if it is oscillating in a potential of the form $V \propto |\phi|^q$, with $q < 1$, as we have discussed in Sec. 8.5. We will not consider this possibility in this chapter, largely because it has been shown that oscillating quintessence leads to excessive inhomogeneities and is therefore excluded.
†Like multiple springs feeding the same river.

the positive λ is decreasing with time ($\dot{\lambda} < 0$) when $\Gamma > 1$. If $V' > 0$ then $\dot{\phi} < 0$ as the field moves to smaller values to minimise V, while Eq. (9.9) suggests that $\lambda < 0$. In this case, Eq. (9.11) suggests that the negative λ is increasing with time ($\dot{\lambda} > 0$) when $\Gamma > 1$. In both cases, $\lambda \to 0$ when $\Gamma > 1$, which means that the potential becomes shallower as time passes. This is not enough for quintessence though. We also need to have the field eventually dominating the Universe. We now show that the decrease of $|\lambda|$ when $\Gamma > 1$ can make sure of this too.

From Eq. (9.3) it is straightforward to obtain

$$1 + w_\phi = \frac{\dot{\phi}^2}{\rho \Omega_\phi} \,, \tag{9.12}$$

and

$$1 - w_\phi = \frac{2V}{\rho \Omega_\phi} \,, \tag{9.13}$$

where we used $\rho \Omega_\phi = \rho_\phi = \frac{1}{2}\dot{\phi}^2 + V$, with $\Omega_\phi \equiv \rho_\phi/\rho$ being the quintessence density parameter in a flat Universe. Taking the time derivative of Eq. (9.3) we find

$$\begin{aligned}
\dot{w}_\phi &= \frac{2V}{\rho \Omega_\phi} \frac{\dot{\phi}\ddot{\phi} + \frac{1}{2}\lambda\dot{\phi}^3/m_P}{\rho_\phi} \\
&= (1 - w_\phi)\dot{\phi}\left(-\frac{3H\dot{\phi}}{\rho_\phi} + \frac{\lambda}{m_P}\right) \\
&= (1 - w_\phi)[-3H(w_\phi + 1) + \lambda\dot{\phi}/m_P] \Rightarrow \\
\dot{w}_\phi &= (1 - w_\phi)\left[-3(w_\phi + 1) + \lambda\sqrt{3\Omega_\phi(w_\phi + 1)}\right]H \,, \tag{9.14}
\end{aligned}$$

where in the first line of the above we have used Eq. (9.9), in the second line Eqs. (9.4) and (9.13), in the third and fourth lines Eq. (9.12) which suggests $\dot{\phi}^2 = 3H^2 m_P^2 \Omega_\phi(w_\phi + 1)$ and we used the Friedman equation (6.2). Attractor solutions feature $-1 < w_\phi \simeq \text{constant} < 1$, which means $\dot{w}_\phi = 0$. In this case, Eq. (9.14) suggests

$$\Omega_\phi = \frac{3}{\lambda^2}(w_\phi + 1) \,. \tag{9.15}$$

The above shows that, as $|\lambda|$ diminishes with time (when $\Gamma > 1$), Ω_ϕ grows, which makes sure that quintessence eventually dominates the Universe as required. This is why

$$\Gamma > 1 \,, \tag{9.16}$$

is called the *tracking condition*.

We can calculate the barotropic parameter of quintessence on the attractor as follows. Consider that the Universe contains only two substances: quintessence with density ρ_ϕ and barotropic parameter w_ϕ and the background substance (matter or radiation, for example) with density ρ_B and barotropic parameter w_B. These are independent substances, which satisfy the continuity equations

$$\dot{\rho}_\phi + 3(1 + w_\phi)H\rho_\phi = 0 \quad \text{and} \quad \dot{\rho}_B + 3(1 + w_B)H\rho_B = 0 \,. \tag{9.17}$$

Then we have

$$\dot{\rho} = \dot{\rho}_\phi + \dot{\rho}_B = -3H[(1 + w_B)\rho + (w_\phi - w_B)\rho_\phi] \,. \tag{9.18}$$

We now take the time derivative of the quintessence density parameter $\Omega_\phi = \rho_\phi/\rho$. We get

$$\begin{aligned}
\dot{\Omega}_\phi &= \frac{1}{\rho}\left(\dot{\rho}_\phi - \rho_\phi\frac{\dot{\rho}}{\rho}\right) \\
&= \frac{1}{\rho}\left\{\dot{\rho}_\phi + 3H\rho_\phi\left[(1 + w_B) + (w_\phi - w_B)\Omega_\phi\right]\right\} \Rightarrow \\
\dot{\Omega}_\phi &= 3H(w_\phi - w_B)\Omega_\phi(\Omega_\phi - 1) \,, \tag{9.19}
\end{aligned}$$

where in the second line we considered Eq. (9.18), and in the third line we employed the continuity equation (9.17) for quintessence. Eq. (9.15) suggests $\dot{\Omega}_\phi/\Omega_\phi = -2\dot{\lambda}/\lambda$. In view of Eqs. (9.11) and (9.19), this becomes

$$3(w_\phi - w_B)(\Omega_\phi - 1) = 2\sqrt{3\Omega_\phi(1 + w_\phi)}(\Gamma - 1)\lambda\,, \tag{9.20}$$

where we also considered Eq. (9.12). Eliminating λ with the use of Eq. (9.15) we end up with

$$(w_\phi - w_B)(\Omega_\phi - 1) = 2(1 + w_\phi)(\Gamma - 1)\,. \tag{9.21}$$

Considering that quintessence is subdominant, we have $\Omega_\phi \ll 1$. Then, the above results in the barotropic parameter

$$w_\phi = \frac{w_B - 2(\Gamma - 1)}{2\Gamma - 1}\,. \tag{9.22}$$

We see that when $\Gamma = 1$ we have $w_\phi = w_B$. This solution satisfies also Eq. (9.21), i.e., it does not depend on the assumption $\Omega_\phi \ll 1$. As we show in the next section, this is the *scaling* attractor. If quintessence is dominant, then $\Omega_\phi \simeq 1$, in which case Eq. (9.21) suggests that $w_\phi \simeq -1$. This means that when a scalar field is dominant and has approximately constant barotropic parameter, then this barotropic parameter is $w_\phi \simeq -1$. This is the case of the inflaton field, for example. We put all the above to use by considering two specific examples, which have been popular in the literature, namely exponential and inverse-power-law quintessence.

9.2.3 Exponential quintessence

This case considers $\lambda = \text{constant}$, which implies

$$V(\phi) = V_0 \exp(-\lambda\phi/m_P)\,, \tag{9.23}$$

where V_0 is some constant density scale [CLW98]. We have already looked into this potential when discussing power-law inflation in Sec. 8.2. We have discussed the attractor solution for the dominant field, shown in Eq. (8.2) repeated here

$$V = \frac{2(6 - \lambda^2)}{\lambda^2}\left(\frac{m_P}{\lambda t}\right)^2 \quad \text{and} \quad \rho_{\text{kin}} \equiv \frac{1}{2}\dot{\phi}^2 = 2\left(\frac{m_P}{\lambda t}\right)^2\,. \tag{9.24}$$

As shown in Eq. (8.4), on this attractor $H = 2/\lambda^2 t$ and the corresponding barotropic parameter is $w_\phi = -1 + \lambda^2/3$. Then, accelerated expansion demands $\lambda < \sqrt{2}$. It is straightforward to see that steady roll is maintained during this attractor. Indeed, noting that $V'' = \lambda^2 V/m_P^2$ and using Eq. (9.24) we find

$$\frac{V''}{H^2} = \frac{1}{2}\lambda^2(6 - \lambda^2) = \text{constant}\,. \tag{9.25}$$

Thus, for $\lambda \sim 1$, $|V''| \sim H^2$ is maintained during the attractor, as required. For example, in the limiting case for dark energy $\lambda = \sqrt{2}$ we have $V'' = 4H^2$.

The above regard the scalar field as dominant. If it is subdominant, however, there is a different attractor solution. This attractor is not a tracker because $\Gamma = 1$ as suggested by Eq. (9.10) (or (9.11) for that matter). Putting this into Eq. (9.22), we find $w_\phi = w_B$ and the subdominant attractor is

$$V = 2\frac{1 - w_B}{1 + w_B}\left(\frac{m_P}{\lambda t}\right)^2 \quad \text{and} \quad \rho_{\text{kin}} \equiv \frac{1}{2}\dot{\phi}^2 = 2\left(\frac{m_P}{\lambda t}\right)^2\,. \tag{9.26}$$

It can be readily checked that the above is an exact solution to the Klein-Gordon equation (9.4). Note that the kinetic density in the dominant and subdominant attractors is the same. This

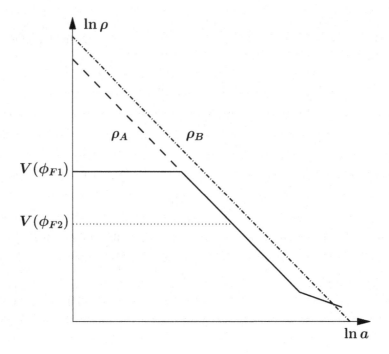

FIGURE 9.1 Log-log plot of the evolution of the quintessence density and the background density in the scaling-freezing scenario with respect to the scale factor of the Universe. The background density ρ_B is depicted with the dash-dotted line. The scaling attractor density ρ_A is depicted with the dashed line. Two possible initial conditions are considered, leading to the same evolution. In both, the quintessence field starts frozen at a value ϕ_{F1} (solid line) or ϕ_{F2} (dotted line) with $\phi_{F2} > \phi_{F1}$. Initially, the field has constant density $V(\phi_{F1})$ or $V(\phi_{F2})$, respectively, with $V(\phi_{F2}) < V(\phi_{F2})$. The field remains frozen until the attractor density becomes comparable to the field's potential density, in which case it unfreezes and follows the attractor. Once the quintessence field follows the attractor, its subsequent evolution is common regardless of its initial frozen value. The scaling attractor keeps the quintessence density at a constant ratio to the background density ρ_B, until the second exponential in the scaling-freezing scenario kicks in, which terminates the scaling behaviour and leads to the eventual domination of quintessence.

means that $\phi(t)$ is also given by Eq. (8.3). The potential and the kinetic density of the attractor are kept at a constant ratio

$$\frac{V}{\rho_{\text{kin}}} = \frac{1 - w_B}{1 + w_B} \,, \tag{9.27}$$

where $\rho_{\text{kin}} = \frac{1}{2}\dot{\phi}^2$. Also, note that the subdominant attractor coincides with the dominant one if $w_B = w_\phi = -1 + \lambda^2/3$, i.e., $\lambda = \sqrt{3(1 + w_B)}$. The density parameter of the subdominant attractor is

$$\Omega_\phi = \frac{3(1 + w_B)}{\lambda^2} = \text{constant} < 1 \,, \tag{9.28}$$

where $\rho_\phi = \rho_{\text{kin}} + V = \frac{4}{1+w_B}(m_P/\lambda t)^2$ (cf. Eq. (9.26)) and $\rho = \frac{4}{3(1+w_B)^2}(m_P/t)^2$ (cf. Eq. (2.41)). The above coincides with Eq. (9.15).

It is straightforward to see that steady roll is maintained during this attractor as well. Indeed, noting again that $V'' = \lambda^2 V/m_P^2$ and using Eq. (9.26), we find

$$\frac{V''}{H^2} = \frac{9}{2}(1 - w_B^2) = \text{constant} \,. \tag{9.29}$$

Thus, $|V''| \sim H^2$ is maintained during the attractor, as required. For example, during the radiation era of the Hot Big Bang, when $w_B = \frac{1}{3}$, we have $V'' = 4H^2$.

Therefore, when subdominant, exponential quintessence follows the subdominant attractor and its density remains a constant fraction of the total. This is why this attractor is called *scaling attractor*. It is evident that a scalar field rolling along the scaling attractor of its exponential potential can never dominate, thus it cannot become quintessence. Also, because $w_\phi = w_B$ on the scaling attractor, this scalar field does not act as dark energy, even subdominant dark energy, because $w_B > -\frac{1}{3}$. Thus, a modification is needed to break the scaling of the rolling scalar field and render it quintessence.

A possible way out was presented by Tiago Barreiro, Edmund J. Copeland, and Nelson J. Nunes, in 2000 [BCN00]. The idea was to consider two exponential functions of the form

$$V(\phi) = V_1 e^{-\lambda_1 \phi / m_P} + V_2 e^{-\lambda_2 \phi / m_P}, \tag{9.30}$$

where V_1, V_2 are constant density scales and $\lambda_1 > \lambda_2$. This condition ensures that the first term in $V(\phi)$ reduces faster than the second term as quintessence rolls towards larger values. As a result, originally only the first term is important, while eventually the second term takes over. Thus, in the beginning, the system assumes the scaling attractor corresponding to the first exponential. Consequently, its density is kept at a constant ratio to the overall density. How large this ratio is allowed to be is primarily constrained by BBN, because the scalar field contribution to the density budget during BBN has to be small enough to not disturb the delicate process of the formation of light nuclei. The constraint is $\Omega_\phi < 0.045$, which (via Eq. (9.28)) translates to the bound $\lambda_1 > 9.4$. Simultaneously, we can obtain a bound on λ_2 considering that the scaling attractor is abandoned when the second exponential term in Eq. (9.30) becomes important, thereby freezing the rolling quintessence field. Thus, at present we expect that the field unfreezes and is about to follow the dominant exponential attractor, for which $w_\phi = -1 + \lambda_2^2 / 3$. Barring phantom dark energy, observations demand that $w_\phi < -0.95$ at present, which produces the bound $\lambda_2 < 0.4$. This scenario is called *scaling-freezing quintessence*, and it is shown in Fig. 9.1.

The problem with this scenario is that V_1 has to be fine-tuned to almost the dark-energy scale $(10^{-3}\,\mathrm{eV})^4 \sim 10^{-120}\, m_P^4$ in order for the freezing to occur at present with the residual potential density comparable to the density of the Universe today. This begs the question, if one is willing to live with such fine-tuning why not stick with ΛCDM in the first place.

Exponential quintessence could work if $\lambda \ll \sqrt{2}$, and the field never assumes the scaling solution but, instead, remains frozen until the present, when it begins to dominate and unfreezes to slow-roll down its potential approaching the dominant attractor in Eq. (9.24). For this scenario to work, the field needs to be frozen at the correct value ϕ_F such that its density when it unfreezes today is given by $\Omega_\Lambda \rho_0$, where $\Omega_\Lambda \simeq 0.687$. In this case, we have classic thawing quintessence, which does not enjoy the benefits of attractor behaviour and needs to have its initial conditions (e.g., the value ϕ_F) explained.

9.2.4 Inverse-power-law quintessence

We have seen that, although there is ample theoretical justification for exponential potentials, the scaling attractor cannot lead to successful quintessence, while the scenario of scaling-freezing quintessence seems to require comparable tuning to ΛCDM, so it is not that appealing. A different suggestion, which is also supported by some supersymmetric theories, considers a truly tracking attractor, and is called *inverse-power-law* (IPL) quintessence. It was first considered by Paul J. Steinhardt, Li-Min Wang, and Ivaylo Zlatev in 1999 [SWZ99].

The potential is

$$V(\phi) = \frac{M^{q+4}}{\phi^q}, \tag{9.31}$$

where M is some constant energy scale and $q > 0$. From Eq. (9.10) we find

$$\Gamma = \frac{q+1}{q} > 1 . \tag{9.32}$$

We see that the tracking condition in Eq. (9.16) is always satisfied. The tracker is

$$\phi(t) = \left[\frac{q(q+2)^2(1+w_B)}{4(q+2) - 2q(1+w_B)} \right]^{\frac{1}{q+2}} M^{\frac{q+4}{q+2}} t^{\frac{2}{q+2}} . \tag{9.33}$$

The above looks like a horrible expression, but it reduces to a much more reasonable form when specific values for q and w_B are considered. For example, if $q = 4$ then the tracker solution in the matter era ($w_B = 0$) is simply $\phi = (3M^4 t)^{1/3}$. For the potential and kinetic density of the attractor we find

$$V = \left[\frac{4(q+2) - 2q(1+w_B)}{q(q+2)^2(1+w_B)} \right]^{\frac{q}{q+2}} M^{2(\frac{q+4}{q+2})} t^{-\frac{2q}{q+2}} \tag{9.34}$$

and

$$\frac{1}{2}\dot{\phi}^2 = \frac{2}{(q+2)^2} \left[\frac{4(q+2) - 2q(1+w_B)}{q(q+2)^2(1+w_B)} \right]^{-\frac{2}{q+2}} M^{2(\frac{q+4}{q+2})} t^{-\frac{2q}{q+2}} . \tag{9.35}$$

It can be easily checked that the above are an exact solution to the Klein-Gordon equation (9.4). Notice also that the kinetic and potential densities stay at a constant ratio given by

$$\frac{V}{\rho_{\rm kin}} = \frac{2(q+2)}{q(1+w_B)} - 1 . \tag{9.36}$$

The above coincides with Eq. (9.27) in the limit $q \to \infty$ because the exponential is the limit of power-law with infinite power. For the same reason, in this limit Eq. (9.32) suggests $\Gamma \to 1$ as in exponential quintessence.

From Eqs. (9.34) and (9.35), we see that $\rho_\phi = \frac{1}{2}\dot{\phi}^2 + V \sim (M^{q+4}t^{-q})^{2/(q+2)}$. We require that $\rho_\phi(t_0) = \Omega_\Lambda \rho_0 = (2.25 \times 10^{-3}\,{\rm eV})^4$, where $t_0 = 13.8\,{\rm Gy}$ is the present time, $\rho_0 = \rho(t_0) = 0.864 \times 10^{-29}\,{\rm g/cm^3}$ is the current density of the Universe and $\Omega_\phi(t_0) = \Omega_\Lambda \simeq 0.687$. Putting all the above together, we find

$$M \sim 10^{\frac{19q-46}{q+4}}\,{\rm GeV} . \tag{9.37}$$

For example, taking $q = 4$ we get $M \sim 10^{3.75}\,{\rm GeV} = ({\rm a\ few}) \times {\rm TeV}$, which is near the electroweak energy scale. Also, the requirement that $V(\phi_0) = M^{q+4}/\phi_0^q \simeq \rho_\phi(t_0)$, where $\phi_0 = \phi(t_0)$ is the value of the tracker quintessence today, suggests

$$\phi_0 \sim 10^{19+1/q}\,{\rm GeV} \sim 10\,m_P . \tag{9.38}$$

This satisfies the condition for acceleration when $q \sim 1$, which is $\lambda < \sqrt{2}$. In view of Eq. (9.9) we find that this condition demands $\phi > (q/\sqrt{2})m_P$.

For the potential in Eq. (9.31) we have $V'' = q(q+1)V/\phi^2$. Evaluating this on the tracker, employing Eqs. (9.33) and (9.34), we obtain

$$V'' = \frac{(q+1)[4(q+2) - 2q(1+w_B)]}{(q+2)^2(1+w_B)} t^{-2} . \tag{9.39}$$

Using this, it is straightforward to see that steady roll is maintained during the tracker. Indeed, we find

$$\frac{V''}{H^2} = \frac{9(q+1)[4(q+2) - 2q(1+w_B)](1+w_B)}{4(q+2)^2} = {\rm constant} . \tag{9.40}$$

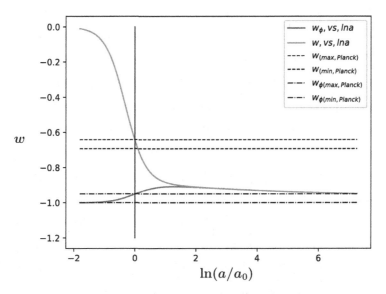

FIGURE 9.2 Behaviour of the barotropic parameter of quintessence w_ϕ (lower solid curve) and of the whole Universe w (upper solid curve) as a function of the logarithm of the scale factor $\ln(a/a_0)$, with a_0 being the value of the scale factor today, for the model $V = M^8/\phi^4$ ($q = 4$), for which $M = 6.25\,\mathrm{TeV}$. We see that originally the Universe is in the matter era with $w = 0$ and the quintessence field is frozen at value ϕ_F =constant with constant density, such that $w_\phi = -1$. However, when approaching the present time (depicted by the vertical solid line) the quintessence unfreezes and $w_\phi(t) > -1$, while it also begins to dominate the Universe so that $w(t) < 0$. We find numerically that successful quintessence is achieved when $\phi_F \geq 6.80\,m_P$. Choosing the limiting case $\phi_F = 6.80\,m_P$ the present values of w_ϕ and w just about satisfy the Planck bounds, depicted by the horizontal lines (w must lie between the horizontal dashed lines and w_ϕ must line between the horizontal dash-dotted lines). In the future, quintessence becomes fully dominant so $w \approx w_\phi$, while it slow-rolls down the scalar potential, ever more slowly, approximating $w = w_\phi \to -1$ in agreement with Eq. (9.43). It is clear that both w_ϕ and w are running at present. Using the CPL parameterisation of Eq. (2.37), we find $w_{\mathrm{DE}}^0 = -0.95$ and $w_a \equiv -dw_\phi/da|_{a=a_0} = -0.0659$.

Thus, for $q \sim 1$, $|V''| \sim H^2$ is maintained during the tracker, as required. The above looks ugly, but it drastically improves if specific values are chosen. For example, taking $q = 4$ during the matter era of the Hot Big Bang, when $w_B = 0$, we have $V'' = 5H^2$.

So far, all seems well with this model. Things turn sour, however, when one considers just what kind of dark energy this model would produce. Indeed, combining Eqs. (9.22) and (9.32) we find that the barotropic parameter of quintessence on the tracker is

$$w_\phi = \frac{qw_B - 2}{q + 2}\,. \tag{9.41}$$

This is no good for reasonable values of q. For example, taking $q = 4$ in the matter era, Eq. (9.41) suggests $w_\phi = -\frac{1}{3}$. This is borderline for dark energy. It does suggest that $\rho_\phi \propto a^{-3(1+w_\phi)} = a^{-2}$, which means that quintessence would eventually come to dominate the Universe as required, but it miserably fails the observational demand that $w_\phi < -0.95$ at present. For this to be true, Eq. (9.41) requires $q \leq 0.1$, which demands $M \lesssim 10^{-11}\,\mathrm{GeV}$ in view of Eq. (9.37). Such values are not reasonable on particle physics grounds.

Not all is lost though. Recall that the tracker becomes invalid when quintessence ceases to be subdominant. Instead, we would expect the dominant quintessence to slow-roll down the potential in Eq. (9.31). For this potential, the slow-roll equation (6.19) results in

$$3H\dot{\phi} \simeq -V' \quad \Rightarrow \quad \phi^{q+2}\dot{\phi}^2 \simeq \frac{q^2}{3}m_P^2 M^{q+4}, \tag{9.42}$$

where we have also used the slow-roll Friedman equation (6.20). Employing the above, we may estimate the barotropic parameter of quintessence using Eq. (9.3). We find

$$w_\phi = \frac{q^2 m_P^2 - 6\phi^2}{q^2 m_P^2 + 6\phi^2}. \tag{9.43}$$

The requirement for dark energy, namely $w_\phi < -\frac{1}{3}$, demands that $\phi > (q/\sqrt{3})m_P$, which is comparable to the requirement for the tracking condition ($\lambda < \sqrt{2}$) to hold, found above. Demanding that $w_\phi < -0.95$ at present requires $\phi_0 > 2.55\, q\, m_P$. This is not too far from the value we estimated in Eq. (9.38), for $q = 4$ for example. Put another way, considering $q = 4$ and $\phi_0 = 10\, m_P$ in Eq. (9.43) we obtain that $w_\phi = -0.95$, which is marginally acceptable.

In reality, we are neither in the purely dominant nor in the purely subdominant case, but somewhere in between. Therefore, it is likely that IPL quintessence can work with q of order unity. However, the tracking attractor cannot be utilised unless $q \ll 1$, which is not realistic. This means that this *tracking-freezing* quintessence scenario, where the tracker brings the quintessence field to dominate today, is not very feasible. In contrast, the model may act as thawing quintessence, where the field is originally frozen and unfreezes today, when it becomes dominant and begins to slow-roll down its potential (but the required value of ϕ_F needs explaining). The tracker is never reached. A successful example of this possibility is discussed in Fig. 9.2. As shown in the figure, the barotropic parameter of quintessence w_ϕ is varying today. Such variation will be observable in the near future.

9.2.5   Freezing versus thawing quintessence

In Sec. 9.2.2, we have explained the attractive nature of a tracker. Now, we need to discuss how the system may find itself on the attractor. Basically, there are three possibilities regarding the initial conditions. First, we cannot exclude the unlikely possibility that the initial conditions place the system miraculously on the attractor right away. This means that both ϕ and $\dot{\phi}$ have exactly the right values to satisfy the attractor requirements. Barring this special case, there are two other possibilities.

In one such possibility, the system has initially more potential than kinetic density compared to the attractor values, that is $V \gg \rho_{\text{kin}}$ (recall that on the attractor $V \sim \rho_{\text{kin}}$). If $|V''| \gg H^2$, the situation immediately changes as the system rapidly rolls to engage into oscillations, for which $V \sim \rho_{\text{kin}}$, as on the attractor. This really means that the attractor is adopted right away and the situation is not too different from the "miraculous" possibility discussed in the beginning. If however, $|V''| \ll H^2$, the scalar field freezes, as we have shown. As time passes, the value of V'' remains constant because the field is frozen, that is its location on the potential slope is fixed and so is the curvature of the potential where the field lies. On the other hand, $H^2 = H^2(t)$ is decreasing with time, until it becomes comparable to V'', when the field unfreezes. Recall that on the attractor $|V''| \sim H^2$, as we have shown (e.g., see Eqs. (9.25), (9.29), and (9.40)). Thus, when the field unfreezes it immediately follows the attractor. Turning this around, when the conditions of the attractor are met, meaning when the potential density of the frozen field becomes comparable to the density that the field would have if it were following the attractor, the field unfreezes and follows the attractor.

The final possibility is to begin with a kinetically dominated field, such that $V \ll \rho_{\text{kin}}$. In this case, the field is in freefall and varies oblivious to its potential. Its density decreases as $\rho_\phi \propto a^{-6}$ until it is depleted enough that the field becomes potentially dominated, and we are

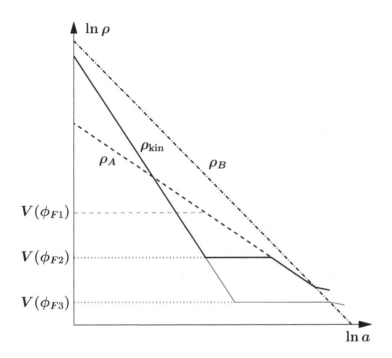

FIGURE 9.3 Log-log plot of the evolution of the quintessence density and the background density in the presence of a tracker attractor solution with respect to the scale factor of the Universe. The background density ρ_B is depicted with the dash-dotted line. The tracker attractor density ρ_A is depicted with the thick dashed line. Three possible initial conditions are considered. In the first one, the quintessence field starts frozen at a value ϕ_{F1} with constant density $V(\phi_{F1})$ and is depicted with the thin dashed line. The field remains frozen until the attractor density becomes comparable to its potential density, when it unfreezes and follows the attractor. The second case is depicted with the thick solid line. In this case, the quintessence field begins kinetically dominated, with $\rho_{\rm kin} \gg V$, in freefall with $\rho_\phi \simeq \rho_{\rm kin} \propto a^{-6}$. Its freefall evolution overshoots the attrator (the thick solid line crosses the dashed line). Eventually, its kinetic density is depleted and the field is frozen at some value ϕ_{F2} with constant density $V(\phi_{F2})$. The field remains frozen until the attractor density becomes comparable to its potential density, when it unfreezes and follows the attractor as with the previous case. The final case is depicted with the thin solid line. In this case too, the quintessence field begins kinetically dominated and in freefall. But, this time it freezes at a value $\phi_{F3} > \phi_{F2}$, such that it has constant density $V(\phi_{F3}) < V(\phi_{F2})$. In this case, the potential density of the field becomes comparable to the background denisty ρ_B before the attractor catches up with it, i.e., $V \sim \rho_B < \rho_A$. The field then unfreezes and begins to slow-roll down its potential without experiencing the attractor. This is thawing quintessence.

back on the previous cases. When the field is in freefall it may "overshoot" the attractor until it freezes and lies dormant waiting for the attractor to catch up with it. This behaviour is shown in Fig. 9.3.

What if the attractor does not catch up with the frozen field until today? Then, there are two possibilities. In one the potential density of the frozen field remains subdominant even today. In this case, the existence of this field is irrelevant (and Occam's razor suggests we dismiss it). The other, is that the potential density of the frozen field is becoming dominant today.* Such a field can be successful quintessence. It unfreezes and starts slow-rolling, as the inflaton field does in inflation. This is a classic thawing quintessence scenario. Thus, we see that, even when there is in principle an attractor, the system can still behave as thawing quintessence if the attractor has not been reached until the present (as we have discussed for IPL quintessence). However, if the attractor-tracker behaviour is not experienced, the benefits regarding the required tuning of the quintessence initial conditions are lost. If thawing quintessence is preferred over attractor-freezing quintessence, its initial conditions must be explained and should not be assumed "by hand." One such possibility is quintessential inflation, where thawing quintessence has its initial conditions fixed by the inflation attractor, as we discuss in Sec. 9.7.

9.2.6 PNGB quintessence

As we have seen so far, thawing quintessence seems more successful than attractor quintessence, even though it suffers from the severe problem of explaining the otherwise tuned initial conditions. One possibility that does not suffer from the problem of initial conditions despite being thawing quintessence (which does not feature any attractors) considers a pseudo Nambu-Goldstone boson (PNGB) [FHSW95], which has similar theoretical justification to the inflaton field of natural inflation in Sec. 7.2.3. The quintessence potential is

$$V(\phi) = V_0[1 + \cos(\phi/qm_P)]\,, \tag{9.44}$$

which only has a phase difference with Eq. (7.30) ($\phi \to \pi q m_P + \phi$), with $q > 0$. The value of the field is presumably homogenised by inflation.[†] From Eq. (9.44) we find

$$V''(\phi) = -\frac{V_0}{q^2 m_P^2} \cos(\phi/qm_P)\,. \tag{9.45}$$

Thus, generically we expect $m^2 = |V''| \simeq V_0/q^2 m_P^2$, unless we fine-tune $|\cos(\phi/qm_P)| \ll 1$. We consider that originally $|V''| \ll H^2$ so the field is frozen. Assuming that the field comes to dominate at present requires that

$$V_0 \simeq \rho_\phi(t_0) = \Omega_\Lambda \rho_0 \sim m_P^2 H_0^2 \sim (10^{-3}\,\text{eV})^4\,, \tag{9.46}$$

where the Friedman equation (6.2) is $\rho_0 = 3m_P^2 H_0^2$. Keeping the field frozen until today requires $m^2 \lesssim H_0^2$, which implies

$$1 \gtrsim \frac{m^2}{H_0^2} \sim \frac{V_0}{q^2 m_P^2 H_0^2} \sim \frac{1}{q^2}\,. \tag{9.47}$$

Therefore, we need that $q \gtrsim 1$. The above suggest that, for generic initial conditions, the model can be successful quintessence, *provided* $V_0 \sim (10^{-3}\,\text{eV})^4$. This is not much different from requiring $V_2 \sim (10^{-3}\,\text{eV})^4$ in the scaling-freezing scenario in Sec. 9.2.3. There we argued that this amounts to the level of tuning of ΛCDM, which defeats the purpose of considering quintessence.

---

*We can disregard the third possibility that the frozen field dominates the Universe in the past, unless it is the far past, in which case we are talking about an inflaton field.

[†]This is barring inhomogeneities due to particle production during inflation. Such inhomogeneities would be absent if the phase transition that gave rise to the axionic degree of freedom ϕ featured in the potential in Eq. (9.44) occurred after inflation, so that during inflation ϕ did not even exist.

However, there is theoretical justification for such a value of V_0 in the case of PNGB quintessence. Indeed, in the last decade string theory considerations have suggested that there might be a plethora of axion fields characterised by a potential of the form in Eq. (9.44) with a wide range of values of V_0, in the context of a theoretical construction called the *axiverse*. Moreover, there have been attempts to explain the desired value of V_0 for a PNGB through a mechanism similar to the one envisaged for neutrino masses, called the *see-saw* mechanism. This makes use of the curious fact that the electroweak energy scale is roughly the geometrical mean of the Planck scale and the recent dark energy scale. Schematically

$$\text{TeV} \sim \sqrt{\Lambda m_P}, \qquad (9.48)$$

where $\Lambda \sim 10^{-3}\,\text{eV}$ symbolises the dark energy scale here.* Using such a "see-saw" mechanism suggests that $V_0^{1/4} \sim v^2/m_P \sim 10^{-3}\,\text{eV}$, where v here stands for the VEV of the electroweak Higgs field ($v = 246\,\text{GeV}$).

9.2.7 The fifth force problem and beyond

As we have seen, quintessence needs to have effective mass-squared $m^2 = |V''(\phi_0)| \lesssim H_0^2$ in order to unfreeze today and start slow-rolling down its potential. This means that the Compton wavelength ℓ_C of the associated particle (the "quintesson"?) is of the order of the present horizon, because $\ell_C = 1/m \gtrsim 1/H_0$. Thus, virtual quintessence particles may mediate a long-range force, similar to photons (or gravitons), which mediate electromagnetic (or presumably gravitational) interactions. Thus, not only is the quintessence field the fifth element in the Universe content but it may correspond to a fifth force beyond the four fundamental forces of gravity, electromagnetism and the weak and strong nuclear interactions. Such a force could lead to the violation of the weak equivalence principle of general relativity, which equates inertial with gravitational mass, since the gravitational mass would be affected by the quintessence force. Indeed, in the nonrelativistic limit, the correction to the Newtonian gravitational potential is of the form

$$V(r) = -G\frac{m_1 m_2}{r}\left(1 + \beta e^{-mr}\right), \qquad (9.49)$$

where m is the mass of the quintessence particle and $\beta \ll 1$ is a constant parametrising the strength of the interaction.

Interactions between quintessence and the standard model fields would take the form [Car98]

$$e^{-\beta_i \Delta\phi/m_P}\mathcal{L}_i \simeq \mathcal{L}_i - \beta_i \frac{\Delta\phi}{m_P}\mathcal{L}_i, \qquad (9.50)$$

where $\mathcal{L}_i$ corresponds to every allowed dimension-4 Lorentz invariant operator (labelled by the index 'i') in the Lagrangian density of the theory, and $\Delta\phi$ is the total excursion of the quintessence field, while the β_i are dimensionless constants parametrising the strength of the corresponding interactions. In the above equation we assumed that $\beta_i \Delta\phi \ll m_P$, which is what we need in order for the the second term in the right hand side to be negligible, so that we preserve the weak equivalence principle. Bounds on the values of the β_i constants are set by Eötvös-type experiments.[†] For example, for electromagnetism we have $\mathcal{L}_{\text{em}} = \frac{1}{4}|F|^2$, where $|F|^2$ is the norm-squared of the Faraday tensor, comprised from the components of the electric and

---

*More accurately, the dark energy scale is $\rho_\Lambda^{1/4} = \sqrt{\Lambda m_P}$, so that Eq. (9.48) is really $\text{TeV} \sim (\Lambda m_P^3)^{1/4}$.

[†]Named from Baron Loránd Eötvös de Vásárosnamény, who was a 19th-century Hungarian physicist known for his important work on gravitation and the invention of the torsion pendulum.

the magnetic field.* In fact, modulation of the electromagnetic term in the Lagrangian density amounts to variation of the fine-structure constant of electromagnetism $\alpha \approx 1/137$. Since $\alpha \propto e^{\beta_{\mathrm{em}}\Delta\phi/m_P}$, we have $\Delta\alpha/\alpha = \beta_{\mathrm{em}}\Delta\phi/m_P$. Thus, the roll of the quintessence field might cause a variation of the fine-structure "constant", which is tentatively observable.

As we have discussed, we expect $\Delta\phi \sim m_P$ for quintessence. To avoid a sizeable fifth force the excursion of the scalar field cannot be super-Planckian (meaning $\Delta\phi \lesssim m_P$), unless all the β_i constants are fine-tuned to be exponentially small. Additionally, the effective field theory behind the form of the scalar potential V is not really trustworthy over super-Planckian distances in field space, because degrees of freedom (other fields) which were assumed heavy and were integrated out (and so, disregarded) could become light and their effect on the form of the scalar potential could not be ignored any more. This is equivalent to saying there could be enhanced symmetry points (ESPs) encountered over super-Planckian excursions, which would deform V. A similar problem exists with radiative corrections, which are contributions to V from loop diagrams. An exception to the above problems is considering PNGB quintessence, as in the previous section, where V is protected by the remnant of an internal symmetry. We briefly return to the fifth force problem in Sec. 9.5.1.

9.3   Phantom and quintom

Observations allow dark energy to have a phantom equation of state, with barotropic parameter $w_{\mathrm{DE}} < -1$. We have briefly discussed phantom dark energy in Sec. 2.8.3. At first sight, Eq. (9.3) suggests that a scalar field cannot feature $w_\phi < -1$. However, phantom dark energy could be modelled as a scalar field with a kinetic term of the "wrong" sign [Cal02]. Thus, the Lagrangian density of a phantom scalar field would look like

$$\mathcal{L} = -\frac{1}{2}(\partial\phi)^2 - V(\phi)\,, \qquad (9.51)$$

where the kinetic density is given by $\rho_{\mathrm{kin}} = -\frac{1}{2}(\partial\phi)^2$. Then the homogeneous scalar field barotropic parameter is again given by Eq. (6.6) which now takes the form

$$w_\phi = \frac{\frac{1}{2}\dot{\phi}^2 + V(\phi)}{\frac{1}{2}\dot{\phi}^2 - V(\phi)}\,, \qquad (9.52)$$

which may be compared with Eq. (9.3) for quintessence. From the above we see that, when a phantom scalar field is dominated by the kinetic density its barotropic parameter is $w_\phi = 1$ as with a "normal" scalar field. However, when a phantom scalar field has $\frac{1}{2}\dot{\phi}^2 \lesssim V$ its barotropic parameter can be $w_\phi \ll -1$. The Klein-Gordon equation of motion is now

$$\ddot{\phi} + 3H\dot{\phi} - V' = 0\,, \qquad (9.53)$$

which can be identified with the "normal" Klein-Gordon in Eq. (6.8) (cf. also Eq. (9.4)) by switching $V \to -V$. As a result, the scalar field slides up the scalar potential. This can be understood if we imagine that the field experiences the mirror image of the potential, similar to the reflection of a landscape on the still surface of a calm lake, as shown in Fig. 9.4. This behaviour means that, if the potential is unbounded from above, the energy density of the scalar field would continue to grow to infinity. It can be shown that, if $V \propto e^{-\lambda\phi/m_P}$, then the field (which eventually dominates the Universe) approaches an attractor with barotropic parameter

---

*We have $|F|^2 = 2(B^2 - E^2/c^2)$, with E and B being the norms of the electric and magnetic field, respectively, and we have reinstated the speed of light here.

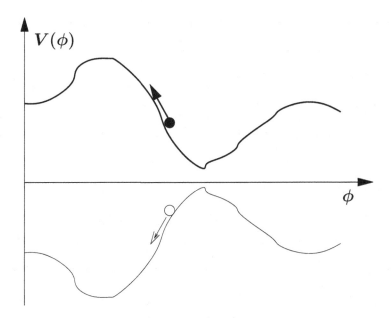

FIGURE 9.4 Schematic plot depicting the dynamics of a phantom scalar field, which is climbing the potential (thick curve) because it experiences the mirror image of the potential (thin curve), as if the potential is a landscape reflected on the still surface of a calm lake at $V = 0$.

$w_\phi = -1 - \lambda^2/3 < -1$, which is the equivalent of the dominant exponential quintessence attractor in Eq. (9.24), which features $w_\phi = -1 + \lambda^2/3$. In contrast, if the potential is bounded from above (i.e., it has a maximum), e.g., $V \propto e^{-\lambda(\phi/m_P)^2}$, then the phantom scalar field, starting from a phantom barotropic parameter approaches $w_\phi = -1$ as the field settles in the potential maximum with $V \gg \frac{1}{2}\dot\phi^2 \simeq 0$ as suggested by Eq. (9.52).

There is a no-go theorem, which specifies that the "phantom divide" $w = -1$ cannot be crossed by a single minimally-coupled (to gravity) scalar field. However, observations might hint that dark energy does cross the phantom divide going from $w_{\mathrm{DE}}(t_1) > -1$ to $w_{\mathrm{DE}}(t_2) < -1$ at different times $t_1 \neq t_2$ in the past. To achieve this with scalar fields one can consider combinations of quintessence and phantom scalar fields in what is called *quintom* dark energy [FWZ05]. The simplest quintom model considers two scalar fields, one quintessence field ϕ_1 and one phantom field ϕ_2 with Lagrangian density

$$\mathcal{L} = \frac{1}{2}(\partial\phi_1)^2 - \frac{1}{2}(\partial\phi_2)^2 - V(\phi_1, \phi_2)\,. \tag{9.54}$$

The quintessence field equation is Eq. (9.4), while the phantom field equation is Eq. (9.53). The barotropic parameter of dark energy is then

$$w_{\mathrm{DE}} = \frac{p_{\mathrm{DE}}}{\rho_{\mathrm{DE}}} = \frac{\frac{1}{2}\dot\phi_1^2 - \frac{1}{2}\dot\phi_2^2 - V}{\frac{1}{2}\dot\phi_1^2 - \frac{1}{2}\dot\phi_2^2 + V}\,, \tag{9.55}$$

from which we can see that crossing the phantom divide could be possible. For a simple potential like $V = V_1 e^{-\lambda_1 \phi_1/m_P} + V_2 e^{-\lambda_2 \phi_2/m_P}$ it can be shown that the evolution of the system begins with quintessence-like dark energy with $w_{\mathrm{DE}} > -1$ and ends with phantom like dark energy with $w_{\mathrm{DE}} < -1$ (in fact, with $w_{\mathrm{DE}} = -1 - \lambda_2^2/3 < -1$). This is so, even if $V(\phi_1, \phi_2)$ includes interaction between the two scalar fields. However, the opposite behaviour is also possible with a careful choice of V, starting with phantom dark energy with $w_{\text{DE}} < -1$ and ending up with quintessence with $w_{\text{DE}} > -1$.

All this is fine, but there is a big elephant in the room that we cannot ignore. A scalar field with kinetic energy featuring the "wrong" sign is also known as a *ghost* field. The existence of a ghost field in the theory is very problematic because the vacuum becomes unstable and decays into a never-ending production of ghost particles. The problem is impossible to ignore, rendering theories with minimally coupled phantom scalar fields seriously sick and best avoided. Fortunately, there are other ways to cross the phantom divide, for example k-essence discussed next.

9.4 k-essence

Leaving behind the spooky world of phantoms and ghosts we come back to the realm of essences. It is not just quintessence this time; it is kinetically-driven quintessence, introduced by Takeshi Chiba, Takahiro Okabe, and Masahide Yamaguchi in 2000 [COY00]. In short, it is called *k-essence*, a name coined by Christian Armendariz-Picon, Viatcheslav F.F. Mukhanov, and Paul J. Steinhardt in 2000 [APMS00]. k-essence is an incarnation of k-inflation, which we discussed in Sec. 7.4, but addressing the current dark energy instead of the early Universe one.

As with k-inflation, in the case of k-essence the pressure and density of the scalar field are given by Eq. (7.103) repeated here

$$p_\varphi = \mathcal{L}_\varphi(\varphi, X) \quad \text{and} \quad \rho_\varphi = 2X\partial_X p_\varphi - p_\varphi, \tag{9.56}$$

where $X \equiv \frac{1}{2}\dot{\varphi}^2$ (cf. Eq. (7.102)) and $\partial_X \equiv \partial/\partial X$. The barotropic parameter is

$$w_\varphi = \frac{p_\varphi}{\rho_\varphi} = \frac{p_\varphi}{2X\partial_X p_\varphi - p_\varphi} = \frac{-1}{1 - 2X\partial_X p_\varphi/p_\varphi}. \tag{9.57}$$

Thus, if $2X|\partial_X \ln p_\varphi| \ll 1$ we have $w_\varphi \simeq -1$. From the above, we see that demanding $\rho_\varphi > 0$ we have $w_\varphi > 0$ if $p_\varphi > 0$, while in the case when $p_\varphi < 0$ then $w_\varphi > -1$ if $\partial_X p_\varphi > 0$ but $w_\varphi < -1$ if $\partial_X p_\varphi < 0$. Thus, in principle, k-essence can behave as a phantom fluid. Is this fluid stable?

To investigate stability we consider the sound speed of perturbations defined in Eq. (2.16). We find

$$c_s^2 = \frac{\partial p_\varphi}{\partial \rho_\varphi} = \frac{\partial_X p_\varphi}{\partial_X \rho_\varphi} = \frac{\partial_X p_\varphi}{\partial_X p_\varphi + 2X\partial_X^2 p_\varphi} = \frac{(\partial_X p_\varphi)^2}{\partial_X[X(\partial_X p_\varphi)^2]}, \tag{9.58}$$

where $\partial_X^2 p \equiv \partial_X(\partial_X p)$. The theory is classically unstable if $c_s^2 < 0$ because an imaginary speed of sound results in the perturbation Fourier modes growing exponentially, instead of just being oscillations, i.e., waves of the field travelling with c_s as is the case of $c_s^2 \geq 0$. Thus, perturbative stability demands $c_s^2 \geq 0$. Quantum stability (no ghosts), on the other hand, demands separately that $\partial_X \rho_\varphi > 0$ (meaning $\partial_X p_\varphi > -2X\partial_X^2 p_\varphi$) and $\partial_X p_\varphi > 0$ (meaning $w_\varphi > -1$ as we have discussed). Also, avoiding superluminal mode propagation demands $c_s^2 \leq 1$, which suggests $\partial_X^2 p \geq 0$ (cf. Eq. (9.58)). This condition implies that $\partial_X \rho_\varphi > 0$ is guaranteed if $\partial_X p_\varphi > 0$.

The case when $p_\varphi = X - V(\varphi)$ is simple quintessence, while when $p_\varphi = -X - V(\varphi)$ we have a phantom field that violates the above stability conditions. It is interesting to consider $p_\varphi = p_\varphi(X)$ (sometimes referred to as *kinetic k-essence*). In this case, taking $w_\varphi = $ constant, Eq. (9.57) can be readily integrated

$$X\partial_X p_\varphi = \frac{w_\varphi + 1}{2w_\varphi}p_\varphi \quad \Rightarrow \quad p_\varphi \propto X^{\frac{w_\varphi+1}{2w_\varphi}}. \tag{9.59}$$

From the above, we can consider several cases of interest. Firstly, when $w_\varphi = -1$ then $p_\varphi = $ constant. This is so even in $p_\varphi = p_\varphi(\varphi, X)$, as suggested by Eq. (9.57). The case $w_\varphi = 0$ (matter like) is not well defined because $\mathcal{L}_\varphi = p_\varphi = 0$ and there is no k-essence. If $w_\varphi = 1$ then

$p_\varphi \propto X$ and we have a canonical free field with $V = 0$. Finally, when $w_\varphi = 1/3$, the density of k-essence scales as radiation and corresponds to $p_\varphi \propto X^2$. We look into this possibility below.

In general, it can be shown that there is a scaling solution in which k-essence mimics the background if $p_\varphi = Xg(Y)$, where g is an arbitrary function and $Y \equiv Xe^{\lambda\varphi/m_P}$, with $\lambda = $ constant. For example, the choice $g(Y) = 1 - V_0/Y$ results in $\mathcal{L}_\varphi = p_\varphi = X - V_0 e^{-\lambda\varphi/m_P}$, which is exponential quintessence that features a scaling attractor as we discussed in Sec. 9.2.3.

In the rest of this section, we implement some of the discussion above into a specific and popular model, to demonstrate how k-essence works. Recall that, in Sec. 9.3, to achieve a phantom barotropic parameter, we considered a scalar field with negative kinetic density $-X$. This gave rise to ghost instabilities. The hope is that a higher-order X^2 in the kinetic density can stabilise the vacuum of the theory. This is the reasoning behind the model

$$p_\varphi = \mathcal{L}_\varphi = f(\varphi)\left(-X + \frac{X^2}{M^4}\right), \qquad (9.60)$$

where M is an energy scale and f is dimensionless. This model is called *ghost condensate* and can be motivated by string theory. Eq. (9.56) suggests that the density is

$$\rho_\varphi = f(\varphi)\left(-X + \frac{3X^2}{M^4}\right). \qquad (9.61)$$

Therefore, from Eq. (9.57), the barotropic parameter of k-essence in this model is

$$w_\varphi = \frac{M^4 - X}{M^4 - 3X}. \qquad (9.62)$$

We see that, if $X = $ constant, then $w_\varphi = $ constant. In fact, k-essence mimics the cosmological constant with $w_\varphi = -1$ when $X/M^4 = \frac{1}{2}$. From the above it is easy to see that

$$\frac{1}{2} \leq \frac{X}{M^4} < \frac{2}{3} \quad \Leftrightarrow \quad -1 \leq w_\varphi < -\frac{1}{3}, \qquad (9.63)$$

$$\frac{1}{3} < \frac{X}{M^4} < \frac{1}{2} \quad \Leftrightarrow \quad -\infty < w_\varphi < -1. \qquad (9.64)$$

When $X/M^4 \geq 2/3$ then $w_\varphi \geq -\frac{1}{3}$ and k-essence cannot be dark energy. When $X/M^4 < 1/3$, Eq. (9.62) suggests $w_\varphi > 1$. This implies superluminal mode propagation because, in this model Eq. (9.58) gives

$$c_s^2 = \frac{M^4 - 2X}{M^4 - 6X} = \frac{1 + w_\varphi}{5 - 3w_\varphi}, \qquad (9.65)$$

which results in $c_s^2 > 1$ when $1 < w_\varphi < 5/3$. The absence of ghosts requires $\partial_X p_\varphi > 0$. This can be satisfied when $w_\varphi < -1$, provided $f(\varphi) < 0$. However, perturbative stability $c_s^2 \geq 0$ requires $w_\varphi \geq -1$, which means that stable phantom k-essence cannot be achieved after all.[*]

Before concluding, it is useful to discuss the scaling properties of the ghost condensate model. Assuming k-essence is originally subdominant, the Hubble scale is $H = 2/3(1 + w_B)t$ (cf. Eq. (2.40)), where w_B is the barotropic parameter of the dominant background substance. Then the continuity equation (6.4) for k-essence can be written as[†]

$$\dot\rho_\varphi + 3H(1 + w_\varphi)\rho_\varphi = 0 \quad \Rightarrow \quad \frac{\mathrm{d}\ln\rho_\varphi}{\mathrm{d}\ln t} = -2\frac{1 + w_\varphi}{1 + w_B}. \qquad (9.66)$$

---

[*]To reinforce the stabilising effect of the X^2 term, the model has been augmented as $p_\varphi \propto -X + e^{\lambda\varphi/m_P}X^2/M^4$, which is called *dilatonic ghost condensate*, but even this model cannot give rise to stable phantom k-essence.

[†]For simplicity, normalisation inside the logarithms is ignored, see Sec. 5.3.2.

We now assume that $X = $ constant. Since $X = \frac{1}{2}\dot{\varphi}^2$, this implies $\varphi \propto t$. In view of Eq. (9.62), it also suggests that $w_\varphi = $ constant. Then, Eqs. (9.61) and (9.66) suggest

$$\frac{\mathrm{d}\ln f}{\mathrm{d}\ln\varphi} = \frac{\mathrm{d}\ln\rho_\varphi}{\mathrm{d}\ln t} = -2\frac{1+w_\varphi}{1+w_B} \;\Rightarrow\; f(\varphi) \propto \varphi^{-2\frac{1+w_\varphi}{1+w_B}} . \tag{9.67}$$

A scaling solution is such that the density of k-essence mimics the background density so that $w_\varphi = w_B$. Thus, the above suggests that, when $X = $ constant, the ghost condensate k-essence model features a scaling solution when $f(\varphi) \propto \varphi^{-2}$. If $w_\varphi = \frac{1}{3}$, then k-essence mimics the background density during the radiation era. However, when matter takes over, the value of w_B changes so scaling is terminated. Then k-essence has a chance to increase its density parameter and come to dominate the Universe at present. This "graceful exit" from the scaling regime when matter domination begins is a rather appealing feature of k-essence models. Note, however, that if the ghost-condensate k-essence mimics a cosmological constant at late times, then $w_\phi \approx -1$. Consequently, Eq. (9.67) suggests that $f \simeq $ constant and Eq. (9.62) suggests $X \simeq \frac{1}{2}M^4$. Putting the above together, Eq. (9.61) results in $\rho_\varphi \simeq \frac{1}{4}fM^4 \simeq $ constant. This means that M and f must be tuned such that $\rho_\varphi \sim (10^{-3}\mathrm{eV})^4$ today. To avoid introducing the dark energy scale $M \sim 10^{-3}$ GeV (i.e., the fine-tuning of ΛCDM), the value of f must be exponentially suppressed, which does not seem very natural.

The ghost condensate k-essence model was but an example. Other proposals include *tachyon* k-essence and DBI k-essence, similar to DBI-inflation mentioned in Sec. 7.4.

9.5 Coupled dark energy

Many cosmologists have considered the possibility that late dark energy interacts with matter. This may pose problems of the fifth force type, but it might help explain why dark energy is taking over not much after matter took over from radiation at the time of equality, when the radiation era ends and the matter era begins. Recall that events in the Universe's history are logarithmic in time: an awful lot happens in the first few minutes, while not much has changed in the last few billion years, except the appearance of dark energy on the scene. So, cosmologically speaking, the present is right after equality. We have no proof that there is such a connection between matter and dark energy. Nevertheless, we take a brief look into some of the most promising ideas.

9.5.1 Coupled exponential quintessence

Coupling matter to dark energy implies that both cease to be independent fluids as energy can flow from one to the other. The most straightforward coupling provides a source term in the continuity equation for dark energy and background matter. Taking dark energy to be quintessence we end up with [Ame00]

$$\dot{\rho}_\phi + 3H(\rho_\phi + p_\phi) = -\frac{Q}{m_P}(\rho_B - 3p_B)\dot{\phi}, \tag{9.68}$$

$$\dot{\rho}_B + 3H(\rho_B + p_B) = \frac{Q}{m_P}(\rho_B - 3p_B)\dot{\phi}, \tag{9.69}$$

where $Q = Q(\phi)$ is dimensionless. Using that $\rho_\phi = \frac{1}{2}\dot{\phi}^2 + V$ and $p_\phi = \frac{1}{2}\dot{\phi}^2 - V$, Eq. (9.68) becomes

$$\ddot{\phi} + 3H\dot{\phi} + V' = -\frac{Q}{m_P}(\rho_B - 3p_B). \tag{9.70}$$

If the background matter is relativistic (radiation) then $p_B = \frac{1}{3}\rho_B$ and the source terms in the above equations vanish, meaning both quintessence and the background fluid are independent during the radiation era. In the matter era, $p_B = p_m = 0$ and $\rho_B = \rho_m$. Then Eqs. (9.69) and (9.70) become

$$\dot{\rho}_m + 3H\rho_m = \frac{Q}{m_P}\rho_m\dot{\phi}, \tag{9.71}$$

$$\ddot{\phi} + 3H\dot{\phi} + V' = -\frac{Q}{m_P}\rho_m. \tag{9.72}$$

Using the flat Friedmann equation (6.2) $3H^2m_P^2 = \rho = \rho_\phi + \rho_m$, assuming $Q = $ constant and considering exponential quintessence with $V = V_0 e^{-\lambda\phi/m_P}$ the evolution of the system is as follows. Quintessence remains utterly negligible during the radiation era, when the interaction is absent. After equality, the system approaches a fixed point where quintessence is kinetically dominated with $w_\phi = 1$ and at a fixed fraction to the overall density, with density parameter $\Omega_\phi = 2Q^2/3$.* The barotropic parameter of the Universe is

$$w = \frac{p}{\rho} = \frac{p_\phi}{\rho} = \frac{p_\phi}{\rho_\phi}\frac{\rho_\phi}{\rho} = w_\phi\Omega_\phi = \frac{2}{3}Q^2, \tag{9.73}$$

where $p = p_m + p_\phi = p_\phi$ is the total pressure. The above result suggests that the scale factor evolves as $a \propto t^{2/(3+2Q^2)}$ during the matter era (cf. Eq. (2.39)). In order not to inhibit structure formation, we need $0 \leq w \ll 1$, which means that $Q^2 \ll 1$.

This fixed point in the field evolution is unstable (provided $Q(Q + \lambda) > -\frac{3}{2}$), and soon (i.e., in a few billion years only!) the system evolves to one of the two attractors of the exponential quintessence, depending on the values of Q and λ. One of the attractors is the dominant exponential attractor in Eq. (9.24), where eventually $w = w_\phi = -1 + \lambda^2/3$, which leads to acceleration if $\lambda^2 < 2$. The condition to end up on this attractor is $\lambda(Q + \lambda) < 3$. In this scenario, the coupling Q does not make much of a difference. The other possibility is to end up with the equivalent of the scaling attractor in Eq. (9.26) but this time the barotropic parameter of the Universe is $w = -Q/(Q + \lambda)$. The condition to end up on this attractor is $\lambda(Q + \lambda) > 3$. Since structure formation demands $Q^2 \ll 1$, we have that $w \approx 0$, and there is no chance of accelerated expansion, as is the case with the scaling attractor in Sec. 9.2.3. One could have accelerated expansion if Q jumped near the present to some sizeable value. Demanding $w < -\frac{1}{3}$, we see that we have acceleration only if $Q > \lambda/2$ or $Q < -\lambda$. However, arranging for this jump is not too different from arranging for a "graceful exit" from the scaling attractor of exponential quintessence, which as we saw in Sec. 9.2.3, may amount to substantial fine-tuning.

Additionally, a large coupling to matter today would result in a fifth force problem, i.e., a violation of the weak equivalence principle, which is strongly supported by Solar System observations (see Sec. 9.2.7). One way to avoid this is the *chameleon* proposal, in which the effect of the interaction between dark energy and matter is screened near high density objects, so it is not felt within our Solar System. Another way is by considering that dark energy interacts only with dark matter and not with the baryons, which make up us and the rest of the objects of the Solar System that we observe. However, in this scenario dark matter would scale differently (weaker) to baryonic matter with the Universe expansion. As a result, there would be a period of baryonic matter domination in the past. This implies that structure formation would begin not at equality but at decoupling because the primary potential wells would form only after baryonic matter can collapse gravitationally. However, this poses many problems with CMB observations, which require the dark matter potential wells to form before decoupling so baryonic matter falls into these preexisting wells. In fact, this is one of the supporting arguments for the existence of dark matter in the first place, as we explained in Sec. 5.4.2.

---

*Despite being kinetically dominated, ϕ is not in freefall because quintessence is not an independent fluid so $\rho_\phi \not\propto a^{-6}$.

9.5.2 Coupled nonminimal quintessence

It is interesting to discuss the origin of the dark energy - matter interaction. This could be due to particle physics. However, another way is from modified gravity. As we have discussed in Sec. 7.3.3, in curved spacetime, scalar fields are expected to develop a direct coupling to gravity of the form $\Delta \mathcal{L} = \frac{1}{2}\xi R\chi^2$, as shown in Eq. (7.63). In this case, even if the field in question is independent and the background fluid satisfies the usual continuity equation (6.4) in the Jordan frame, when switching to the Einstein frame, a coupling between the scalar field and the background fluid is developed so that the continuity equation features a source term as in Eq. (9.69), with [Ame99]

$$Q = -\frac{1}{2}m_P \, \partial_\chi \ln \Omega^2 \left[\frac{3}{2}(\partial_\chi \ln \Omega^2)^2 + \frac{1}{\Omega^2} \right]^{-1/2}, \tag{9.74}$$

where $\Omega^2 = 1 + \xi\chi^2/m_P^2$ is the conformal factor given in Eq. (7.65). Putting this into the above it is straightforward to find

$$Q = -\frac{\xi\chi}{m_P\sqrt{1 + \xi(6\xi + 1)\chi^2/m_P^2}}. \tag{9.75}$$

With $\xi \gg 1$ as in Higgs inflation, for example, we have $Q \simeq -1/\sqrt{6} \simeq -0.4$. This would mean that $\Omega_\phi = \frac{2}{3}Q^2 \simeq 0.1$, which is challenging for structure formation.

9.5.3 Mass-varying neutrinos

One other interesting possibility is to consider a coupling between quintessence and neutrinos. This is loosely motivated by the fact that the dark energy scale $\rho_\Lambda^{1/4} \sim 10^{-3}$ eV is comparable to the neutrino masses as inferred by solar and atmospheric observations, $m_\nu \sim 10^{-1} - 10^{-3}$ eV. The modified Klein-Gordon for quintessence is [LWFZ02]

$$\ddot{\phi} + 3H\dot{\phi} + V' = -\frac{Q}{m_P}(\rho_\nu - 3p_\nu), \tag{9.76}$$

which is Eq. (9.70) when taking the background fluid to be neutrinos. In the above, we have

$$Q = \frac{\mathrm{d}\ln m_\nu(\phi)}{\mathrm{d}\phi}. \tag{9.77}$$

Eq. (9.76) suggests that when neutrinos are relativistic (in which case $\rho_\nu = 3p_\nu$) the interaction vanishes. However, as we mentioned in Sec. 3.3.1, neutrinos are expected to start turning nonrelativistic near the present time. For nonrelativistic neutrinos $p_\nu \simeq 0$ and $\rho_\nu \simeq n_\nu m_\nu(\phi)$, where n_ν is the neutrino number density. Then the quintessence effective potential becomes

$$V_{\text{tot}}(\phi) = V(\phi) + n_\nu m_\nu(\phi), \tag{9.78}$$

where $V(\phi)$ is the bare potential. Even if $V(\phi)$ does not have a minimum (that is $V(\phi)$ is of the usual runaway type) $V_{\text{tot}}(\phi)$ may feature one instantaneous minimum, which varies with time. At the minimum, $V'_{\text{tot}} = 0$, which means

$$n_\nu = -\frac{V'}{m'_\nu} = -\frac{\partial V}{\partial m_\nu}. \tag{9.79}$$

Because $n_\nu > 0$, the above suggests that V_{tot} features a minimum only if $\partial V/\partial m_\nu < 0$. As quintessence rolls into this temporary minimum it gets stabilised there with negligible kinetic density $\rho_{\text{kin}} \simeq 0$. This means that the total density and pressure of the quintessence-neutrino fluid is $\rho_{\text{tot}} \simeq n_\nu m_\nu + V$ and $p_{\text{tot}} \simeq 0 - V$, so that the corresponding barotropic parameter is

$$w_{\phi+\nu} = \frac{p_{\text{tot}}}{\rho_{\text{tot}}} = -\frac{V(\phi)}{V(\phi) + n_\nu m_\nu(\phi)} = -1 + \frac{m_\nu}{V/n_\nu + m_\nu} . \tag{9.80}$$

Thus, we see that $w_{\phi+\nu} \simeq -1$ if $n_\nu m_\nu \ll V$. One possibility is to consider that the ratio of the neutrino and the scalar field densities is approximately constant after the onset of the acceleration. Then, it can be shown that

$$w_{\phi+\nu} = -1 + \frac{m_\nu(t_0)}{12\,\text{eV}} , \tag{9.81}$$

where $t_o \simeq 13.8\,\text{Gy}$ is the present time. This demonstrates that observing the barotropic parameter of dark energy might provide information on the neutrino masses.

9.6 Other proposals

Moving on from scalar fields, we briefly discuss some of the more fanciful ideas for the nature of late dark energy.

9.6.1 Chaplygin gas

In 2001, Alexander Kamenshchik, Ugo Moschella, and Vincent Pasquier [KMP01] attempted to model dark matter and late dark energy in a single fluid, inspired by the works of Sergey Alexeyevich Chaplygin, a famous Russian and Soviet physicist, mathematician, and mechanical engineer.* Chaplygin's work in aerodynamics suggested that the air-flow near the wings of an aircraft is characterised by an equation of state of the form

$$p = -\frac{A}{\rho} , \tag{9.82}$$

where A is a positive constant. In a cosmological setting, a fluid with this equation of state could act as matter in the past and dark energy afterwards. Such a fluid is called *Chaplygin gas* and can be generated in braneworld cosmologies.

Indeed, inputting Eq. (9.82) into the continuity equation (6.4) we find

$$\dot{\rho} + 3H\left(\rho - \frac{A}{\rho}\right) = 0 \quad \Rightarrow \quad \int \frac{\rho\,d\rho}{\rho^2 - A} = -3 \int \frac{da}{a}$$

$$\Rightarrow \quad \rho = \sqrt{A + Ba^{-6}} , \tag{9.83}$$

where B is an integration constant. The above suggests that

$$a \ll (B/A)^{1/6} \quad \Rightarrow \quad \rho \propto a^{-3} \quad \text{and}$$
$$a \gg (B/A)^{1/6} \quad \Rightarrow \quad \rho = \sqrt{A} = -p = \text{constant} , \tag{9.84}$$

that is the substance interpolates between matter at early times and vacuum density at late times.

---

*Chaplygin, among other distinctions, has a town named after him (the administrative centre of the Russian Chaplyginsky district), as well as a large impact crater in the far side of the Moon. Plus, a cosmological model as well!

However, there is a serious problem when considering perturbations, which do behave like the ones of matter early on but get suppressed near the present. This can be easily understood when considering the Jeans length, given by Eq. (5.44), which (using Eq. (2.38)) is written as

$$\lambda_J = 2\pi\sqrt{\frac{2}{3}}\frac{c_s}{H}\,. \tag{9.85}$$

From Eq. (2.16), the sound speed is

$$c_s^2 = \frac{\partial p}{\partial \rho} = \frac{A}{\rho^2} = -\frac{p}{\rho} = -w. \tag{9.86}$$

Early on, Chaplygin gas behaves as matter and $c_s^2 = -w \approx 0$, meaning $\lambda_J \simeq 0$. Thus, gravitational collapse of overdensities is uninhibited. However, near the present $c_s^2 = -w \approx 1$ and $\lambda_J \sim 1/H$, meaning the Jeans length grows up to horizon size. Consequently, ever larger overdensities are unable to collapse and structure formation is inhibited at large scales, resulting in substantial loss of power in the primordial CMB anisotropy.

To improve the situation, modifications of the model have been put forward. One proposal is called *generalised Chaplygin gas* and considers an equation of state $p \propto -\rho^{-\alpha}$. However, observations impose the constraint $|\alpha| < 10^{-5}$, which makes the model unrealistic. Another proposal is to consider $p = -\rho\left[1 - \frac{\sin(C/\rho)}{C/\rho}\right]$, where C is a constant density scale. At early and late times we have $\rho \gg C$ (thus $p \simeq 0$) and $\rho \ll C$ (thus $p \simeq -\rho$), respectively, so the model exhibits the behaviour in Eq. (9.84). However, it is difficult to see where this equation of state might originate. The problems of the Chaplygin gas model can be ameliorated if it is considered as only a model of dark energy.

9.6.2   Holographic dark energy

Two and a half millennia ago, in classical Greece, the philosopher Plato proposed that what we experience as reality is but a shadow of the ideal forms of all the objects we see, similar to prisoners in a cave who can only see the two-dimensional shadows of the three-dimensional real objects outside the cave, projected on the walls of the cave by the external sunlight. Modern physics contemplates the inverse of Plato's cave and considers that the real world is the wall of the cave, i.e., it can be understood as a hologram, which falsely appears three-dimensional, even though in reality it is two-dimensional only. In physics jargon, the Holographic Principle states that the world with a theory of gravity is dual to (that is, through a mathematical transformation, it can be represented as) a field theory without (dynamical) gravity existing on the boundary, meaning on the surface of one dimension less. As a result of this principle, it can be shown that the number of degrees of freedom (e.g., fields) of a bounded system is finite. This produces a relation between the short distance cutoff of the theory (which may be m_P^{-1}) with the long distance cutoff of the theory, which can only be cosmological. The above relation is thought to link the vacuum density, which is due to zero-point quantum fluctuations with cosmological scales.

One can also understand this as follows. Suppose that there is nonzero vacuum energy density, which gravitates. In order for empty space not to collapse into black holes we must insist that the vacuum energy E contained in a given region of dimension L is less than the mass M of a black hole with Schwarzschild radius L. Thus, we demand

$$E = \rho_{\mathrm{DE}}L^3 \lesssim Lm_P^2 = M\,, \tag{9.87}$$

where $L = 2GM$ with $8\pi G = m_P^{-2}$. Parametrising the strength of the above inequality with some unknown positive constant $c \sim 1$ (not to be confused with the speed of light) we can solve for the vacuum density and we find

$$\rho_{\mathrm{DE}} = 3c^2 m_P^2 L^{-2}\,, \tag{9.88}$$

where the factor of 3 was introduced for convenience and L is a cosmological scale. If $L \sim H_0^{-1}$ then $\rho_{\rm DE} \sim \rho_0 \sim (10^{-3}\,{\rm eV})^4$ as required from dark energy observations. This is called *holographic dark energy* and was first introduced by Miao Li in 2004 [Li04].

But why $L \sim H_0^{-1}$? Is it possible to equate L with the Hubble radius: $L = 1/H$? If this were the case we would have $\rho_{\rm DE} \sim \rho$ all the time, which would not allow the Hot Big Bang to work. One more promising possibility is to identify L with the future event horizon. According to Eq. (2.56), we have

$$L = D_H = -a \int_\infty^t \frac{dt'}{a(t')} = -a \int_\infty^a \frac{da}{Ha^2}\,, \tag{9.89}$$

where we used that $H \equiv \dot{a}/a$. Taking the time derivative of the above, we obtain

$$\dot{D}_H = -\dot{a} \int_\infty^a \frac{da}{Ha^2} - a\frac{\dot{a}}{Ha^2} = HD_H - 1\,. \tag{9.90}$$

Equations (9.88), (9.89), and (9.90) can be combined to calculate the barotropic parameter of the holographic dark energy $w_{\rm DE}$. For simplicity, consider that $w_{\rm DE}$ varies very slowly, so that we can take $w_{\rm DE} \approx$ constant. Then, starting from the continuity equation for dark energy (cf. Eq. (6.4)) we have

$$\dot{\rho}_{\rm DE} + 3(1 + w_{\rm DE})H\rho_{\rm DE} = 0$$
$$\Rightarrow \quad -6c^2 m_P^2 D_H^{-3}(HD_H - 1) + 9(1 + w_{\rm DE})Hc^2 m_P^2 D_H^{-2} = 0$$
$$\Rightarrow \quad w_{\rm DE} = -\frac{1}{3} - \frac{2}{3HD_H} = -\frac{1}{3} - \frac{2\sqrt{\Omega_{\rm DE}}}{3c}\,, \tag{9.91}$$

where the density parameter of holographic dark energy is

$$\Omega_{\rm DE} \equiv \frac{\rho_{\rm DE}}{\rho} = \frac{3c^2 m_P^2 D_H^{-2}}{3m_P^2 H^2} = \left(\frac{c}{HD_H}\right)^2\,, \tag{9.92}$$

and we used Eqs. (6.2) and (9.88) with $L = D_H$.

When the holographic dark energy is subdominant then $\Omega_{\rm DE} \ll 1$ and Eq. (9.91) suggests that $w_{\rm DE} \simeq -\frac{1}{3}$, which means that $\rho_{\rm DE} \propto a^{-3(1+w_{\rm DE})} = a^{-2}$ (cf. Eq. (2.17)). Thus, $\rho_{\rm DE}$ is diluted less rapidly with the Universe expansion compared to matter and radiation. Consequently, eventually the holographic dark energy dominates and the Hot Big Bang ends. When $\rho_{\rm DE}$ is dominant then $\Omega_{\rm DE} \simeq 1$ and Eq. (9.91) suggests that $w_{\rm DE} \simeq -\frac{1}{3} - \frac{2}{3c} < -\frac{1}{3}$, which means that the Universe engages in accelerated expansion and $\rho_{\rm DE}$ truly acts as dark energy. What kind of dark energy we end up with depends on the value of c. For $c = 1$ we have $w_{\rm DE} = -1$ and the Universe expansion becomes quasi-de Sitter, as in ΛCDM. When $c > 1$ we have $-1 < w_{\rm DE} < -\frac{1}{3}$ and $\rho_{\rm DE}$ acts as exponential quintessence, when following the dominant attractor in Eq. (9.24). Finally, when $0 < c < 1$ then $w_{\rm DE} < -1$ and holographic dark energy becomes a phantom substance.

There is some theoretical prejudice in favour of the $c = 1$ case. Indeed, considering the holographic dark energy inside the observable Universe (which is a sphere of radius D_H), we have $E = \frac{4\pi}{3} D_H^3 \rho_{\rm DE}$. The mass of a black hole of Schwarzschild radius equal to that of the observable Universe is $2GM = D_H \Rightarrow M = 4\pi m_P^2 D_H$, where $8\pi G = m_P^{-2}$. Equating the two we arrive at $\rho_{\rm DE} = 3m_P^2 D_H^{-2}$, which is Eq. (9.88) with $L = D_H$ and $c = 1$.

However, there is a conceptual problem with this kind of holographic dark energy. Indeed, it seems that the dark energy presupposes the existence of a finite event horizon, while a finite event horizon is only due to the presence and dominance of a dark energy substance, as we discussed in Sec. 2.10. In other words, which came first, the chicken or the egg? This is why other proposals for the value of L have been put forward.

A prominent possibility considers that $L \sim 1/\sqrt{R}$, where R is the scalar curvature, cf. Sec. 7.3.2 and Appendix A. For a constant w, $R = 3(1 - 3w)H^2$ so that $L \sim 1/H$ as desired. But

w is not necessarily constant. In fact, it can be shown that this model only works when $w = w(t)$. In flat FLRW $R = 6(2H^2 + \dot{H})$ (see Eq. (A.9)) and Eq. (9.88) suggests $\rho_{\text{DE}} \propto L^{-2} \propto 2H^2 + \dot{H}$. In the past $\Omega_{\text{DE}} \to$ constant and $w_{\text{DE}} \to 0$ so that that the holographic dark energy behaves more like holographic dark matter, similar to the Chaplygin gas. This means that the substance undergoes tracker evolution, which ameliorates the coincidence problem, i.e., the tuning of the initial conditions required so that dark energy takes over at present.

The tuning requirement of the initial conditions of dynamical dark energy is completely eliminated (and not only ameliorated) with the quintessential inflation idea discussed next.

9.7   Quintessential inflation

It is only natural to connect early- and late-time dark energy. After all, the inflationary paradigm and quintessence are based on the same idea: that the Universe undergoes accelerated expansion when dominated by the potential density of a scalar field. This was exactly what P. James E. Peebles and Alexander Vilenkin proposed in 1999, and they called it *quintessential inflation* [PV99]. The original potential of quintessential inflation is

$$V(\phi) = \left\{ \begin{array}{ll} \lambda(\phi^4 + M^4) & \text{for} \quad \phi < 0 \\[2mm] \dfrac{\lambda M^8}{\phi^4 + M^4} & \text{for} \quad \phi > 0 \end{array} \right. , \tag{9.93}$$

where $0 < M \ll m_P$ is a suitable energy scale. For negative values of the inflaton field $\phi \ll -M$, the above potential reduces to quartic chaotic inflation, which has been excluded by observations unless it is "warmed up", by considering significant dissipation effects (see Sec. 7.6). During inflation $\phi \sim -m_P$. For positive values of the field $\phi \gg M$ the potential becomes quartic inverse power-law (IPL) quintessence. As we have discussed in Sec. 9.2.4, IPL quintessence features a tracker solution, which however, is too steep to satisfy observations in the case of an inverse quartic potential $V \propto \phi^{-4}$. Nevertheless, the field may not follow the tracker. Instead, after the end of inflation, it may rush down its runaway potential and freeze at a value $\phi_F \sim m_P$ (see why later) with some residual potential density, which explains dark energy. At present, the field would unfreeze and begin slowly rolling down its potential acting as thawing quintessence (see Sec. 9.2.5). Demanding that $V(\phi_F) \simeq \lambda M^8/\phi_F^4 \sim \rho_\Lambda \sim (10^{-3} \text{ eV})^4$ with $\phi_F \sim m_P$ suggests $\lambda^{1/8} M \sim$ TeV, which is the electroweak scale; a rather natural scale in particle physics.

9.7.1   The scenario in general

When the model in Eq. (9.93) was first put forward both quartic chaotic inflation and tracker quartic IPL quintessence were acceptable. Not anymore though. This is why other quintessential inflation models have been constructed, where a scalar field can be both the inflaton and the quintessence fields.

Quintessential inflation has a number of important advantages. Utilising a single degree of freedom (one scalar field) is a very economic proposal, which pairs well with the principal mission of the quintessence idea, which is to overcome the fine-tuning of ΛCDM. It also attempts to treat early and late dark energy in a common theoretical framework. It is heavily constrained, and therefore easily falsifiable because any quintessential inflation model must account for both inflation and dark energy observations. Additionally, the fact that the inflationary attractor fixes the initial conditions of quintessence, implies that one of the main disadvantages of the quintessence idea (the tuning of its initial conditions) is overcome without the need of attractor behaviour, which would only ameliorate the problem anyway. This means that the coincidence requirement, that is the requirement that late accelerated expansion commences near the present time, is facilitated only by means of the model parameters.

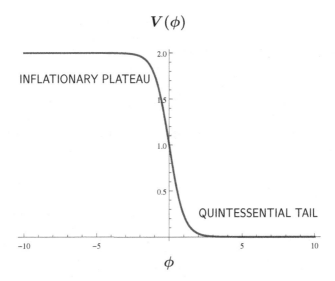

FIGURE 9.5 Typical form of the scalar potential $V(\phi)$ of quintessential inflation. There are two flat regions, the inflationary plateau and the quintessential tail. The drop in density between them is about 110 orders of magnitude. Fiducial units are used in the plot, and without the loss of generality, we assumed that the scalar field ϕ is smaller in the inflationary plateau than in the quintessential tail, such that $\dot{\phi} > 0$.

Typically, a quintessential inflation model considers a scalar potential, which features two flat regions: the *inflationary plateau* and the *quintessential tail*. The original model in Eq. (9.93) did not feature an inflationary plateau, but this was before the recent inflation observations, which clearly favour a plateau in the scalar potential. In contrast, the quintessential tail corresponds to the usual runaway quintessential potential, whose minimum is placed at infinity. The form of the potential is shown in Fig. 9.5. In most inflation models, the energy scale of inflation is slightly lower than the scale of grant unification $\sim 10^{16}\,\mathrm{GeV}$, while the current dark energy scale is $\sim 10^{-12}\,\mathrm{GeV}$. This means that there is an enormous jump in energy density between the inflationary plateau and the quintessential tail; about $\sim (10^{15.5}/10^{-12})^4 = 10^{110}$! It is very hard to construct a model that naturally bridges such a gap in density scales and is based on a theoretical framework which remains valid at both extremes, while managing to simultaneously satisfy inflation and the current dark energy observations. Yes, it is hard but not impossible.

Assuming we have such a model at hand, there is an extra complication to the story, unique to quintessential inflation. In model-building inflation, the usual approach is to consider that, after the end of inflation, the inflaton field ends up oscillating around its VEV, which leads to perturbative reheating or preheating (or both) as we have discussed in Secs. 6.3.1 and 6.3.3 respectively. This happens through the decay of the inflaton condensate. However, in quintessential inflation, the scalar field condensate needs to survive until the present to play the role of quintessence. Thus, the Universe must be reheated by other means. We have already discussed some alternative mechanisms which apply to nonoscillatory inflation models, like the ones used in quintessential inflation. The most minimal is gravitational reheating, mentioned in Sec. 6.3.4. However, as we explain below, this might overproduce gravitational waves which challenge BBN, cosmology's sacred cow. A more popular mechanism is instant preheating, presented in Sec. 6.3.4. This requires the presence of a suitable ESP along the inflaton direction in field space (amounting to an interaction between the inflaton and another scalar field), which goes a little against the minimalist philosophy of quintessential inflation. Another reheating mechanism is called *curvaton reheating* [FL03], and it is based on the presumption that a spectator field (the curvaton) reheats the Universe through its decay after inflation, in accordance to the curvaton scenario, which we talked about in Sec. 6.5.5. This time, there is no need to produce the curvature perturbations through the curvaton field, as the latter's function is primarily to facilitate reheating. This scenario must involve an additional field, as is also the case in instant preheating, but no interaction with the inflaton is needed (i.e., no ESP). Finally, another possibility is that inflation is warm, as we discussed in Sec. 7.6, but dissipation ceases once the field moves away from the inflationary part of its potential. Then the radiation bath is due to dissipation of the inflaton's density, before the latter dies out. This scenario is called *warm quintessential inflation* [DDW19].

One final ingredient to the story is a new epoch in the Universe history, where the dominant substance is the kinetic density of the scalar field. Let us look into this period.

9.7.2   Kination

As shown in Fig. 9.5, at the end of the inflationary plateau the potential becomes curved and very steep and inflation ends. What happens afterwards? After the field rolls off the inflationary plateau, it plunges down the steep potential cliff as its potential density is drastically diminished. As a result, the density of the scalar field is dominated by its kinetic part $\rho_\phi \simeq \rho_{\rm kin}$. According to Eq. (6.7), when the density of a scalar field is dominated by its kinetic density $\rho_{\rm kin} \gg V$, its effective barotropic parameter becomes $w_\phi = 1$. Such matter is said to be subject to a "stiff" equation of state. In view of Eq. (2.17), we see that the density of the scalar field decreases as $\rho_\phi \propto a^{-6}$. This is similar to our findings in Sec. 9.2.1. A kinetically dominated scalar field, whose density decreases as $\rho_\phi \propto a^{-6}$, is said to be in "freefall" as it is oblivious to the scalar potential, which is not featured in its equation of motion and does not affect its dynamics. This is why, the dynamics of the kinetically dominated scalar field studied in this section are model-independent.

Indeed, the contribution of the potential in the Klein-Gordon equation (6.8) practically disappears and the equation becomes

$$\ddot{\phi} + 3H\dot{\phi} \simeq 0 \,, \tag{9.94}$$

which is of the same form as Eq. (8.43) but there is a crucial difference: the field and the Universe are dominated by the kinetic density $\rho_\phi \simeq \rho_{\rm kin} = \frac{1}{2}\dot{\phi}^2$. As a result, the flat Friedmann equation (6.2) suggests

$$3H^2 m_P^2 = \frac{1}{2}\dot{\phi}^2 \;\Rightarrow\; H = \frac{1}{\sqrt{6}}\frac{\dot{\phi}}{m_P} \,, \tag{9.95}$$

where, without loss of generality, we took $\dot{\phi} > 0$, in agreement with Fig. 9.5, which suggests that the field moves towards larger values in field space to minimise $V(\phi)$. Inserting the above into Eq. (9.94) we find

$$\frac{1}{\dot{\phi}} - \frac{1}{\dot{\phi}_{\rm end}} = \frac{3}{\sqrt{6}}\frac{t - t_{\rm end}}{m_P} \;\Rightarrow\; \dot{\phi} \simeq \sqrt{\frac{2}{3}}\frac{m_P}{t} \,, \tag{9.96}$$

where, for simplicity, we assumed that kinetic density domination occurs right after inflation ends, denoted by the subscript 'end'. In the second equation above, we considered that the decrease of the kinetic density is rapid,* such that $\dot{\phi}(t) \ll \dot{\phi}_{\rm end}$ when $t \gg t_{\rm end}$. Integrating Eq. (9.96) we obtain

$$\phi(t) = \phi_{\rm end} + \sqrt{\frac{2}{3}}m_P \ln\left(\frac{t}{t_{\rm end}}\right) \,. \tag{9.97}$$

The period of the Universe after inflation ends, when the dominant content is the kinetic density of the scalar field, is called *kination*, a name coined in 1998 by Michael Joyce and Tomislav

---

*Recall that $\frac{1}{2}\dot{\phi}^2 = \rho_\phi = \rho \propto a^{-6} \Rightarrow \dot{\phi} \propto a^{-3}$.

Prokopec [JP98]. In kination $w = 1$. Consequently, Eqs. (2.17), (2.39) and (2.40) suggest

$$\rho \propto a^{-6}, \quad a \propto t^{1/3} \quad \text{and} \quad H = \frac{1}{3t}. \tag{9.98}$$

Using this, Eq. (9.95) readily results in Eq. (9.96). As we have shown above, during kination the field value increases logarithmically, without end. This is not so when the field ceases to be dominant. Indeed, via one of the numerous mechanisms which we mentioned before, reheating takes place and the Universe becomes radiation dominated. Our scalar field continues to roll in freefall for a while but now its roll is different from Eq. (9.97). Let us calculate it.

During radiation domination $H = 1/2t$ (cf. Eq. (2.40)) and Eq. (9.94) is solved by

$$\dot{\phi} \propto t^{-3/2} \Rightarrow \dot{\phi} = \dot{\phi}_{\text{reh}} \left(\frac{t_{\text{reh}}}{t} \right)^{3/2}, \tag{9.99}$$

where "reh" denotes the moment of reheating, when radiation takes over. Eq. (9.96) is valid at reheating, which is the end of the kination era and the beginning of the radiation era. Thus, $\dot{\phi}_{\text{reh}} = \sqrt{\frac{2}{3}}(m_P/t_{\text{reh}})$. Using this in Eq. (9.99), we arrive at

$$\dot{\phi} = \sqrt{\frac{2}{3}} \frac{m_P \sqrt{t_{\text{reh}}}}{t^{3/2}}, \tag{9.100}$$

during the radiation era. Integrating the above, we obtain

$$\phi(t) = \phi_{\text{reh}} + 2\sqrt{\frac{2}{3}} m_P \left(1 - \sqrt{\frac{t_{\text{reh}}}{t}} \right). \tag{9.101}$$

This equation shows that the roll of the scalar field comes to a halt when $t \gg t_{\text{reh}}$. The reason for this is that Eq. (9.94) ceases to be valid because the kinetic density is no longer dominant and the full Klein-Gordon equation (6.8) applies. We now find ourselves in the quintessential tail of the scalar potential, shown in Fig. 9.5. This is a very flat region, which means that the curvature of the potential is very small and $|V''| \ll H^2(t)$. Our scalar field is light and subdominant to the radiation bath. In this case, our discussion in Sec. 9.2.1 suggests that the field is frozen. We conclude, therefore, that the field freezes at the value

$$\phi_F = \phi_{\text{reh}} + 2\sqrt{\frac{2}{3}} m_P. \tag{9.102}$$

The value of the frozen field is of great importance because, if it is to become thawing quintessence, it determines how much residual potential density is stored, which should be comparable to the observed dark energy today $V(\phi_F) \sim \rho_\Lambda \sim (10^{-3} \text{ eV})^4$. For the field to remain frozen until today we also require $|V''(\phi_F)| \lesssim H_0^2$. Of course, in principle, the field could unfreeze before the present and follow some suitable tracker solution, but designing a reasonable model with a tracker solution that satisfies dark energy observations is difficult, as we have discussed in Sec. 9.2.4. However, Eq. (9.102) is not enough to estimate the value of ϕ_F because ϕ_{reh} is unknown and model dependent.

The value of ϕ_{reh}, when radiation dominates, depends on the mechanism that creates the radiation, which is destined to become the thermal bath of the Hot Big Bang. Let us denote the moment when this happens as $t_\star$. This could be the moment that the rolling scalar field crosses an ESP and generates radiation through instant preheating. It also could be the time when a curvaton field decays into radiation, when subdominant.[*] If the radiation bath were created

---

[*] A curvaton may also dominate before it decays because its density scales as $\rho_\sigma \propto a^{-3}$ when it is oscillating, which is diluted less drastically than the background density in kination $\rho \propto a^{-6}$. For simplicity we do not consider this possibility here.

through gravitational reheating then $t_\star = t_{\rm end}$. The same is true if inflation is warm and the thermal bath is generated during inflation through dissipation.

Evaluating Eq. (9.97) at $t_\star$ and at $t_{\rm reh}$ it is easy to show that

$$\phi_{\rm reh} = \phi_\star + \sqrt{\frac{2}{3}} m_P \ln\left(\frac{t_{\rm reh}}{t_\star}\right). \tag{9.103}$$

We can find the ratio $t_{\rm reh}/t_\star$ in terms of the efficiency of the radiation-producing mechanism as follows. Using the fact that radiation scales as $\rho_r \propto a^{-4}$, while the background density scales as $\rho = \rho_\phi \propto a^{-6}$ during kination, the radiation density parameter scales as $\Omega_r = \rho_r/\rho \propto a^2$. Thus,

$$1 \simeq \Omega_r^{\rm reh} = \Omega_r^\star \left(\frac{a_{\rm reh}}{a_\star}\right)^2 \Rightarrow \frac{t_{\rm reh}}{t_\star} = (\Omega_r^\star)^{-3/2}, \tag{9.104}$$

where we used that $a \propto t^{1/3}$ during kination (cf. Eq. (9.98)), and we employed the fact that, at reheating, radiation takes over so its density parameter is maximised to unity. Inserting this into Eq. (9.103) we obtain

$$\phi_{\rm reh} = \phi_\star - \sqrt{\frac{3}{2}} m_P \ln \Omega_r^\star. \tag{9.105}$$

Combining the above with Eq. (9.102), we arrive at

$$\phi_F = \phi_\star + \sqrt{\frac{2}{3}} \left(2 - \frac{3}{2} \ln \Omega_r^\star\right) m_P. \tag{9.106}$$

The quintessential inflation story can be visualised in Fig. 9.6.

Considering that a period of kination follows the end of inflation, increases the value of N_*, the remaining number of e-folds of inflation when the cosmological scales exit the horizon. Indeed, from Eq. (6.93), taking $w = 1$ after inflation and until reheating, we see that N_* is increased by the amount

$$\Delta N_* = \frac{1}{3} \ln\left(\frac{V_{\rm end}^{1/4}}{T_{\rm reh}}\right). \tag{9.107}$$

Therefore, the increase is maximal when we consider gravitational reheating, discussed in Sec. 6.3.4, because this results in the lowest possible value of $T_{\rm reh}$ as the gravitationally produced radiation is unavoidable. We may estimate this as follows. According to Eq. (6.62), we have $(\rho_r^{\rm gr})_{\rm end} \sim 10^{-2} H_{\rm end}^4$, while the Friedmann equation (6.2) suggests $\rho_{\rm end} \sim H_{\rm end}^2 m_P^2$. Therefore, the density parameter of gravitationally produced radiation at the end of inflation is

$$(\Omega_r^{\rm gr})_{\rm end} \sim 10^{-2} \left(\frac{H_{\rm end}}{m_P}\right)^2. \tag{9.108}$$

Now, Eq. (9.104) suggests that $(a_{\rm end}/a_{\rm reh})^2 \sim (\Omega_r^{\rm gr})_{\rm end}$, where $a_\star = a_{\rm end}$ in the case of gravitational reheating. Thus, we find $a_{\rm end}/a_{\rm reh} \sim 10^{-1} H_{\rm end}/m_P$. Using this, we obtain

$$(\rho_r^{\rm gr})_{\rm reh} = (\rho_r^{\rm gr})_{\rm end} \left(\frac{a_{\rm end}}{a_{\rm reh}}\right)^4 \sim 10^{-6} H_{\rm end}^4 \left(\frac{H_{\rm end}}{m_P}\right)^4, \tag{9.109}$$

where we used that $\rho_r \propto a^{-4}$. From Eq. (6.1), we have

$$\rho_r^{\rm gr} = \frac{\pi^2}{30} g_*^{\rm gr} T^4 \Rightarrow (\rho_r^{\rm gr})_{\rm reh} \sim 10^2 (T_{\rm reh}^{\rm gr})^4, \tag{9.110}$$

where we have taken $g_*^{\rm gr} \sim 10^2$. Combining Eqs. (9.109) and (9.110) we find

$$T_{\rm reh}^{\rm gr} \sim 10^{-2} H_{\rm end}^2/m_P. \tag{9.111}$$

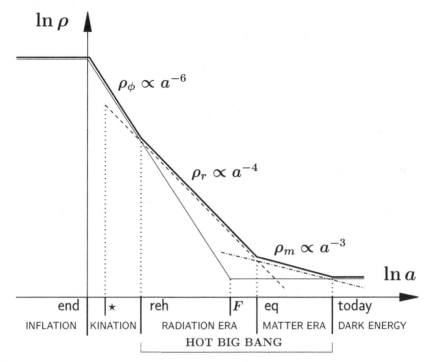

FIGURE 9.6   Schematic log-log plot of the density of the Universe and its individual components with respect to the scale factor according to quintessential inflation. The thin solid line depicts the density of the scalar field, which is roughly constant during inflation (i.e., until inflation ends, denoted by "end") and after freezing (denoted by F) but scales as $\rho_\phi \propto a^{-6}$ in between, regardless of whether it is dominant or not. The dashed line depicts the density of radiation (formed at the moment denoted by "$\star$"), which scales as $\rho_r \propto a^{-4}$. Radiation dominates at reheating (denoted by "reh"), which marks the onset of the Hot Big Bang. The dot-dashed line depicts the density of matter, which scales as $\rho_m \propto a^{-3}$ and takes over from radiation at the time of equality (denoted by "eq"). Today is the end of the Hot Big Bang, when the density of the scalar field takes over again accounting for dark energy. The thick solid line depicts the total density.

The typical value of the inflationary Hubble scale is $H_{\rm end} \sim 10^{13}\,{\rm GeV}$, which suggests $T_{\rm reh}^{\rm gr} \sim 10^6\,{\rm GeV}$. Considering $V_{\rm end}^{1/4} \sim \sqrt{H_{\rm end} m_P} \sim 10^{15.5}\,{\rm GeV}$, Eq. (9.107) suggests $\Delta N_* \simeq 7$. As we have seen, for oscillatory inflation near the GUT-scale we have $N_* \simeq 60$. This means that kination may increase N_* up to $N_* \simeq 67$. Because there is no faster decrease of the inflaton field than "freefall" and because the energy scale of inflation cannot be larger than the GUT-scale (cf. Eq. (6.91)) we can regard this number as the maximum value of N_* (recall that N_* increases with the inflation scale, as evident from Eq. (6.93)).

The result in Eq. (9.106) demonstrates that the mechanism which generates the radiation of the Hot Big Bang fully determines the value ϕ_F of the frozen field. It also shows that, even if radiation is produced right away at $t_\star = t_{\rm end}$ with maximum efficiency $\Omega_r^\star \simeq 1$ the field still travels distance $2\sqrt{\frac{2}{3}}\,m_P \simeq 1.6\,m_P$ in field space.* If one adds to this the excursion of the field during inflation we see that there is a danger that the scalar field traverses a super-Planckian

---

*This is because it becomes kinetically dominated at the end of inflation as it falls off the potential cliff.

span in field space. This means that quintessential inflation generically suffers from the fifth-force problem we discussed in Sec. 9.2.7. For the same reason, there is a danger that radiative corrections lift the flatness of the plateaus and require fine-tuning to remain under control. Finally, the validity of effective field theory itself (upon which our treatment is based) becomes questionable when traversing super-Planckian distances in field space.

All these issues are problems of quintessence that plague quintessential inflation as well. But there is an additional danger, special to quintessential inflation. This has to do with the possible overproduction of gravitational waves if the period of kination is too large.

9.7.3   Spike of gravitational waves

As we have discussed in Sec. 6.5.4, through the process of particle production, inflation generates gravitational waves (GWs). Such a GW background, if observed, will confirm a definite prediction of inflation. Typically, inflation-generated GWs are too weak to affect post-inflation cosmology. However, a period of kination results in a spike in the energy density of GWs, which are otherwise scale-invariant. When kination lasts too long, this spike is so substantial that it threatens the delicate process of BBN, which is one of the main pillars supporting the Hot Big Bang.

Let us estimate the contribution of the inflationary GWs to the density budget of the Universe. From our discussion in Sec. 6.5.4, we have that the metric perturbations are $h_{\rm GW} = \frac{1}{2}\sqrt{16\pi G}(H/2\pi)$. In natural units we have

$$h_{\rm GW}^2 = \frac{H_{\rm end}^2}{8\pi^2 m_P^2} \, . \tag{9.112}$$

The inflationary GW generation is not too different from the gravitational particle production of light, nonconformally invariant fields considered in Sec. 6.3.4. Mirroring Eq. (6.62), the density of the produced GWs is

$$\rho_{\rm GW}^{\rm end} = \frac{\hat{q} H_{\rm end}^4}{240\pi^2} \, , \tag{9.113}$$

where $\hat{q} \sim 1$ and we considered that GWs are bosonic and have two degrees of freedom (the two polarisation states). Combining Eqs. (9.112) and (9.113) and using the flat Friedmann equation (6.2), the above is rendered

$$\rho_{\rm GW}^{\rm end} = \frac{\hat{q}}{90} h_{\rm GW}^2 \rho_{\rm end} \;\Rightarrow\; \Omega_{\rm GW}^{\rm end} = \frac{\hat{q}}{90} h_{\rm GW}^2 \, , \tag{9.114}$$

where $\Omega_{\rm GW} = \rho_{\rm GW}/\rho$ is the GW density parameter. Similarly to electromagnetic radiation, the density of gravitational radiation scales as $\rho_{\rm GW} \propto a^{-4}$. During kination, the background density scales as $\rho \propto a^{-6}$. This means that $\Omega_{\rm GW} \propto a^2$. Thus, in view of Eq. (9.114), we find

$$\Omega_{\rm GW} = \frac{\hat{q}}{90} h_{\rm GW}^2 \left(\frac{a}{a_{\rm end}}\right)^2 \, . \tag{9.115}$$

After the production of radiation by one of the mechanisms we have discussed, the radiation density parameter also scales as $\Omega_r \propto a^2$, because $\rho_r \propto a^{-4}$. Then, we have (cf. Eq. (9.104))

$$1 \simeq \Omega_r^{\rm reh} = \Omega_r^\star \left(\frac{a_{\rm reh}}{a_\star}\right)^2 \;\Rightarrow\; \left(\frac{a_{\rm reh}}{a_{\rm end}}\right)^2 = \frac{1}{\Omega_r^\star}\left(\frac{a_\star}{a_{\rm end}}\right)^2 \, . \tag{9.116}$$

Using the above, we write Eq. (9.115) as

$$\Omega_{\rm GW}^{\rm reh} = \frac{\hat{q}}{90} h_{\rm GW}^2 \frac{1}{\Omega_r^\star}\left(\frac{a_\star}{a_{\rm end}}\right)^2 \, . \tag{9.117}$$

$$\log(f/\mathrm{Hz})$$

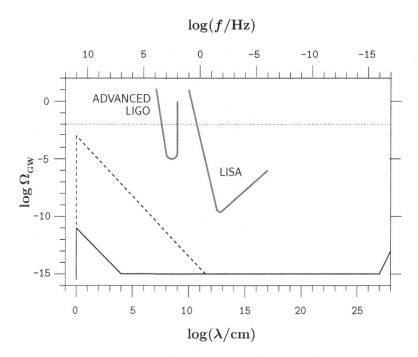

FIGURE 9.7 Log-log plot of the spectral GW density parameter $\Omega_{\mathrm{GW}}(\lambda)$ as a function of the comoving wavelength λ and the associated frequency $f = c/\lambda$. The horizontal branch corresponds to the radiation era, while there is a small spike for very large λ (small f) corresponding to the matter era, when $\Omega_{\mathrm{GW}}(\lambda) \propto \lambda^2$. The much more substantial spike at lower values of λ (higher values of f) corresponds to an era of kination when $\Omega_{\mathrm{GW}}(\lambda) \propto \lambda^{-1}$. We have assumed $H_{\mathrm{end}} \sim 10^{13}$ GeV. The slanted dashed line depicts the case of gravitational reheating, when $T_{\mathrm{reh}} \sim 10^6$ GeV and the duration of kination is maximal, while the slanted solid line depicts the case of $T_{\mathrm{reh}} \sim 10^{14}$ GeV, when the duration kination is significantly smaller. We see that the GW spike, in the case of gravitational reheating, might challenge BBN constraints depicted as the horizontal dotted line at $\Omega_{\mathrm{GW}} \sim 10^{-2}$. The parameter space expected to be probed by future surveys of Advance LIGO and LISA is also shown.

After reheating, the Hot Big Bang begins, and the background density is radiation. Because the densities of both the radiation background and GWs scale in the same way ($\rho_r, \rho_{\mathrm{GW}} \propto a^{-4}$), the GW density parameter Ω_{GW} remains constant. The GW density cannot be too large, for otherwise GWs would disturb BBN. The requirement is

$$\Omega_{\mathrm{GW}}^{\mathrm{reh}} = \Omega_{\mathrm{GW}}^{\mathrm{BBN}} \lesssim 10^{-2}. \tag{9.118}$$

Eq. (9.117) suggests that this requirement produces a lower bound to $\Omega_r^\star$, i.e., to the efficiency of the radiation generating mechanism.

In the case of gravitational reheating, discussed in Sec. 6.3.4, we have that the mechanism operates at the end of inflation so $a_\star = a_{\mathrm{end}}$ and $\Omega_r^\star = \Omega_r^{\mathrm{end}}$. Then, Eqs. (9.117) and (9.118) result in the bound

$$\Omega_r^{\mathrm{end}} \gtrsim \frac{100\,\hat{q}}{90} h_{\mathrm{GW}}^2 \;\Rightarrow\; \rho_r^{\mathrm{end}} \gtrsim \frac{5\,\hat{q}}{12\pi^2} H_{\mathrm{end}}^4\,, \tag{9.119}$$

where we used Eq. (9.112) and the flat Friedmann equation (6.2). Comparing the above with Eq. (6.62), which gives the density of the graviationally produced radiation at the end of inflation, we see that the bound is satisfied only when the number of effective relativistic degrees of freedom, which are also nonconformally invariant is $g_*^{\mathrm{gr}} \gtrsim 200(\hat{q}/q)$. Given that in current fundamental theories we expect $g_*^{\mathrm{gr}} = \mathcal{O}(100)$, we see that the BBN bound is challenged. This casts a shadow on the gravitational reheating mechanism, which otherwise would be the most minimal proposal, and this is why it was assumed in the original quintessential inflation model in Eq. (9.93).

But why do we talk about a "spike" of GWs? This has to do with the spectrum of the GWs. As we have discussed in Sec. 6.5.4, the spectrum of the inflation-generated GWs is almost scale invariant. Strictly speaking, this is only true for GWs whose wavelengths correspond to lengthscales which reenter the horizon during the radiation era of the Hot Big Bang. For larger wavelengths, which correspond to lengthscales entering the horizon during the matter era, there is some growth in amplitude, but because the matter era does not last very long (recall that the timeline of events in cosmology is logarithmic) this amplification is not that important. However, growth may occur also for very short wavelengths, which correspond to lengthscales that enter the horizon during kination. This results in a spike, which is larger the longer kination lasts. If kination lasts very long, the spike threatens BBN.

This is easy to understand when plotting the spectral GW density parameter defined as $\Omega_{\rm GW}(\lambda) = d\Omega_{\rm GW}/d\ln\lambda$, where λ is the comoving GW wavelength,[*] as shown in Fig. 9.7. The scaling of the spectral GW density parameter with respect to λ is [SSS02]

$$\Omega_{\rm GW}(\lambda) \propto \begin{cases} \lambda^2 & \lambda_{\rm eq} < \lambda \leq \lambda_H & \text{matter era} \\ \text{constant} & \lambda_{\rm reh} < \lambda \leq \lambda_{\rm eq} & \text{radiation era} \\ 1/\lambda & \lambda_{\rm end} \leq \lambda \leq \lambda_{\rm reh} & \text{kination} \end{cases} , \qquad (9.120)$$

where "eq" denotes the time of matter-radiation equality in the Hot Big Bang and $\lambda_H = D_H(t_0)$ is the horizon distance at present. Since GWs travel with the speed of light[†] the frequency of the GWs is related to their wavelength as $f = c/\lambda$, where we have reinstated c. Thus, a period of kination results in a spike of high frequency GWs, which may be observable in the near future, rendering the quintessential inflation proposal very interesting.

9.7.4 Quintessential inflation with α-attractors

To showcase how quintessential inflation works, it is useful to present a concrete example. For this, we combine exponential quintessence, discussed in Sec. 9.2.3, with the α-attractors proposal in Sec. 7.3.5. The kinetic poles introduced by α-attractors "flatten" the scalar potential of the canonical inflaton field because the poles are transposed to infinity. As such, two plateaus appear in the scalar potential, which is exactly what quintessential inflation requires for the inflationary plateau and the quintessential tail. We only need to consider a monotonic scalar potential for the noncanonical field. An excellent choice is an exponential potential, which has ample theoretical justification. Thus, we may consider the model [DO17]

$$\mathcal{L} = \frac{1}{2} \frac{(\partial\varphi)^2}{\left[1 - \frac{1}{6\alpha}\left(\frac{\varphi}{m_P}\right)^2\right]^2} - V_0 e^{-\kappa\varphi/m_P} - V_\Lambda , \qquad (9.121)$$

where α and κ are dimensionless positive constants, V_0 is a constant density scale and $V_\Lambda = \Lambda m_P^2$ is the cosmological constant. In the above, the noncanonical kinetic term of the field features poles at $\varphi = \pm\sqrt{6\alpha}\,m_P$, and has the standard form of α-attractor models motivated by supergravity (see Sec. 7.3.5). The effect of the exponential potential is to drive φ to large values.

---

[*]This is given by $\lambda = D_H(t)(a_0/a)$, where a_0 is the scale factor today and $D_H(t) \simeq ct$ is the particle horizon (c is reinstated here).

[†]This was spectacularly confirmed by the recent (2017) observation of a neutron star merging event, which transmitted both a GW pulse and a gamma-ray burst that were independently observed only about 1.7 seconds apart, even though the merger took place 43 Mpc away and 140 million years ago.

Hang on, you might say. Quintessence was introduced to explain the dark energy observations without making use of the cosmological constant. Well, as was standard practice until the observation of dark energy, we assume that, due to some *unknown* symmetry, the vacuum density in the Universe is zero. However, because of the positive pole present in the model, φ cannot go to infinity in the vacuum; it is capped at $\varphi = \sqrt{6\alpha}\, m_P$. This means that zero vacuum energy density requires $V_0 e^{-\kappa\sqrt{6\alpha}} = -V_\Lambda$, that is a negative value of Λ is needed to ensure that the vacuum density is zero. We assume that this is provided by the mechanism which eliminates vacuum density. Note that string theory (one of the most popular theories of everything) favours a negative Λ. Substituting this back into Eq. (9.121), the Lagrangian density becomes

$$\mathcal{L} = \frac{1}{2}\frac{(\partial\varphi)}{\left[1 - \frac{1}{6\alpha}\left(\frac{\varphi}{m_P}\right)^2\right]^2} - V_0 e^{-n}\left\{e^{n\left[1 - \frac{1}{\sqrt{6\alpha}}\left(\frac{\varphi}{m_P}\right)\right]} - 1\right\}, \tag{9.122}$$

where $n \equiv \kappa\sqrt{6\alpha}$. Because of the poles, the potential for the canonical field becomes stretched at $\varphi \to \pm\sqrt{6\alpha}\, m_P$. Hence, we find two plateaus in the model, with $V \to V_0 e^{-n}(e^{2n} - 1)$ or $V \to 0$. If the scalar field dominates the Universe, the two plateaus can result in early and late periods of accelerated expansion. Thus, they can be the inflationary plateau and the quintessential tail.

As in Sec. 7.3.5, to assist our intuition and help with studying the model, we make the field redefinition in Eq. (7.87) to obtain a canonical kinetic term for the scalar field. Our canonical field, ϕ, can take any value whilst our noncanonical φ may remain sub-Planckian at all times, as long as $\alpha < \frac{1}{6}$. The potential becomes now

$$V(\phi) = e^{-2n}M^4\left\{\exp\left[n\left(1 - \tanh\frac{\phi}{\sqrt{6\alpha}\, m_P}\right)\right] - 1\right\}, \tag{9.123}$$

where we have defined $M^4 \equiv e^n V_0$, which stands for the inflation energy scale because $V \to (1 - e^{-2n})M^4 \simeq M^4$ when $\phi \to -\infty$ (we show later on that $n \gg 1$). Note, also, that $V_\Lambda = -e^{-2n}M$. The potential has the desired form of a quintessential inflation potential, shown in Fig. 9.5.

As $\tanh(\phi/\sqrt{6\alpha}\, m_P)$ approaches a constant value when $|\phi|$ is very large, the potential becomes asymptotically constant, featuring plateaus. At the locations of these plateaus the field slow rolls and accelerated expansion occurs. In what follows, we examine these two periods; that of inflation when $\phi \to -\infty$ ($\varphi \to -\sqrt{6\alpha}\, m_P$) and that of quintessence when $\phi \to +\infty$ ($\varphi \to \sqrt{6\alpha}\, m_P$).

Consider inflation first, when $\phi \to -\infty$. In this limit the potential in Eq. (9.123) becomes[*]

$$V(\phi) \simeq M^4 \exp\left(-2n e^{\frac{2\phi}{\sqrt{6\alpha}m_P}}\right) \simeq M^4\left(1 - 2n e^{\frac{2\phi}{\sqrt{6\alpha}m_P}}\right). \tag{9.124}$$

This is a standard α-attractors potential, which leads to inflationary observables, common to α-attractors models, given by $n_s \simeq 1 - 2/N_*$ and $r \simeq 12\alpha/N_*^2$ (cf. Eq. (7.90)). As we have discussed in Sec. 7.3.5 these predictions hit the sweet spot of Planck observations, for $N_* \simeq 60$ and $\alpha \lesssim 1$. But even when N_* is larger, as is the case in quintessential inflation, the success of inflation is preserved. Indeed, if we take $N_* = 67$, which is the maximum value of N_*, we find $n_s = 0.970$ and $r = 2.67 \times 10^{-3}\alpha$, which is still in excellent agreement with the observations, if $\alpha < 26$.

Now, consider quintessence. In the limit $\phi \to +\infty$, the potential in Eq. (9.123) becomes

$$V \simeq 2n e^{-2n}M^4 e^{-\frac{2\phi}{\sqrt{6\alpha}m_P}}. \tag{9.125}$$

---

[*]We have considered $e^{-2n} \ll 1$ because we show later that $n \sim 100$ or so.

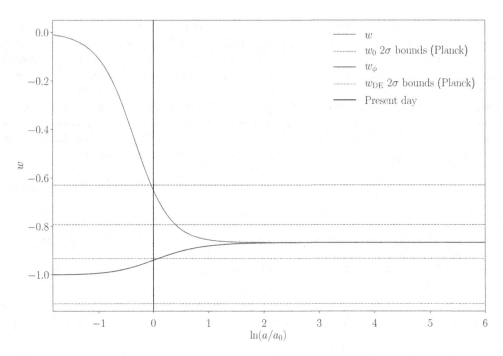

FIGURE 9.8   Behaviour of the barotropic parameter of quintessence w_ϕ (lower solid curve) and of the whole Universe w (upper solid curve) as a function of the logarithm of the scale factor $\ln a$, with a_0 being the value of $a(t)$ today, taking $\alpha = \frac{5}{3}$ ($\lambda = \sqrt{0.4}$). We see that originally the Universe is in the matter era with $w = 0$ and the quintessence field is frozen at value $\phi_F =$ constant with constant density, such that $w_\phi = -1$. However, when approaching the present time (depicted by the vertical solid line) the quintessence unfreezes and $w_\phi(t) > -1$, while it also begins to dominate the Universe so that $w(t) < 0$. We find numerically that successful quintessence is achieved and the present values of w_ϕ and w satisfy the Planck bounds, depicted by the horizontal dashed lines. In the future, quintessence becomes fully dominant so $w \approx w_\phi$, while it slow-rolls down the scalar potential with constant barotropic parameter $w_\phi = -1 + \lambda^2/3 = -0.87$. It is clear that both w_ϕ and w are running at present. Using the CPL parametrisation of Eq. (2.37), we find $w_{\mathrm{DE}}^0 = -0.94$ and $w_a \equiv -dw_\phi/da|_{a=a_0} = -0.013$.

Thus, the potential features a classic exponential quintessential tail, of the form shown in Eq. (9.23) $V(\phi) = V_Q \exp(-\lambda\phi/m_P)$ where $V_Q = 2ne^{-2n}M^4$ and $\lambda = 2/\sqrt{6\alpha} = (2/n)\kappa$. As we have discussed in Sec. 9.2.3, accelerated expansion requires $\lambda < \sqrt{2}$, which implies $\alpha > \frac{1}{3}$. Then, the dominant attractor evolution applies, which when attained leads to $\lambda^2 = 3(1 + w_\phi)$. Given that current observations suggest that, for quintessence $-1 < w_\phi \le -0.95$, we find the bound $\lambda < 0.4$, which translates into the requirement $\alpha \gtrsim 4$. This is too strong though, because we expect that quintessence is thawing, that is it unfreezes today and approaches the attractor, but the latter has not been attained yet. As a result, w_ϕ is variable and smaller than the attractor value. This means that, for a given w_ϕ, λ can be larger, and therefore α can be smaller. In fact, the model works fine with $\alpha \sim 1$; see Fig. 9.8.

Next we find the parameter space for n and κ. To do this, we enforce the coincidence requirement, which demands that the density of quintessence (which is similar to $V_F \equiv V(\phi_F)$) must be comparable with the density of the Universe at present ρ_0. We have

$$\frac{\rho_{\mathrm{inf}}}{\rho_0} \simeq \frac{V_{\mathrm{inf}}}{V_F} \simeq \frac{e^{\lambda\phi_F/m_P}}{2ne^{-2n}} \sim 10^{110} \quad , \tag{9.126}$$

where for the inflation energy scale we have $\rho_{\text{inf}} \simeq V_{\text{inf}} = (1 - e^{-2n}) M^4 \simeq M^4 \sim (10^{15.5} \text{ GeV})^4$ (we find later that $n \gg 1$), $\rho_0 \sim (10^{-3} \text{ eV})^4$, $V_F = V_Q \exp(-\lambda \phi_F/m_P)$, and $V_Q = 2ne^{-2n}M^4$. The above leads to

$$2n - \ln(2n) \simeq 110 \ln 10 - \frac{2}{\sqrt{6\alpha}} \frac{\phi_F}{m_P}, \qquad (9.127)$$

where we used $\lambda = 2/\sqrt{6\alpha}$ and ϕ_F is given by Eq. (9.106). The value of ϕ_F is largely determined by the mechanism, which produces the radiation bath of the Hot Big Bang. However, we can still deduce some generic findings. The potential cliff in Eq. (9.123) is near the origin meaning that we expect $|\phi_{\text{end}}| < \phi_F$ because this is the edge of the inflationary plateau. Consequently, if radiation is produced at the end of inflation (as with gravitational reheating or warm quintessential inflation) then we can probably ignore $\phi_\star = \phi_{\text{end}}$ in Eq. (9.106). Similarly, if we consider instant preheating, the efficiency of the radiation production is proportional to the "velocity" $\dot{\phi}$ of the field in field space, cf. Eq. (6.58). Because $\dot{\phi} \propto a^{-3}$ during kination, this is larger the earlier in kination we are. Thus, even though there may be several ESPs along the inflaton trajectory, it is the one encountered earliest that produces the most radiation. Thus, $\phi_\star$ is probably not too different from ϕ_{end}, and therefore $|\phi_\star| < \phi_F$. This means that Eq. (9.127) can be written as

$$2n - \ln(2n) \simeq 110 \ln 10 - \frac{1}{\sqrt{\alpha}} \left(\frac{4}{3} - \ln \Omega_r^\star \right), \qquad (9.128)$$

The last term in the above becomes significant only when the production of radiation is very inefficient, so that $\Omega_r^\star \ll 1$. This makes sense because, when little radiation is produced, it takes time for it to dominate the Universe even though $\Omega_r \propto a^2$ during kination. As a result, reheating occurs late, and the field manages to reach a large value before freezing at ϕ_F. The least efficient mechanism is gravitational reheating, for which $\Omega_r^\star$ is given by Eq. (9.108). Taking $H_{\text{end}} \sim 10^{13}$ GeV we estimate $\Omega_r^\star \sim 10^{-12}$. Using this and $\alpha \sim 1$, the last term in Eq. (9.128) is about 30. Thus, in all cases the dominant term in Eq. (9.128) is $110 \ln 10 \simeq 250$. Therefore, we find that $n \sim 100 \gg 1$. Consequently, $\kappa = n/\sqrt{6\alpha}$ is also similar but slightly smaller. This means that the energy scale which suppresses the exponent in the original exponential potential in Eq. (9.121) is near the GUT-scale $m_P/\kappa \sim 10^{16}$ GeV. No incredibly fine-tuned, tiny number is required. From the above, we see that the coincidence requirement is determined by the model parameters (including the reheating mechanism which decides ϕ_F) and there is no need to worry about the initial conditions of quintessence. Finally, it can be argued that, even though $\phi_F \gg m_P$ (but $\varphi(\phi_F) \sim m_P$), in the case of α-attractors radiative corrections and interactions are exponentially suppressed.

9.8   The end of time

At the end of this chapter, we take a brief look at what modern cosmology can say about the ultimate fate of the Universe, based on what we know so far and what we can conjecture from fundamental theory. In Sec. 2.5.1, we saw that geometry determines destiny. The Friedmann universes suggest that the fate of the Universe is sealed by the sign and value of the curvature parameter k, which reflects the overall geometry of the cosmological FLRW spacetime. However, as we explained later on in Sec. 2.6.1, things change when a nonzero cosmological constant is introduced. In particular, a positive cosmological constant results in an eternally expanding Universe, even if closed (i.e., with $k > 0$), given that the observations suggest a total density parameter very close to unity. The Friedmann universes are eventually overwhelmed by the exponential expansion, because ultimately the cosmological constant term dominates. Conversely, if the cosmological constant is negative the Universe recollapses, even if open/flat (i.e., $k \leq 0$) and ends in a Big Crunch. The same is true if the late-time dark energy gives rise to a positive or negative constant vacuum density, which is indistinguishable from a pure cosmological constant.

For example, this may be the value of the potential density of quintessence at its VEV. Given that a cosmological constant can be also thought as a feature of the geometry (see Appendix A), we may still claim that geometry determines destiny, albeit in a more complicated way than what Friedmann had envisioned.

9.8.1 Open-ended fate

Suppose that the Universe does not have a cataclysmic end into a Big Crunch or a Big Rip at some finite time in the future and continues to expand forever. What do we expect to happen? Well, in enough time all peculiar motions of objects revolving around each other would end, because of energy loss due to emission of electromagnetic or gravitational radiation, which would mean that revolving bodies would come close to each other and merge into one. For our galaxy, this means that everything will eventually merge with the supermassive black hole at the galactic centre. Furthermore, the galaxies in the Local Group would also merge with ours in a giant black hole. This is so with other galaxy groups in the nearby galactic cluster. The end result of this process is a single black hole inside the horizon, absorbing what is left from the CMB. In time, the CMB will be redshifted so much that it will become subdominant to the Hawking radiation emitted through particle production on the event horizon of this cosmic black hole. Slowly, the cosmic black hole will shrink and eventually evaporate. Radiation from this black hole will continue to fade along with the CMB in an ever-expanding Universe. This scenario is called *heat death* or Big Chill (or Big Freeze). The rationale behind heat death is that the Universe content has reached maximum thermal equilibrium, which would not allow any new processes to take place.[*]

The above scenario is more complex when the Universe dynamics are considered in more detail. If we disregard dark energy, the cosmological horizon is a particle horizon, which expands with lightspeed, faster than the Universe expansion itself (see Sec. 2.10). This means that, even if there is enough time for all matter in our observable Universe to merge in a cosmic black hole, as the horizon expands, other neighbouring cosmic black holes, which were hitherto beyond causal contact, will enter our observable Universe and try to merge with "our" cosmic black hole, a process that will continue to happen indefinitely. Such mergers might keep the cosmic black holes from evaporating. Thus, barring the dimming CMB background, the Universe will reach a semi-stationary condition and the end of time corresponds to nothing new happening.

Now, let us take into account dark energy and suppose that it is constant, as in ΛCDM. This dark energy is taking over at present. As a result, the cosmological horizon is turning from a particle horizon into an event horizon, which grows slower that the superluminal expansion of the Universe. For constant dark energy the expansion approaches de Sitter, and the horizon size becomes constant too. We expect the Universe expansion to drive out of our horizon volume all the unbound and semi-bound structures such as galactic superclusters. In about $\Delta t \sim 100\,\mathrm{Gy}$, everything beyond the local group of galaxies (so everything at a distance of more than a few Mpcs from us) will be gone beyond contact forever, leaving behind a group of about 100 galaxies as a kind of "island universe" (out of the two trillion galaxies inside the observable Universe today). These will eventually merge to a cosmic black hole, much smaller than the one we discussed above. However, there are two important differences in the ΛCDM scenario, compared to the stationary no dark energy scenario. One is that, because of the expansion and the fact that the horizon size remains constant, no other cosmic black hole will ever enter our observable

---

[*]The idea of heat death has been considered since the 19th century, well before the Universe expansion or black holes were known. It was originally put forward by William Thomson, 1st Baron Kelvin, who was an Irish-Scottish mathematical physicist and engineer (it was possible to be all these things in those days). He is the reason absolute temperature is measured in units of Kelvin.

Universe. We will be truly alone (apart from being long gone). The second difference is the fact that, although the CMB will be redshifted away, the constant event horizon will fill the observable Universe with Hawking radiation of temperature $T_H = H_0/2\pi$. This will dominate the fading CMB and will become a source for feeding the cosmic black hole. We should compare this temperature with the Hawking temperature of the cosmic black hole to see whether the net effect on the latter will be to gain or lose mass. The black hole Hawking temperature is given by $T_{\mathrm{BH}} = \kappa/2\pi$, where $\kappa = 1/4GM$ is called the surface gravity of the black hole. The mass of the observable Universe at present is*

$$M_H = \frac{4\pi}{3} H_0^{-3} \rho_0 = \frac{1}{2GH_0} \,, \tag{9.129}$$

where we used the flat Friedmann equation (2.38). However, the mass of the cosmic black hole M is comprised only by the mass of the local group of galaxies, while the majority of the mass in M_H corresponds to the vacuum energy driving the de Sitter expansion. Taking that today the mass in galaxies and in the vacuum density are comparable (but they will not be in the future, since most galaxies will be inflated away) we expect

$$M \sim \frac{\text{Local Group}}{\text{two trillion galaxies}} M_H \sim \frac{100}{2 \times 10^{12}} M_H \sim \tfrac{1}{2} \times 10^{-10} M_H \,. \tag{9.130}$$

This implies that

$$T_{\mathrm{BH}} = \frac{1}{2\pi} \frac{1}{4GM} \sim \frac{1}{2\pi} \frac{10^{10}}{2GM_H} = 10^{10} T_H \gg T_H \,. \tag{9.131}$$

Thus, the radiation emitted by the de Sitter horizon would not be nearly enough to stop the eventual evaporation of the cosmic black hole, the reason being that the cosmic black hole would emit much more radiation than it receives, leading in a runaway manner to its eventual evaporation. After this, the remaining radiation, either from the evaporation of the cosmic black hole or from the redshifting CMB, will continue to be diluted until they are overwhelmed by the Hawking radiation due to the cosmological event horizon. From then on, nothing will ever happen and the Universe will reach a static (not stationary) state which deserves to be considered the end of time. The only caveat in the above is the possibility of the particle production eventually destabilising de Sitter expansion. At the moment, it is not clear whether this happens and cosmologists' opinions are (sometimes bitterly) divided.

Another interesting issue is the stability of the electroweak vacuum. In the timescale of $\Delta t \sim 100 \, \mathrm{Gy}$, the electroweak Higgs field might tunnel through to its true vacuum. As a result, its expectation value may jump from 246 GeV, which is its current value, to about m_P. If there is still "normal" matter around (so not everything is yet eaten by the cosmic black hole) the mass of each particle will grow by a factor of 10^{16}, which may enlarge the mass of the observable Universe considerably, possibly leading to a Big Crunch. However, experiment is still inconclusive at to whether the electroweak vacuum is metastable or not.

Variants of the above scenarios correspond to decaying dark energy, for example quintessence which asymptotes to zero vacuum density, depending how fast the decay of the dark energy is, which in turn determines the growth of the cosmological event horizon. However, there is another, much more exciting range of possibilities, which might lead to catastrophic future singularities, like the Big Crunch or the Big Rip, that may occur in finite time in the future. We take a look into some of them next.

---

*Note that the Schwarzschild radius of the Universe is $2GM_H = 1/H_0$, which is equal to its Hubble radius. Does this mean that we live inside a black hole?

9.8.2   Cataclysmic end

We have discussed the Big Crunch in Sec. 2.5.1 and the Big Rip in Sec. 2.10. However, there are a number of other finite-time singularities, which may appear in the future. The most important are (for a review see Ref. [CST06]):

Type 0:	Big Crunch	$t \to t_\star$	$\Rightarrow$	$a \to 0,$	$\rho \to +\infty,$ $\vert p \vert \to +\infty$ or 0
Type I:	Big Rip	$t \to t_\star$	$\Rightarrow$	$a \to +\infty,$	$\rho \to +\infty,$ $\vert p \vert \to +\infty$
Type II:	sudden	$t \to t_\star$	$\Rightarrow$	$a \to a_\star,$	$\rho \to \rho_\star,$ $\vert p \vert \to +\infty$
Type III:	Big Freeze	$t \to t_\star$	$\Rightarrow$	$a \to a_\star,$	$\rho \to +\infty,$ $\vert p \vert \to +\infty$
Type IV:	generalised sudden	$t \to t_\star$	$\Rightarrow$	$a \to a_\star,$	$\rho \to \rho_\star,$ $\vert p \vert \to \vert p_\star \vert$

higher derivatives of H diverge

Type V: w − singularity $t \to t_\star$ $\Rightarrow$ $a \to a_\star,$ $\rho \to 0,$ $\vert p \vert \to 0$
but $w = p/\rho \to \infty$

Type VI: quiescent $t \to t_\star$ $\Rightarrow$ $a \to +\infty,$ $\rho \to \rho_\star,$ $\vert p \vert \to \vert p_\star \vert$
higher derivatives of H diverge

where the subscript "$\star$" denotes a finite value at the time of the singularity.

The origin of the above finite-time future singularities is related to the violation of all or part of the energy conditions, which are:

$$
\begin{aligned}
\text{null} \quad &: \quad \rho + p \geq 0 \\
\text{weak} \quad &: \quad \rho \geq 0 \ \& \ \rho + p \geq 0 \\
\text{dominant} \quad &: \quad \rho \geq \vert p \vert \\
\text{strong} \quad &: \quad \rho + p \geq 0 \ \& \ \rho + 3p \geq 0
\end{aligned}
\tag{9.132}
$$

For example, as we have discussed in Sec. 2.8.3, the Big Rip singularity (Type I) is realised in a flat Universe dominated by phantom dark energy with $p < -\rho$, which evidently violates the null energy condition. Below we briefly discuss some dark energy toy-models which may lead to finite-time future singularities of other types, assuming dark energy domination.

One convenient way to design a model which may lead to a finite-time future singularity is by considering the following barotropic equation of state:

$$
p = -\rho - f(\rho) \,,
\tag{9.133}
$$

where the function $f(\rho)$ is the deviation from ΛCDM, for which $p = -\rho$. Using the continuity equation (6.4), we find

$$
\dot{\rho} = 3H f(\rho) \ \Rightarrow \ \frac{da}{a} = \frac{1}{3} \frac{d\rho}{f(\rho)} \ \Rightarrow \ a \propto \exp\left(\frac{1}{3} \int \frac{d\rho}{f(\rho)} \right) .
\tag{9.134}
$$

Using the above, we also find

$$
\dot{\rho} = 3H f(\rho) \ \Rightarrow \ dt = \frac{d\rho}{3H f(\rho)} \ \Rightarrow \ t = \frac{m_P}{\sqrt{3}} \int \frac{d\rho}{\sqrt{\rho}\, f(\rho)} \,,
\tag{9.135}
$$

where we used the Friedmann equation (6.2).

To move on we introduce the ansatz

$$
f(\rho) = \frac{C}{(\rho_\star - \rho)^\gamma} \,,
\tag{9.136}
$$

where C is a dimensionful constant and γ is a parameter. From Eq. (9.133), we see that the dark energy equation of state is

$$
w = \frac{p}{\rho} = -1 - \frac{f(\rho)}{\rho} \,.
\tag{9.137}
$$

In order to arrange for a finite-time future singularity we would prefer the dark energy to be phantom.* Thus, assuming that before the moment of the singularity the density was $\rho < \rho_\star$ so that $\rho_\star$ is the maximum density, we see that dark energy is phantom only when $f(\rho) > 0$, which means $C > 0$. We are interested in the possibility that $\rho \to \rho_\star$, which may lead to future singularities of type II, IV and VI.

Inputting Eq. (9.136) into Eq. (9.134) we find

$$a = a_\star \exp\left(-\frac{1}{3}\frac{(\rho_\star - \rho)^{\gamma+1}}{C(\gamma + 1)}\right), \tag{9.138}$$

where $\gamma \neq -1$. When $\rho \to \rho_\star$, the above suggests that $a \to a_\star$, which leaves only types II and IV. Also, from Eq. (9.136), we see that $f(\rho)$ diverges, which means that p (as given by Eq. (9.133)), diverges too, leaving only type II as a possibility.

Near the singularity, when $\rho \sim \rho_\star$, Eq. (9.135) results in

$$t = \frac{m_P}{\sqrt{3}}\int\frac{(\rho_\star - \rho)^\gamma}{C\sqrt{\rho}}d\rho \simeq \frac{m_P}{\sqrt{3\rho_\star}\,C}\int(\rho_\star - \rho)^\gamma d\rho \Rightarrow t \simeq t_\star - \frac{m_P}{\sqrt{3\rho_\star}\,C}\frac{(\rho_\star - \rho)^{\gamma+1}}{\gamma + 1}. \tag{9.139}$$

Thus, the above shows that the limit $\rho \to \rho_\star$ is reached in finite time, when $t \to t_\star$. Note also that, because $C > 0$, we have $t < t_\star$ when $\rho < \rho_\star$, confirming that this is a future singularity, provided also that $\gamma > -1$. Thus, in this case we have a sudden singularity of type II.

Now consider the possibility that $\rho \to +\infty$ at the singularity, corresponding to types 0, I, and III. By virtue of Eq. (9.133) we see that $|p| \to +\infty$ too, regardless of the value of $f(\rho)$. To model this case we modify Eq. (9.136) by scrapping $\rho_\star$ to get

$$f(\rho) = C\rho^{-\gamma}, \tag{9.140}$$

where we continue to demand that $C > 0$ so that $f(\rho) > 0$ and dark energy is phantom, $p < -\rho$. For $\gamma = -1$ we have $p = -(1 + C)\rho$, which is the usual phantom dark energy, which leads to the Big Rip (singularity type I). For $\gamma \neq -1$ Eq. (9.134) gives

$$a \propto \exp\left(\frac{1}{3}\frac{\rho^{\gamma+1}}{C(\gamma + 1)}\right). \tag{9.141}$$

In the limit $\rho \to +\infty$, the above suggests that $a \to a_\star$ (finite) if $\gamma < -1$ and $a \to +\infty$ (diverges) if $\gamma > -1$. Using Eq. (9.135), we find

$$t = t_\star + \frac{m_P}{\sqrt{3}\,C}\frac{\rho^{\gamma+\frac{1}{2}}}{\gamma + \frac{1}{2}} \qquad \text{when} \quad \gamma \neq -\frac{1}{2},$$

$$\tag{9.142}$$

$$t = t_\star + \frac{m_P}{\sqrt{3}\,C}\ln\left(\frac{\rho}{\rho_\star}\right) \qquad \text{when} \quad \gamma = -\frac{1}{2}.$$

This suggests that in the limit $\rho \to +\infty$ we have $t \to t_\star$ if $\gamma < -\frac{1}{2}$, while $t \to +\infty$ if $\gamma \geq -\frac{1}{2}$.

Thus, when $\gamma < -1(< -\frac{1}{2})$ we have $a \to a_\star$ and $t \to t_\star$, while $\rho, |p| \to +\infty$. This means that we have singularity type III (Big Freeze). However, when $-1 \leq \gamma < -\frac{1}{2}$ we have $a \to +\infty$ but still $t \to t_\star$, while $\rho, |p| \to +\infty$. This means that we have singularity type I (Big Rip). In both cases, because $\gamma + \frac{1}{2} < 0$ we have $t < t_\star$ before the singularity is reached (cf. Eq. (9.142)), which confirms that it is a future singularity. Finally, when $\gamma \geq -\frac{1}{2}$ we have $t \to +\infty$ when $\rho, |p| \to +\infty$, which means that divergence occurs in infinite time in the future.

---

*Although this is not strictly necessary.

Transposing the Big Rip singularity to the infinite future is the motivation of the Little Rip idea. In the Little Rip scenario, the density increases with phantom barotropic parameter $w < -1$, which however is also increasing, tending to $w = -1$ asymptotically. The scenario still leads to the gradual dissolution of bound structures but it takes infinite time for the Big Rip to occur. As an example of Little Rip we can represent the Hubble parameter as

$$H(t) = H_\star \exp(\xi t), \tag{9.143}$$

where $H_\star$ and ξ are constants. Related to the Little Rip is the Pseudo Rip proposal, where H approaches a constant in infinite time. Cosmology is phantom but asymptoptes to de Sitter. For example, in this case, we may represent the Hubble parameter as

$$H(t) = H_\star \tanh(\xi t), \tag{9.144}$$

which means that $H \to H_\star$ as $t \to +\infty$. For the barotropic parameter we find

$$w = \frac{p}{\rho} = -1 + \frac{\rho + p}{\rho} = -1 - \frac{\dot{\rho}}{3H\rho} = -1 - \frac{2\dot{H}}{3H^2}, \tag{9.145}$$

where we considered the continuity equation (6.4) and also that $\dot{\rho} = 6H\dot{H}m_P^2$, which is obtained by taking the time derivative of the Friedmann equation (6.2). Using Eq. (9.144), the above gives

$$w = -1 - \frac{2\xi}{3H_\star} \frac{1}{\sinh^2(\xi t)}. \tag{9.146}$$

This result shows that $w < -1$ always and $w \to -1$ when $t \to +\infty$ as with Little Rip.

When considering the divergence of ρ or p or both we should note that the scalar curvature also diverges. Indeed, we have (see Appendix A)

$$R = 6\left(\frac{\ddot{a}}{a} + H^2\right) = m_P^{-2}(\rho - 3p), \tag{9.147}$$

where in the last step we used Eqs. (6.2) and (6.3). Thus, as R grows, we would expect higher order corrections of gravity to come into play, see for example Secs. 7.3.2 and 7.3.4.

The discussion in this section is but a glimpse of the various possibilities and the alternative scenarios for the ultimate fate of our Universe. We may well never be able to determine the future of the Universe with certainty. In that sense, the fate of our Universe is truly dark, that is unknown and possibly unknowable. One thing is undeniable though. This future is shaped by the nature and the characteristics of dark energy.

EXERCISES

1. In supergravity scalar fields typically obtain masses of order the Hubble scale H (this is the source of the η−problem), i.e., for the potential V, we have $V'' \sim H^2$, where the prime denotes derivative with respect to the scalar field.
   Investigate the fate of a minimally coupled spectator scalar field with potential

   $$V(\phi) = \tfrac{1}{2}cH^2\phi^2\,,$$

   after inflation, where $H = 2/3(1+w)t$ is the Hubble parameter, with w being the barotropic parameter of the Universe and $c \geq 0$ is a constant, with $c \sim 1$. (Note that supergravity suggests that $c = 0$ in the radiation era.) In order for the contribution of the field to the density budget of the Universe to be negligible, we require $|\phi| < \sqrt{\tfrac{6}{c}}\,m_p$ so that $V < \rho = 3H^2m_P^2$.
   In particular:

   (a) By changing variable as $dt = (1+w)t\,d\tau$, show that the Klein-Gordon equation for the scalar field becomes

   $$\frac{\partial^2\phi}{d\tau^2} + (1-w)\frac{d\phi}{d\tau} + \frac{4c}{9}\phi = 0\,.$$

   By solving this equation, show that the spectator field undergoes oscillations if $c > c_{\mathrm{x}}$ or follows a scaling solution if $c \leq c_{\mathrm{x}}$, where $\sqrt{c_{\mathrm{x}}} \equiv \tfrac{3}{4}(1-w)$.

   (b) Find the oscillating and the scaling solution mentioned in part (a) by assuming that initially the field begins as stationary $\dot{\phi} = 0$.

   (c) By calculating how the density parameter $\Omega_\phi = \rho_\phi/\rho$ scales with the scale factor $\Omega_\phi = \Omega_\phi(a)$ demonstrate that, in all cases, the spectator field can never dominate the Universe.

   (d) With the help of a diagram depicting how the field rolls down a varying potential, briefly discuss the physical interpretation of the scaling solution.

2. The potential of the original quintessential inflation model is

   $$V(\phi) = \begin{cases} \lambda(\phi^4 + M^4) & \text{for} \quad \phi < 0 \\ \dfrac{\lambda M^8}{\phi^4 + M^4} & \text{for} \quad \phi > 0 \end{cases}\,,$$

   where $\lambda > 0$ is a dimensionless constant and M is mass scale with $0 \ll M \ll m_P$. In view of current observations, the above is not viable because quartic chaotic inflation as well the tracker of quartic inverse-power-law quintessence have both been excluded. However, the model can work if inflation is assumed warmed up and quintessence is considered thawing instead of tracking.
   Analyse warm quintessential inflation assuming $Q \ll 1$, where $Q \equiv \Upsilon/3H$, with H being the Hubble scale and Υ being the dissipative coefficient. You may also take $Q = $ constant during inflation for simplicity. In particular:

   (a) Show that, in warm inflation with $Q \ll 1$ the radiation density during inflation is $\rho_r \simeq \tfrac{1}{2}\varepsilon QV$, where $\varepsilon \equiv \tfrac{1}{2}m_P^2(V'/V)^2$ is the usual slow-roll parameter with the prime denoting derivative with resoect to the inflaton field ϕ.

   Using this, show that the temperature of the subdominant thermal bath during inflation is

   $$T = \left(\tfrac{45}{\pi^2 g_*}\right)^{1/4}(\varepsilon Q)^{1/4}\sqrt{Hm_P}\,,$$

where g_* is the effective relativistic degrees of freedom.

(b) Show that, in warm inflation with $Q \ll 1$ the power-spectrum of the curvature perturbation is written as

$$\mathcal{P}_\zeta = \frac{1}{4\pi^2} \left(\frac{45}{\pi^2 g_*} \right)^{1/4} \varepsilon^{-3/4} Q^{1/4} \left(\frac{H}{m_P} \right)^{3/2} .$$

(c) During inflation $\phi \ll -M$ and the model becomes $V \simeq \lambda \phi^4$.

Calculate the slow-roll parameters and show that $\eta = \frac{3}{2}\varepsilon$. Considering that inflation ends when $\varepsilon \simeq 1$, find the value of the inflaton field $\phi(N)$ as a function of the remaining number of e-folds of inflation N.

Hence, show that $\varepsilon = 1/(N+1)$.

[Hint: Recall that in warm inflation $T > H$.]

(d) Show that during inflation $H = 8\sqrt{\lambda/3}\,(N+1)m_P$.

Using this, obtain that

$$\mathcal{P}_\zeta = \frac{2\sqrt{8}}{\pi^2} \left(\frac{45}{\pi^2 g_*} \right)^{1/4} Q^{1/4} (\lambda/3)^{3/4} (N+1)^{9/4} .$$

Choosing the numbers $Q = 0.002$, $N_* + 1 = 60$, and $g_* = 106.75$ (corresponding the the standard model at high energies) and employing that the observations suggest $\mathcal{P}_\zeta = 2.10 \times 10^{-9}$, estimate the values of λ, H, and T/H when the cosmological scales exit the horizon.

(e) The scalar spectral index in warm inflation is given by

$$n_s - 1 = -\frac{17 + 9Q}{4(1+Q)^2} \varepsilon + \frac{3}{2(1+Q)} \eta - \frac{1 + 9Q}{4(1+Q)^2} \beta ,$$

where $\beta \equiv m_P^2 \frac{\Upsilon' V'}{\Upsilon V}$.

Considering that the dissipative coefficient does not depend on the inflaton field $\Upsilon \neq \Upsilon(\phi)$, show that in our case

$$n_s = 1 - \frac{2}{N+1} .$$

The tensor to scalar ratio is given by the standard expression

$$r = \frac{2}{\pi^2 \mathcal{P}_\zeta} \left(\frac{H}{m_P} \right)^2 .$$

Using the numbers given in part (d), obtain the values of n_s and r. Briefly comment on how they compare with the observations.

(f) Find the density parameter of radiation at the end of inflation Ω_r^{end}.

The field is briefly kinetically dominated, but this is enough for it to be propelled down the quintessential tail. The expectation value of the field changes by several factors of m_P. This suggests that the dissipative effects present during inflation are no more important so they can be safely ignored after inflation.

After the field ceases to be kinetically dominated, it freezes at a value ϕ_F on the quintessential tail. Using the obtained value of Ω_r^{end}, estimate ϕ_F when $Q = 0.002$ during inflation.

(g) Consider that the field remains frozen until today, when it is about to dominate the Universe and start slow-rolling down the quintessential tail. This is thawing quintessence.

By making the crude approximation that, after unfreezing, the field has not moved much so that $\phi \simeq \phi_F$ today, use the coincidence requirement to estimate the value of M in the potential. Comment whether this is a "reasonable" result.

[Hint: You may employ that the dark energy parameter at present is $\Omega_\Lambda \simeq 0.692$ and the current density of the Universe is $\rho_0 = 0.864 \times 10^{-29} \frac{\text{g}}{\text{cm}^3} = 3.72 \times 10^{-47} \, \text{GeV}^4$.]

3. Gravitational production of light nonconformal fields during accelerated expansion results in the generation of a bath of particles with density

$$\rho_{\text{gr}} = \frac{q g_*^{\text{gr}}}{480\pi^2} H^4 \,,$$

where $q \sim 1$ is some numerical coefficient, H is the Hubble parameter and g_*^{gr} is the effective relativistic and nonconformal degrees of freedom. Because of this, the Friedmann equation obtains an additional term of the form

$$H^2 = \frac{8\pi G}{3} \rho + C H^4 \,,$$

where ρ is the density of the substance which causes the accelerated expansion and

$$C = \frac{8\pi G \rho_{\text{gr}}}{3 H^4} = \frac{qG}{180\pi} g_*^{\text{gr}} \,.$$

This contribution is typically negligible in quintessence but it might become important when dark energy is phantom and H is growing with time.

Study the backreaction of this extra term in the Friedmann equation in the case of phantom dark energy and investigate how it may affect the Big Rip. In particular:

(a) By solving the modified Friedmann equation (recall that H is a growing function of ρ), show that there is an upper limit of the density

$$\rho_{\max} = \frac{3}{32\pi GC} = (6\pi)^2 \frac{30}{q g_*^{\text{gr}}} m_P^4$$

Find the corresponding maximum value $H_{\max}$ of the Hubble scale.

Obtain a lower bound on g_*^{gr} such that $\rho_{\max}$ is sub-Planckian and quantum gravity corrections are avoided.

(b) If backreaction from gravitational production becomes important then the continuity equation must be augmented to account for this. As a result

$$\dot{\rho} + 3(1+w)H(\rho - \rho_{\text{gr}}) = 0$$

$$\Rightarrow \quad \dot{\rho} - 3|1+w|H\rho = -3|1+w|H\rho_{\text{gr}} = -\frac{9|1+w|C}{8\pi G} H^5 \,,$$

where for phantom dark energy $1 + w < 0$. Thus, the backreaction introduces a negative source term on the right-hand side of the continuity equation. The form of this term is given by $\delta\rho/\delta t$, where $\delta\rho = -\rho_{\text{gr}} \propto H^4$ and $\delta t \sim H^{-1}$ is the Hubble time. This is because the relativistic particles produced gravitationally in a Hubble time are diluted by the accelerated expansion and replenished by the particles produced in the following Hubble time.

By defining $u \equiv \rho/\rho_{\max}$ show that the augmented continuity equation can be written as

$$\dot{u} = 6|1 + w|H_{\max}\sqrt{1 - u}\,(1 - \sqrt{1 - u})^{3/2}\,.$$

Discuss what this equation suggests when $\rho = \rho_{\max}$.

Show that the equation reduces to the usual continuity equation in the limit $\rho \ll \rho_{\max}$.

[Hint: The solution to the modified Friedmann equation suggests $H = H_{\max}$ $(1 - \sqrt{1 - u})^{1/2}$]

(c) Verify that the solution to the augmented continuity equation is

$$\frac{\rho}{\rho_{\max}} = u = 1 - \left\{1 - \left[\tfrac{3}{2}|1 + w|H_{\max}(t_{\max} - t) + 1\right]^{-2}\right\}^2\,.$$

(d) Sketch this solution and comment on its implications with respect to the Big Rip. Discuss what happens when $t > t_{\max}$.

10

Epilogue

Observations suggest that we live on a planet (one of eight) of an average star, belonging (one of a few hundred billions) to a common spiral galaxy, which is part of a group of galaxies (about 50) that is embedded inside a giant supercluster of galaxies (called Virgo, with at least a hundred other galaxy clusters and groups), which is a fraction of a bubble-wall/filament (the Pisces-Cetus supercluster complex) in the large scale structure that corresponds to many thousands of similar bubbles (containing about 10 million galaxy superclusters) inside the observable Universe. Our civilisation is about ten thousand years old, which is one millionth of the age of the Universe. This begs the question: Could we be more insignificant?

Yet, in the last hundred years, we believe that we have discovered the equations that describe the Universe evolution. Our observations verify the validity of the inferred history of the Universe as far back as Big Bang nucleosynthesis (BBN), about 14 billion years ago and only a few seconds after the beginning of our Universe itself. We also have evidence of processes that took place even earlier than BBN in the form of the primordial density perturbations. In fact, we suspect that our description of the laws of physics in our Universe may be valid as early as the emergence of spacetime as we know it, i.e., as far back as the Planck time: $t_P \sim 10^{-43}$ sec. Is this not a tremendous achievement?[*]

The story of the Hot Big Bang, at least after BBN, is an undeniable fact, not much different from the fact that the Milky Way is a spiral galaxy or even that the Earth is round. Our fundamental theories may change in the future, as Newtonian physics gave way to Einstein's Relativity (and also to Quantum Mechanics), but this did not change the fact that the Earth is a globe and the Milky Way is a spiral galaxy. In the same way, the Hot Big Bang is here to stay. Oh sure, there are glitches: the Lithium abundance in BBN is about 2–3 times smaller, there is a tension regarding the value of the Hubble constant H_0 (CMB says it is 68 km/sec Mpc while late Universe measurements suggest 73 km/sec Mpc at the time of writing), and we still have no clue what dark matter is (is it axions, is it primordial black holes?). Of course it took 26 years for neutrinos to be discovered (they were proposed in 1930 and finally observed in 1956). These glitches are similar to open questions about the spiral structure of the Milky Way or about the convection currents in the Earth's mantle and outer core. Such questions do not mean that the Earth is not round, or the Milky Way is not a spiral galaxy. In the same way, the lithium underabundance or the Hubble tension do not mean that the Hot Big Bang is questionable.

---

[*]However, before congratulating ourselves too much, we need to be reminded that we still do not seem to have a clue of what makes up 95% of the Universe at present.

Things are less robust at the edges of the Hot Big Bang, either the very end, i.e., the present time (meaning the last billion years or so), or the very beginning. Mounting evidence suggests that the Universe underwent a phase of accelerated expansion in both cases. General relativity, which has managed to survive intense scrutiny for a century, suggests that the Universe must be dominated by some exotic substance, with pressure negative enough, in order to undergo accelerated expansion. This substance has been named dark energy.

The phase of the early universe accelerated expansion is called cosmic inflation. The idea of cosmic inflation has been around for 40 years now. The main observational support for inflation is based on the fact that the observed primordial density perturbations exhibit acausal (super-horizon) correlations. It is also supported by the fact that our observable Universe is spatially flat and uniform. The latter observation can be overcome when considering pre–big-bang extensions, proposed by some alternatives to inflation. It should be noted however, that alternatives to inflation are typically based on speculative theories (like string theory) which are not supported yet by observation or experiment. In contrast, inflation uses general relativity and quantum field theory. Both these frameworks have been thoroughly scrutinised by observation and experiment.

Inflation is a paradigm and not a specific model. The increasing precision of cosmological observations suggests that inflation is simple; driven by a single degree of freedom. Modelling this as a scalar field, the inflaton field, then most probably the inflaton slow-rolls down a plateau scalar potential. Observations have already ruled out many otherwise well-motivated specific inflationary models, even families of models. However, a number of models of inflation still survive. This does not mean that inflation does not have predictive power or that it is unfalsifiable, as some of its critics suggest. Indeed, inflation has predicted the peak structure in the CMB. When this structure was observed it was the rival theory of structure formation, that of cosmic strings, which collapsed. Were it the other way around and no acoustic peaks were observed, then this would have falsified inflation. Even beyond the density perturbations, inflation is possible to falsify if, for example, a global rotation of the Universe were observed (inflation would dilute this similarly to the curvature of space). Additionally, if nontrivial topology were observed at subhorizon scales today, it would contradict the way that inflation solves the horizon and flatness problems by inflating a causally connected patch to size larger than our observable Universe. However, neither rotation nor nontrivial topology have been observed, even though they could have been. The above indicate that inflation is indeed falsifiable and this does not contradict the fact that the number of viable inflation models is still large.

Will we however be able to ever prove inflation? Here, we should remember that, as Karl Popper put it, "no amount of data can really prove a theory, but even a single key data point can potentially disprove it." This means that, in principle a theory is only possible to disprove. No matter how tight the observational constraints become, as long as inflation is not falsified, there will be a number of inflation models which would still work. Observations cannot home in to a single model. If there is ambiguity, the blame should be put to fundamental theory that sets the framework upon which inflationary model-building is based. The situation is similar to questioning the constancy of Newton's gravitational constant G, or of the fine-structure constant α, or of the speed of light c.* One can only produce ever more stringent bounds of the variability of G, α, and c, but cannot prove they are indeed constant. In fact, there has been a lot of effort to construct theories which allow the variability of such constants and rightfully so. This is how science works. In a similar manner, it is important to explore alternatives to inflation. But this

---

*Interestingly, the variable speed of light (VSL) hypothesis is one possible alternative to inflation, as suggested by Andreas Albrecht and João Magueijo in 1998 [AM99]. VSL can accommodate causally produced perturbations which appear acausal at late times when the speed of light is reduced.

does not mean that there is a problem with inflation. The fact that there is intense work on modified gravity does not mean that general relativity is in trouble.

Jumping ahead to the near present, dark energy seems to be taking over our Universe again, leading to accelerated expansion. Late-accelerated expansion is virtually undisputed. Its main support comes from supernovae type-Ia and BAO observations, the resolution of the age problem of the Universe and the fact that the observable Universe is flat but filled with matter only by about 30%. Many cosmologists resisted this unexpected finding for almost a decade, hoping that it would go away due to "systematics" in observations. But we have been forced to concede.

The easiest way to explain away the observed dark energy is by assuming the existence of a positive cosmological constant Λ. This introduces no extra degrees of freedom and no new physics, as such a term is already part of the general relativity package. This proposal is called ΛCDM and has been regarded as 'the concordance model' of cosmology. However, despite ΛCDM being so convenient with astronomers, cosmologists (especially theoretical cosmologists like myself) are not at all happy with it. As we have explained, this is because Λ must be fine-tuned by about a hundred and twenty orders of magnitude to match the observations. Thus, many cosmologists decided to put the cosmological constant back in its drawer and consider that the late dark energy is due to some other dark energy source, similar to inflation. If this is modelled like a scalar field, it is called quintessence, the fifth element after baryonic matter, cold dark matter (CDM), photons and neutrinos.

Rendering late dark energy dynamical introduces a new tuning problem, that of its initial conditions. As we have discussed, one way around this issue was to consider systems which exhibit attractor behaviour, such that their eventual evolution is the same for multiple initial conditions. If the attractor is such that the dark energy substance is led to eventual domination, then the attractor is called a tracker. Unfortunately, for the simplest proposals, trackers cannot seem to satisfy observations any more, now that the observational bounds on late dark energy have tightened. We are forced to consider 'freezing' or 'thawing' models, whose success however depends strongly on initial conditions. One way out of this problem is to link late with early dark energy, allowing the initial conditions for late dark energy to be determined by inflation, which has its own attractor behaviour. Modelling dark energy as a scalar field, this quintessential inflation is severely constrained by both inflation and quintessence physics. Additionally, it faces the challenge of treating early and late dark energy, whose densities differ by more than a hundred orders of magnitude, in a common, hopefully not much fine-tuned, theoretical framework.

Forthcoming observations are expected to constrain late dark energy even more. It would be really interesting if ΛCDM was disproved and late dark energy had to be dynamical. What if future data suggest that late dark energy is phantom? As we have discussed, a phantom scalar field is pathological because it suffers from a ghost instability. What kind of dark energy would this be then? Regarding early dark energy, meaning inflation, future observations might discover the primordial tensor background. This is a definite prediction of inflation, but so were the CMB acoustic peaks. The question of whether cosmic inflation will lead to a Nobel prize, even in this case, is still open, unlike the case of late dark energy, whose discovery resulted in a Nobel prize in 2011.

Modified gravity is becoming a popular alternative to scalar fields to model both early and late dark energy. However, modified gravity models are hopelessly speculative at the moment. In contrast, scalar fields are possible to consider in the context of effective quantum field theory, a framework which is verified every day in accelerators around the world. Some modest extensions of general relativity, which are behind R^2 or nonminimal inflation (e.g., Higgs inflation), are reasonable to consider. Late dark energy, however, requires ad hoc constructions which are not particularly inviting and, in many cases, suffer from the same degree of fine-tuning as ΛCDM. This is why I have chosen not to delve into modified gravity as late dark energy in this book. There is also a mathematical degeneracy between modified gravity and scalar fields at the level of background dynamics. Indeed, as we have discussed, a conformal transformation can take us from a modified gravity Jordan frame to a scalar field theory in the Einstein frame. This degeneracy can in principle be broken at the level of perturbations, which correspond to the growth of structure, that proceeds differently in the two pictures. However, the data are not precise enough at the moment to distinguish between the two. Maybe this will change in the future and vindicate or condemn modified gravity proposals for late dark energy. Additionally, advances in the theories of modified gravity may suggest compelling avenues for modelling late dark energy that way. But we are not there yet.

Finally, once in a while we have touched upon philosophical matters, especially when talking about spacetime foam, eternal inflation, the beginning and the end of time. It is impossible not to contemplate existential questions, e.g., about causality or the nature of reality, given the subject matter. I have tried to keep myself on the ground and only hint at speculative directions, which beckon cosmologists to explore. However, it seems to me that dark energy holds the key to the origin and the ultimate fate of our Universe. As such, its study is directly linked with the fundamental reasons human beings have been investigating the world around them since the dawn of our existence, making philosophical contemplation unavoidable.

A Taste of General Relativity

A.1 The dynamics of the Universe...................... 257
A.2 The cosmological constant 259

A.1 The dynamics of the Universe

This is not a book on general relativity (GR) and we will not review this beautiful theory here, not even briefly. The reader is invited to look into other sources to develop an understanding of the theory. Here we will only present a few elements of GR directly related with cosmology.

We are based on the FLRW metric as given by Eq. (2.51) repeated here:

$$ds^2 = -(c\,dt)^2 + a^2(t)\left[\frac{(dr)^2}{1-kr^2} + r^2(d\theta)^2 + r^2\sin^2\theta(d\phi)^2\right].\tag{A.1}$$

This metric contains all the information of the geometry of a homogeneous and isotropic space-time and GR reveals the dynamics of such a spacetime. How does the theory do this? Well, the recipe is outlined as follows:

First we start with the metric of spacetime. In general, a patch of spacetime with a notion of distance can be described by a metric $g_{\mu\nu}$ corresponding to the line element:

$$ds^2 = \sum_\mu \sum_\nu g_{\mu\nu}\,dx^\mu\,dx^\nu\,,\tag{A.2}$$

where $\mu,\nu = 0,1,2,3$, and x^μ are the four spacetime coordinates (one temporal and three spatial), with x^0 the temporal coordinate and x^1, x^2, x^3 the spatial ones, by convention.[*] The metric is symmetric $g_{\mu\nu} = g_{\nu\mu}$ because spacetime is parity invariant, meaning that the mirror image of a given spacetime is also acceptable as a solution in GR. This suggests that, from the total of 16 components of the 4×4 matrix of the metric, only 10 are actually independent.

The metric in Eq. (A.1) is diagonal with nonzero components

$$g_{00} = -1\,,\quad g_{11} = a^2/(1-kr^2)\,,\quad g_{22} = a^2r^2 \quad\text{and}\quad g_{33} = a^2r^2\sin^2\theta\,.\tag{A.3}$$

From the metric we calculate the Christoffel symbols $\Gamma^\rho_{\mu\nu}$. They are determined by the metric and its derivatives as

$$\Gamma^\rho_{\mu\nu} = \frac{1}{2}\sum_\sigma g^{\rho\sigma}[\partial_\mu g_{\sigma\nu} + \partial_\nu g_{\sigma\mu} - \partial_\sigma g_{\mu\nu}]\,,\tag{A.4}$$

---

[*]Einstein summation is not assumed to avoid confusion.

where $\partial_\mu \equiv \partial/\partial x^\mu$ and $g^{\mu\nu}$ is the inverse metric defined as $\sum_\rho g_{\mu\rho} g^{\rho\nu} = \delta^\nu_\mu$, where δ^ν_μ is the Kronecker's delta, for which

$$\delta^\nu_\mu = \begin{cases} 1 & \text{if } \mu = \nu \\ 0 & \text{if } \mu \neq \nu \end{cases}. \tag{A.5}$$

Thus, Kronecker's delta corresponds really to the identity matrix and the inverse metric is the inverse 4×4 matrix of the metric. Mirroring the symmetry of the metric, the Christoffel symbols are symmetric in the lower (covariant) indexes $\Gamma^\rho_{\mu\nu} = \Gamma^\rho_{\nu\mu}$. Now, for the FLRW metric in Eq. (A.1) the nonzero Christoffel symbols are:

$$\Gamma^0_{11} = c^{-1} a\dot{a}/(1 - kr^2) \; ; \qquad \Gamma^1_{11} = kr/(1 - kr^2) \; ; \qquad \Gamma^2_{33} = -\sin\theta \, \cos\theta$$

$$\Gamma^0_{22} = c^{-1} r^2 a\dot{a} \; ; \qquad \Gamma^1_{22} = -r(1 - kr^2) \; ; \qquad \Gamma^3_{23} = \cot\theta$$

$$\Gamma^0_{33} = c^{-1} r^2 a\dot{a} \sin^2\theta \; ; \qquad \Gamma^1_{33} = -r(1 - kr^2)\sin^2\theta \; ; \tag{A.6}$$

$$\Gamma^1_{01} = \Gamma^2_{02} = \Gamma^3_{03} = c^{-1}\dot{a}/a \; ; \quad \Gamma^2_{12} = \Gamma^3_{13} = 1/r \, .$$

From the Christoffel symbols, we can obtain the components of the Ricci tensor, which are given by

$$R_{\mu\nu} = \sum_\rho \partial_\rho \Gamma^\rho_{\mu\nu} - \partial_\nu \sum_\rho \Gamma^\rho_{\mu\rho} + \sum_\rho \sum_\sigma \Gamma^\rho_{\mu\nu} \Gamma^\sigma_{\rho\sigma} - \sum_\rho \sum_\sigma \Gamma^\rho_{\mu\sigma} \Gamma^\sigma_{\nu\rho} \, . \tag{A.7}$$

As with the metric, the Ricci tensor is also symmetric $R_{\mu\nu} = R_{\nu\mu}$, which means that it has 10 independent components. For the FLRW metric in Eq. (A.1) the non-vanishing components of the Ricci tensor are

$$R_{00} = -3\frac{\ddot{a}}{a} \quad \text{and} \quad R_{ii} = \frac{a\ddot{a} + 2\dot{a}^2 + 2kc^2}{a^2} g_{ii} \, , \tag{A.8}$$

where $i = 1, 2, 3$ and the components R_{11}, R_{22}, R_{33} are determined by the components of the metric g_{11}, g_{22}, g_{33} respectively. The final piece of the puzzle is the Ricci scalar, also called the scalar curvature, which is defined as $R \equiv \sum_\mu \sum_\nu g^{\mu\nu} R_{\mu\nu}$. Using the above, the Ricci scalar for the FLRW metric is (cf. Eq. (6.9))

$$R = 6\frac{a\ddot{a} + \dot{a}^2 + kc^2}{a^2} = 6\left(\dot{H} + 2H^2 + \frac{kc^2}{a^2}\right) \, . \tag{A.9}$$

All the above elements are now introduced in the Einstein equations

$$R_{\mu\nu} - \frac{1}{2}g_{\mu\nu}R = \frac{8\pi G}{c^2} T_{\mu\nu} \, , \tag{A.10}$$

which are 10 independent equations, corresponding to the independent components of $g_{\mu\nu}$ and $R_{\mu\nu}$. In the above, we have discussed the left-hand side quantities only, which have to do with the geometry of spacetime and are ultimately determined by the metric, which is a kind of potential for the gravitational field. The right-hand side of the Einstein equations is determined by the the material filling spacetime, i.e., its content. The Einstein equations express a balance between content and geometry, which was put eloquently by John Archibald Wheeler as "Spacetime tells matter how to move and matter tells spacetime how to curve."

The content of spacetime is quantified through the energy-momentum tensor $T_{\mu\nu}$. You guessed it, $T_{\mu\nu}$ is also symmetric; it has to be since both $g_{\mu\nu}$ and $R_{\mu\nu}$ are. For cosmology we model the Universe content as a perfect fluid (whose particles can be the galaxies, for example). The energy-momentum tensor of a perfect fluid with density ρ and pressure p is

$$T_{\mu\nu} = (\rho c^2 + p)u_\mu u_\nu + p g_{\mu\nu} \, , \tag{A.11}$$

where u_μ is the four-velocity. Considering comoving observers who follow the Hubble flow in the expanding Universe, the four-velocity is $u_\mu = (1, 0, 0, 0)$, i.e., it has no spatial components.* For a homogeneous and isotropic Universe, we also have $\rho = \rho(t)$ and $p = p(t)$. Considering the above energy-momentum tensor in an FLRW Universe described by the metric in Eq. (A.1) implies that $T_{\mu\nu}$ is also diagonal with nonzero components

$$T_{00} = \rho c^2 \quad \text{and} \quad T_{ii} = p\, g_{ii} \,. \tag{A.12}$$

Now we may derive the evolution equations of an FLRW Universe. Considering the 00-component of the Einstein equations we readily obtain the Friedman equation in Eq. (2.10). Considering the ii-component of the Einstein equations we obtain the Raychadhuri equation

$$2\frac{\ddot{a}}{a} + H^2 + \frac{kc^2}{a^2} = -8\pi G \frac{p}{c^2} \,. \tag{A.13}$$

Combining this with the Friedman equation we arrive at the acceleration equation in Eq. (2.21). Further, by combining the acceleration equation with the Friedman equation again, it is straightforward to derive the continuity equation in Eq. (2.12). This is the reverse order compared to our discussion in Chapter 2, where we derived the acceleration equation by combining the Friedman and the continuity equations. In fact, the continuity equation in Eq.(2.12) is obtained independently by requiring the (covariant) conservation of the energy-momentum tensor $T_{\mu\nu}$, so that the fact that one can get it by combining the Einstein equations (i.e., the Friedman and acceleration equations in this case) is a nice consistency test of GR.

A.2   The cosmological constant

In the previous subsection, we ignored the cosmological constant. However, the symmetries of GR admit a contribution to the Einstein equations of the form:

$$R_{\mu\nu} - \frac{1}{2}g_{\mu\nu}R + g_{\mu\nu}c^2\Lambda = \frac{8\pi G}{c^2}\,T_{\mu\nu}\,, \tag{A.14}$$

where Λ is an unspecified constant. Then, following the procedure outlined in the previous section, it is straightforward to derive Eqs. (2.28) and (2.29).

As we have explained in Sec. 2.6.2, there is a dual origin of the cosmological constant; it may be a dimensionful constant Λ_C introduced by the theory of GR, similar to Newton's gravitational constant G, which appears in the Einstein equations, or it can have quantum-mechanical origin, Λ_Q, corresponding to the zero-point energy introduced by all quantum fields. These contributions can be distinct, so one could write the Einstein equations as

$$R_{\mu\nu} - \frac{1}{2}g_{\mu\nu}R + g_{\mu\nu}c^2\Lambda_C = \frac{8\pi G}{c^2}\,T_{\mu\nu} - g_{\mu\nu}c^2\Lambda_Q\,, \tag{A.15}$$

considering the classical Λ_C as a feature of the geometry (so in the left-hand side) and the quantum Λ_Q as inherently related (introduced) by the content (so in the right-hand side).

The cosmological constant problem is why these two contributions seem to cancel each-other $\Lambda_C + \Lambda_Q \approx 0$ at an exponentially high precision. ΛCDM cosmology, however, assumes that this cancellation is not exact and that there is a net cosmological constant $|\Lambda_C + \Lambda_Q| \sim 10^{-120}\, c^5/\hbar G$. Considering that the natural values of $|\Lambda_C|$ and $|\Lambda_Q|$ are indeed $\sim G\rho_P = c^5/\hbar G$, we see that this approximate cancellation must be incredibly fine-tuned, by 120 orders of magnitude!

---

*This means that the particles of the fluid (e.g., the galaxies) are not moving through space but only carried apart by the Hubble flow, i.e., peculiar velocities are ignored.

B

Correlators of the Curvature Perturbation

Here we calculate the two-point correlators of the curvature perturbation. First, we analyse the latter in Fourier components, which are complex functions given by Eq. (5.24) repeated here

$$\zeta_{\mathbf{k}} \equiv \zeta(\mathbf{k}) = \int \zeta(\mathbf{x})\, e^{i\mathbf{k}\, \cdot\, \mathbf{x}} d^3x \,. \tag{B.1}$$

Using this, the two-point correlator in momentum space is

$$
\begin{aligned}
\langle \zeta_{\mathbf{k}} \zeta_{\mathbf{k}'} \rangle &= \int d^3x\, d^3x' \langle \zeta(\mathbf{x})\zeta(\mathbf{x}') \rangle e^{i(\mathbf{k}\cdot\mathbf{x}+\mathbf{k}'\cdot\mathbf{x}')} \\
&= \int d^3x\, d^3x' \langle \zeta(\mathbf{x})\zeta(\mathbf{x}') \rangle e^{i(\mathbf{k}+\mathbf{k}')\cdot\mathbf{x}} e^{i\mathbf{k}'\cdot(\mathbf{x}'-\mathbf{x})} \\
&= (2\pi)^3 \delta^{(3)}(\mathbf{k}+\mathbf{k}') \int d^3x' \langle \zeta(\mathbf{x})\zeta(\mathbf{x}') \rangle e^{i\mathbf{k}'\cdot(\mathbf{x}'-\mathbf{x})} \,,
\end{aligned} \tag{B.2}
$$

where we used $\mathbf{k} \cdot \mathbf{x} + \mathbf{k}' \cdot \mathbf{x}' = (\mathbf{k} + \mathbf{k}')\cdot\mathbf{x} + \mathbf{k}'\cdot(\mathbf{x}' - \mathbf{x})$ and the definition of the 3-D Dirac delta function

$$\delta^{(3)}(\mathbf{q}) = \frac{1}{(2\pi)^3} \int e^{i\mathbf{q}\, \cdot\, \mathbf{x}} d^3x \,, \tag{B.3}$$

with $\mathbf{q} = \mathbf{k} + \mathbf{k}'$. Because of translation invariance we can set $\mathbf{x} = 0$. Then the above gives

$$\langle \zeta_{\mathbf{k}} \zeta_{\mathbf{k}'} \rangle = (2\pi)^3 \delta^{(3)}(\mathbf{k} + \mathbf{k}') P_\zeta(k) \,, \tag{B.4}$$

where we have defined

$$P_\zeta(k) = P_\zeta(\mathbf{k}) = \int d^3x \langle \zeta(0)\zeta(\mathbf{x}) \rangle e^{i\mathbf{k}\, \cdot\, \mathbf{x}} \tag{B.5}$$

and we used $P_\zeta(k) = P_\zeta(\mathbf{k})$ because of rotational invariance and defined $k \equiv |\mathbf{k}|$. We also removed the primes: $\mathbf{x}' \to \mathbf{x}$.

Using the above, we can also find the two-point spatial correlator. From Eq. (5.26), repeated here, we have

$$\zeta(\mathbf{x}) = \frac{1}{(2\pi)^3} \int \zeta_{\mathbf{k}} e^{-i\mathbf{k}\, \cdot\, \mathbf{x}} d^3k \,. \tag{B.6}$$

Thus, we find

$$
\begin{aligned}
\langle \zeta(\boldsymbol{x})\zeta(\boldsymbol{x}')\rangle
&= \frac{1}{(2\pi)^6} \int d^3k\, d^3k'\, \langle \zeta_{\mathbf{k}}\zeta_{\mathbf{k}}\rangle e^{-i(\boldsymbol{k}\cdot\boldsymbol{x}+\boldsymbol{k}'\cdot\boldsymbol{x}')} \\
&= \frac{1}{(2\pi)^3} \int d^3k\, d^3k'\, \delta^{(3)}(\boldsymbol{k}+\boldsymbol{k}') P_\zeta(k) e^{-i(\boldsymbol{k}\cdot\boldsymbol{x}+\boldsymbol{k}'\cdot\boldsymbol{x}')} \\
&= \frac{1}{(2\pi)^3} \int d^3k\, P_\zeta(k) e^{-i\boldsymbol{k}\cdot(\boldsymbol{x}-\boldsymbol{x}')} ,
\end{aligned}
\tag{B.7}
$$

where we used $\boldsymbol{k}\cdot\boldsymbol{x}+\boldsymbol{k}'\cdot\boldsymbol{x}' = (\boldsymbol{k}+\boldsymbol{k}')\cdot\boldsymbol{x} - \boldsymbol{k}'\cdot(\boldsymbol{x}-\boldsymbol{x}')$ and Eq. (B.3) with $\boldsymbol{q} = \boldsymbol{k}+\boldsymbol{k}'$. Setting $\boldsymbol{x}' = \boldsymbol{x}$ we get

$$
\langle \zeta^2(\boldsymbol{x})\rangle = \frac{1}{(2\pi)^3} \int P_\zeta(k) d^3k = \int_0^\infty \mathcal{P}_\zeta(k) \frac{dk}{k} ,
\tag{B.8}
$$

which is Eq. (5.27) and we have defined the *spectrum* of the curvature perturbation as

$$
\mathcal{P}_\zeta(k) \equiv \frac{k^3}{2\pi^2} P_\zeta(k) .
\tag{B.9}
$$

Thus, in view of Eqs. (B.4) and (B.9) we find

$$
\langle \zeta_{\mathbf{k}}\zeta_{\mathbf{k}'}\rangle = (2\pi)^3 \delta^{(3)}(\boldsymbol{k}+\boldsymbol{k}') \frac{2\pi^2}{k^3} \mathcal{P}_\zeta(k) ,
\tag{B.10}
$$

which is Eq. (5.25).

From Eq. (B.1) we find the reality condition $\zeta_{\mathbf{k}}^* = \zeta_{-\mathbf{k}}$. This is the only correlation between the Fourier components of the curvature perturbation. Thus, the curvature perturbation $\zeta(\boldsymbol{x})$ is an infinite sum of uncorrelated quantities. As such, statistics says that $\zeta(\boldsymbol{x})$ has a Gaussian probability distribution. Using the reality condition we find

$$
\langle |\zeta_{\mathbf{k}}|^2 \rangle = \langle \zeta_{\mathbf{k}}\zeta_{\mathbf{k}}^* \rangle = \langle \zeta_{\mathbf{k}}\zeta_{-\mathbf{k}} \rangle = (2\pi)^3 \delta^{(3)}(0) P_\zeta(k) = P_\zeta(k) = \frac{2\pi^2}{k^3} \mathcal{P}_\zeta(k) ,
\tag{B.11}
$$

which is Eq. (5.28).

With Gaussian statistics, all odd-order higher correlators are exactly zero. However, the possibility is entertained that the statistics of the curvature perturbation are not exactly Gaussian. The level of non-Gaussianity is explored by considering the three-point correlator, which is written as

$$
\langle \zeta_{\mathbf{k_1}}\zeta_{\mathbf{k_2}}\zeta_{\mathbf{k_3}} \rangle = (2\pi)^3 \delta^{(3)}(\boldsymbol{k}_1+\boldsymbol{k}_2+\boldsymbol{k}_3) B_\zeta(k_1, k_2, k_3) ,
\tag{B.12}
$$

where $B_\zeta(k_1, k_2, k_3)$ is called the *bispectrum* and is quantified as

$$
B_\zeta(k_1, k_2, k_3) = \frac{6}{5} f_{\mathrm{NL}}(k_1, k_2, k_3)[P_\zeta(k_1)P_\zeta(k_2) + P_\zeta(k_2)P_\zeta(k_3) + P_\zeta(k_3)P_\zeta(k_1)] .
\tag{B.13}
$$

The quantity f_{NL} is called the non-linearity parameter and depends on the configuration of the $\boldsymbol{k}_1$, $\boldsymbol{k}_2$ and $\boldsymbol{k}_3$ momentum vectors. Stringent observational constraints on the value of f_{NL} confirm that the curvature perturbation is highly Gaussian with f_{NL} constrained almost to the limit of observability.

Light Scalar Field Superhorizon Spectrum

Here we briefly outline the procedure to calculate the spectrum of the superhorizon perturbations of a light scalar field during quasi-de Sitter inflation.

The equation of motion of a scalar field in a flat FLRW spacetime is given by Eq. (7.7) repeated here

$$\ddot{\phi} + 3H\dot{\phi} - a^{-2}\nabla^2\phi + V'(\phi) = 0 \,, \tag{C.1}$$

where $a^{-2}\nabla^2$ is the Laplacian in physical coordinates.

We consider a perturbation $\delta\phi(\boldsymbol{x}, t)$ over the homogeneous scalar field $\bar{\phi}(t)$ of the form $\phi(\boldsymbol{x}, t) = \bar{\phi}(t) + \delta\phi(\boldsymbol{x}, t)$ with $|\delta\phi/\bar{\phi}| \ll 1$. Inserting this into Eq. (C.1) and noting that $\bar{\phi}(t)$ satisfies the Klein-Gordon equation (6.8) we obtain

$$\ddot{\delta\phi} + 3H\dot{\delta\phi} - a^{-2}\nabla^2\delta\phi + m^2\delta\phi = 0 \,, \tag{C.2}$$

where we also Taylor approximated $V'(\bar{\phi} + \delta\phi) \simeq V'(\bar{\phi}) + V''(\bar{\phi})\delta\phi$ with $m^2 = V''(\bar{\phi})$. Note that when considering superhorizon scales the Laplacian term is exponentially suppressed (because $a(t)$ grows enormously) so we end up with Eq. (6.72).

Next, we Fourier expand the scalar field in momentum space, using Eq. (6.69). Then, Eq. (C.2) becomes[*]

$$\ddot{\delta\phi}_{\mathbf{k}} + 3H\dot{\delta\phi}_{\mathbf{k}} + [m^2 + (k/a)^2]\delta\phi_{\mathbf{k}} = 0 \,, \tag{C.3}$$

where the comoving momentum is $k \equiv |\boldsymbol{k}|$. Further, we promote the Fourier components of the field perturbations into quantum mechanical operators (second quantisation) so that

$$\delta\hat{\phi}(\boldsymbol{x}, t) = \frac{1}{(2\pi)^3} \int \left[\delta\varphi_{\mathbf{k}}(t)\hat{a}_{\mathbf{k}} + \delta\varphi_{\mathbf{k}}^*(t)\hat{a}_{\mathbf{k}}^\dagger \right] e^{i\boldsymbol{k}\cdot\boldsymbol{x}} d^3k \,, \tag{C.4}$$

where $\hat{a}_{\mathbf{k}}^\dagger$ and $\hat{a}_{\mathbf{k}}$ are creator and annihilator operators respectively ($\hat{a}_{\mathbf{k}}^\dagger$ creates particles and $\hat{a}_{\mathbf{k}}$ destroys particles), which satisfy the algebra

$$[\hat{a}_{\mathbf{k}}, \hat{a}_{\mathbf{k}'}^\dagger] = (2\pi)^3\delta^{(3)}(\boldsymbol{k} - \boldsymbol{k}') \quad \text{and} \quad [\hat{a}_{\mathbf{k}}, \hat{a}_{\mathbf{k}'}] = 0 \; \& \; [\hat{a}_{\mathbf{k}}^\dagger, \hat{a}_{\mathbf{k}'}^\dagger] = 0 \,. \tag{C.5}$$

---

[*]Recall that $a^{-1}\nabla e^{i\boldsymbol{k}\cdot\boldsymbol{x}} = i(k/a)e^{i\boldsymbol{k}\cdot\boldsymbol{x}}$.

The mode functions $\delta\varphi_{\mathbf{k}}$ satisfy Eq. (C.3). We now assume quasi-de Sitter inflation with $H =$ constant and $a \propto e^{Ht}$. In this case, the generic solution of this equation is

$$\delta\varphi_{\mathbf{k}} = \left(\frac{k}{aH}\right)^{3/2} \left[c_1 J_\nu\left(\frac{k}{aH}\right) + c_2 J_{-\nu}\left(\frac{k}{aH}\right)\right], \tag{C.6}$$

where c_1 and c_2 are complex integration constants and $J_\nu(z)$ denotes the Bessel function of the first kind, of order

$$\nu \equiv \frac{3}{2}\sqrt{1 - \left(\frac{2m}{3H}\right)^2}. \tag{C.7}$$

The Bessel function interpolates between a harmonic function and a power-law function. Deep inside the horizon $\delta\varphi_{\mathbf{k}}$ is oscillating harmonically. The oscillating field corresponds to particles, which may be identified with the virtual particles which appear in the vacuum. This is because deep inside the horizon the curvature of space is not felt. Thus, we may identify the oscillating $\delta\varphi_{\mathbf{k}}$ mode with the Bunch-Davies vacuum [BD78] given by*

$$\delta\varphi_{\mathbf{k}} = \frac{H}{\sqrt{2k^3}} \left(i + \frac{k}{aH}\right) e^{ik/aH}. \tag{C.8}$$

This is the boundary condition, which serves to evaluate the integration constants in Eq. (C.6). In order to come into contact with the Bunch-Davis boundary condition, we need to consider early times, when the perturbation is deep inside the horizon. This means that the wavelength of the fluctuation is $a/k \ll H^{-1}$, meaning we need to consider the limit $k/aH \to +\infty$. In this limit, the Bessel functions approximate harmonic functions

$$\lim_{z\to+\infty} J_{\pm\nu}(z) = \sqrt{\frac{2}{\pi z}} \cos\left[z - \left(\frac{1}{2} \pm \nu\right)\frac{\pi}{2}\right]. \tag{C.9}$$

Using this expression in the solution in Eq. (C.6) and after a little algebra we obtain[†]

$$\lim_{k/aH\to+\infty} \delta\varphi_{\mathbf{k}} = \frac{1}{\sqrt{2\pi}} \left(\frac{k}{aH}\right) \left[\left(c_1 e^{-i\pi\nu/2} + c_2 e^{i\pi\nu/2}\right)\frac{1-i}{\sqrt{2}} e^{ik/aH}\right.$$
$$\left. + \left(c_1 e^{i\pi\nu/2} + c_2 e^{-i\pi\nu/2}\right)\frac{1+i}{\sqrt{2}} e^{-ik/aH}\right]. \tag{C.10}$$

Comparing the above with Eq. (C.8) it is evident that the coefficient of $e^{-ik/aH}$ must be zero, which implies the relation $c_2 = -c_1 e^{i\pi\nu}$. Using this, Eq. (C.10) reduces to

$$\lim_{k/aH\to+\infty} \delta\varphi_{\mathbf{k}} = \frac{1-i}{2\sqrt{\pi}} \left(\frac{k}{aH}\right) e^{-i\pi\nu/2}\left(1 - e^{i2\pi\nu}\right) c_1 e^{ik/aH}. \tag{C.11}$$

---

*Eq. (C.8) can be also written as

$$\delta\varphi_{\mathbf{k}} = \frac{e^{-ik\tau}}{a\sqrt{2k}} \left(1 - \frac{i}{k\tau}\right)$$

if we consider conformal time $dt = ad\tau$, in which case the scale factor during quasi-de Sitter inflation is $a = -1/H\tau$ with $-\infty < \tau < -1/H$. In the limit $|k\tau| \to \infty$, we are deep inside the horizon and the expansion may be ignored, so we can set $a = 1$ and $t = \tau$. In this limit, the vacuum becomes $\delta\varphi_{\mathbf{k}} = e^{-ikt}/\sqrt{2k}$. This is the solution of the free-field mode equation in flat (Minkowski) spacetime $\ddot{\delta\varphi}_{\mathbf{k}} + k^2\delta\varphi_{\mathbf{k}} = 0$, which is obtained by Eq. (C.3), when $H, m \to 0$ and $a = 1$.

[†]Recall that $\cos x = \frac{1}{2}(e^{ix} + e^{-ix})$ and also $e^{ix} = \cos x + i\sin x$ so that $e^{\pm i\pi/4} = \frac{1\pm i}{\sqrt{2}}$.

Equating the above with Eq. (C.8) we find c_1 and then $c_2 = -c_1 e^{i\pi\nu}$. Using these values Eq. (C.6) can now be written as

$$\delta\varphi_{\mathbf{k}} = a^{-3/2}\sqrt{\frac{\pi}{2H}}\frac{(1+i)e^{i\pi\nu/2}}{1-e^{i2\pi\nu}}\left[J_\nu\left(\frac{k}{aH}\right) - e^{i\pi\nu}J_{-\nu}\left(\frac{k}{aH}\right)\right]. \qquad (C.12)$$

We now follow the mode evolution as time passes and it is taken out of the horizon by the expansion. To find the ultimate value of $\delta\varphi_{\mathbf{k}}$ in the superhorizon limit we consider that the wavelength of the fluctuation is $a/k \gg H^{-1}$, meaning we need to consider the limit $k/aH \to 0^+$. In this limit the Bessel functions approximate power-law functions

$$\lim_{z\to 0} J_{\pm\nu}(z) = \frac{1}{\Gamma(1\pm\nu)}\left(\frac{z}{2}\right)^{\pm\nu}, \qquad (C.13)$$

where $\Gamma(x)$ is the gamma function, which is a generalisation of the factorial, in that $\Gamma(n) = (n-1)!$ for positive integer numbers, $n > 0$.[*] For half-integers, we have $\Gamma(\frac{1}{2}+n) = \frac{(2n)!}{4^n n!}\sqrt{\pi}$ and $\Gamma(\frac{1}{2}-n) = \frac{(-4)^n n!}{(2n)!}\sqrt{\pi}$. For example, if our field is exactly massless then Eq. (C.7) suggests that $\nu = \frac{3}{2}$. Therefore. $\Gamma(1\pm\nu) = \Gamma(5/2) = \frac{3}{4}\sqrt{\pi}$ or $\Gamma(-1/2) = -2\sqrt{\pi}$. Using Eq. (C.13), the expression in Eq. (C.12) becomes

$$\lim_{k/aH\to 0^+}\delta\varphi_{\mathbf{k}} = -\frac{a^{-3/2}}{\Gamma(1-\nu)}\sqrt{\frac{\pi}{2H}}\frac{(1+i)e^{i3\pi\nu/2}}{1-e^{i2\pi\nu}}\left(\frac{k}{2aH}\right)^{-\nu}, \qquad (C.14)$$

where we kept only the growing mode because $(k/2aH)^\nu \ll 1$ in this limit (recall that $\nu > 0$).

We are now able to calculate the power spectrum as (cf. Eq. (6.70))

$$\mathcal{P}_{\delta\phi}(k) = \frac{k^3}{2\pi^2}|\delta\varphi_{\mathbf{k}}|^2 = \frac{1}{|\Gamma(1-\nu)|^2}\frac{2/\pi}{1-\cos(2\pi\nu)}\left(\frac{k}{2aH}\right)^{3-2\nu}H^2. \qquad (C.15)$$

In the limit of a light field $m \ll H$, Eq (C.7) suggests $\nu \simeq \frac{3}{2} - \eta$, where $\eta \simeq \frac{1}{3}(m/H)^2$ (cf. Eq. (6.25)). In this limit, Eq. (C.15) becomes

$$\mathcal{P}_{\delta\phi} = \left(\frac{H}{2\pi}\right)^2\left(\frac{k}{2aH}\right)^{2\eta}. \qquad (C.16)$$

In the limit $\eta \to 0$ the above reduces to Eq. (6.71). Notice that $(\delta\phi)^2 \sim \mathcal{P}_{\delta\phi} \propto a^{-2\eta}$ agrees with the finding in Eq. (6.74), where $\delta\phi \propto a^{-\eta}$.[†] Also, the exponent 2η is the origin of the same term in Eq. (6.83) for the spectal index n_s. The ε contribution in that equation is absent here because we have assumed $H = $ constant so that Eq. (6.10) gives $\epsilon = 0$ and $\varepsilon \simeq \epsilon$ in slow-roll.

There is an important issue regarding the origin of vacuum fluctuations. The cosmological scales start of well within the horizon to the point that their wavelength is actually smaller than the Planck length m_P^{-1}. It is possible that quantum gravity may influence the Bunch-Davis vacuum state in Eq. (C.8). Such *transplanckian effects* might result in *enhanced vacuum states*, which could alter the generation of perturbations from inflation. The above transplanckian problem is one of the open issues in inflationary physics [MB01].

---

[*] For $n \leq 0$ we have $1/\Gamma(n) = 0$.
[†] The mode function $\delta\varphi_{\mathbf{k}}$ should not be confused with the field perturbation $\delta\phi$. Even dimensionally they are different since $[\delta\phi] = E$ while $[\delta\varphi_{\mathbf{k}}] = E^{-1/2}$ as evident from, e.g., Eq. (C.8).

D

Field Equation and Energy-Momentum of a Free Scalar Field

Here we briefly outline the procedure to calculate the equation of motion and the energy-momentum tensor in the theory of a free minimally-coupled scalar field. Some knowledge of general relativity is assumed.

The Lagrangian density of the theory is given in Eq. (7.6), which can be written as[*]

$$\mathcal{L} = -\frac{1}{2} \sum_\rho \sum_\sigma g^{\rho\sigma} \partial_\rho \phi \partial_\sigma \phi - V(\phi), \tag{D.1}$$

where $g^{\rho\sigma}$ is the inverse metric (see Appendix A) and $\partial_\mu \equiv \partial/\partial x^\mu$, where x^μ stands for the spacetime coordinates with $\mu = 0, 1, 2, 3$, for example $(x^0, x^2, x^2, x^3) = (t, x, y, z)$ if the coordinate system is Cartesian.[†]

We now employ Eq. (7.5) noting that, for a scalar field $\nabla_\mu \phi = \partial_\mu \phi$. We find

$$\sum_\mu \nabla_\mu \left[\frac{\partial \mathcal{L}}{\partial(\partial_\mu \phi)} \right] = \frac{1}{\sqrt{-g}} \sum_\mu \partial_\mu \left[\sqrt{-g} \frac{\partial \mathcal{L}}{\partial(\partial_\mu \phi)} \right] = \sum_\mu [\partial_\mu + (\partial_\mu \ln \sqrt{-g})] \frac{\partial \mathcal{L}}{\partial(\partial_\mu \phi)}, \tag{D.2}$$

where g is the determinant of the metric $g \equiv \det(g_{\mu\nu})$. Using Eq. (D.1), we find

$$\frac{\partial \mathcal{L}}{\partial(\partial_\mu \phi)} = -\frac{1}{2} \sum_\rho \sum_\sigma g^{\rho\sigma} (\delta^\mu_\rho \partial_\sigma \phi + \delta^\mu_\sigma \partial_\rho \phi) = -\sum_\rho g^{\mu\rho} \partial_\rho \phi, \tag{D.3}$$

where we considered δ^μ_ρ is Kronecker's delta (cf. Eq. (A.5)). Combining the above we obtain

$$\sum_\mu \nabla_\mu \left[\frac{\partial \mathcal{L}}{\partial(\partial_\mu \phi)} \right] = -\sum_\mu \left\{ [\partial_\mu + (\partial_\mu \ln \sqrt{-g})] \sum_\rho g^{\mu\rho} \partial_\rho \phi \right\}. \tag{D.4}$$

---

[*]Einstein summation is not assumed to avoid confusion.
[†]Comparing the above with Eq. (7.6) we see that $(\partial \phi)^2 = -\sum_\rho \sum_\sigma g^{\rho\sigma} \partial_\rho \phi \partial_\sigma \phi$. Note that there is a sign difference with much of the literature.

To continue, consider a flat FLRW Universe with metric $g_{\mu\nu} = \text{diag}(-1, a^2, a^2, a^2)$ [cf. Eq. (2.53)] and inverse metric $g^{\mu\nu} = \text{diag}(-1, a^{-2}, a^{-2}, a^{-2})$. Then the determinant of the metric is $g = -a^6$, such that $\sqrt{-g} = a^3$. Hence, using that the metric and its inverse are diagonal, it is straightforward that

$$\sum_\mu \nabla_\mu \left[\frac{\partial \mathcal{L}}{\partial(\partial_\mu \phi)} \right] = (\partial_t + \partial_t \ln a^3)\partial_t \phi - \sum_i (\partial_i + \partial_i \ln a^3)a^{-2}\partial_i \phi = \ddot{\phi} + 3H\dot{\phi} - a^{-2}\nabla^2 \phi, \quad (D.5)$$

where $\partial_t = \partial/\partial t$ and $\partial_i = \partial/\partial x^i$ with $i = 1, 2, 3$ (runs over spatial components only). In the above we used that $\partial_i \ln a^3 = 0$ because $a = a(t)$ and $\partial_t \ln a^3 = 3H$, with $\sum_i \partial_i^2 \phi \equiv \nabla^2 \phi$. From Eq. (D.1) we also have

$$\frac{\partial \mathcal{L}}{\partial \phi} = -\frac{\partial V}{\partial \phi}. \quad (D.6)$$

In view of Eq. (7.5), combining Eqs. (D.5) and (D.6) we obtain the equation of motion of a minimally-coupled free scalar field in a flat FLRW spacetime

$$\ddot{\phi} + 3H\dot{\phi} - a^{-2}\nabla^2 \phi + V'(\phi) = 0, \quad (D.7)$$

which is Eq. (7.7).

For the energy-momentum tensor, we have

$$T_{\mu\nu} = -\frac{2}{\sqrt{-g}} \frac{\delta S}{\delta g^{\mu\nu}} = -\frac{2}{\sqrt{-g}} \frac{\partial(\sqrt{-g}\mathcal{L})}{\partial g^{\mu\nu}} = -2\frac{\partial \mathcal{L}}{\partial g^{\mu\nu}} + g_{\mu\nu}\mathcal{L}. \quad (D.8)$$

Considering $\mathcal{L}$ as given in Eq (D.1), we find

$$\frac{\partial \mathcal{L}}{\partial g^{\mu\nu}} = -\frac{1}{2}\sum_\rho \sum_\sigma \delta_\mu^\rho \delta_\nu^\sigma \partial_\rho \phi \partial_\sigma \phi = -\frac{1}{2}\partial_\mu \phi \partial_\nu \phi. \quad (D.9)$$

Hence, Eq. (D.8) gives

$$T_{\mu\nu} = \partial_\mu \phi \partial_\nu \phi - g_{\mu\nu}\left(\frac{1}{2}\sum_\rho \sum_\sigma g^{\rho\sigma} \partial_\rho \phi \partial_\sigma \phi + V\right). \quad (D.10)$$

Using this, we obtain for the energy density

$$\rho_\phi = T_{00} = \frac{1}{2}[\dot{\phi}^2 + a^{-2}(\nabla\phi)^2] + V(\phi), \quad (D.11)$$

where $(\nabla\phi)^2 = \sum_i (\partial_i \phi)^2$ with $i, j = 1, 2, 3$ The pressure corresponds to the spatial components of $T_{\mu\nu}$. The off diagonal spatial components are simply $T_{ij} = \partial_i \phi \partial_j \phi$, where $i \neq j$. The diagonal components (associated with the principal pressures) are given by $T_{ii} = (\partial_i \phi)^2 + a^2 p_\phi$ with

$$p_\phi = \mathcal{L} = \frac{1}{2}[\dot{\phi}^2 - a^{-2}(\nabla\phi)^2] - V(\phi). \quad (D.12)$$

If the scalar field configuration is homogeneous then $\nabla\phi = 0$ and the above expressions reduce to Eq. (6.6). Note that the energy-momentum tensor of a scalar field is not diagonal if the field is inhomogeneous. As a result, an inhomogeneous scalar field cannot correspond to a perfect fluid, for which $T_\mu^\rho = \text{diag}(-\rho, p, p, p)$, where $T_\mu^\rho = \sum_\nu g^{\rho\nu} T_{\mu\nu}$.

References

1. P. A. R. Ade et al. Planck 2015 results. XX. Constraints on inflation. *Astron. Astrophys.*, 594:A20, 2016.

2. N. Aghanim et al. Planck 2018 results. VI. Cosmological parameters. 2018. e-Print: 1807.06209 [astro-ph.CO]

3. Y. Akrami et al. Planck 2018 results. X. Constraints on inflation. 2018. e-Print: 1807.06211 [astro-ph.CO]

4. Rouzbeh Allahverdi, Kari Enqvist, Juan Garcia-Bellido, and Anupam Mazumdar. Gauge invariant MSSM inflaton. *Phys. Rev. Lett.*, 97:191304, 2006.

5. Ana Achucarro, Jinn-Ouk Gong, Sjoerd Hardeman, Gonzalo A. Palma, and Subodh P. Patil. Mass hierarchies and non-decoupling in multi-scalar field dynamics. *Phys. Rev.*, D84:043502, 2011.

6. Andreas Albrecht and Joao Magueijo. A time varying speed of light as a solution to cosmological puzzles. *Phys. Rev.*, D59:043516, 1999.

7. Luca Amendola. Scaling solutions in general nonminimal coupling theories. *Phys. Rev.*, D60:043501, 1999.

8. Luca Amendola. Coupled quintessence. *Phys. Rev.*, D62:043511, 2000.

9. C. Armendariz-Picon, T. Damour, and Viatcheslav F. Mukhanov. k-inflation. *Phys. Lett.*, B458:209–218, 1999.

10. C. Armendariz-Picon, Viatcheslav F. Mukhanov, and Paul J. Steinhardt. A dynamical solution to the problem of a small cosmological constant and late time cosmic acceleration. *Phys. Rev. Lett.*, 85:4438–4441, 2000.

11. Mohsen Alishahiha, Eva Silverstein, and David Tong. DBI in the sky. *Phys. Rev.*, D70:123505, 2004.

12. K. G. Begeman, A. H. Broeils, and R. H. Sanders. Extended rotation curves of spiral galaxies: Dark haloes and modified dynamics. *Mon. Not. Roy. Astron. Soc.*, 249:523, 1991.

13. T. Barreiro, Edmund J. Copeland, and N. J. Nunes. Quintessence arising from exponential potentials. *Phys. Rev.*, D61:127301, 2000.

14. C. Brans and R. H. Dicke. Mach's principle and a relativistic theory of gravitation. *Phys. Rev.*, 124:925–935, 1961. [,142(1961)].

15. T. S. Bunch and P. C. W. Davies. Quantum Field Theory in de Sitter Space: Renormalization by Point Splitting. *Proc. Roy. Soc. Lond. A*, A360:117–134, 1978.

16. Arjun Berera. Warm inflation. *Phys. Rev. Lett.*, 75:3218–3221, 1995.

17. Lotfi Boubekeur and David. H. Lyth. Hilltop inflation. *JCAP*, 0507:010, 2005.

18. S. Burles, K. M. Nollett, and Michael S. Turner. Deuterium and Big Bang nucleosynthesis. *Nucl. Phys. A*, 663:861c–864c, 2000.

19. Fedor L. Bezrukov and Mikhail Shaposhnikov. The standard Model Higgs boson as the inflaton. *Phys. Lett.*, B659:703–706, 2008.

20. J. C. Bueno Sanchez, Konstantinos Dimopoulos, and David H. Lyth. A-term inflation and the MSSM. *JCAP*, 0701:015, 2007.

21. Matthew Colless et al. The 2dF Galaxy Redshift Survey: Spectra and redshifts. *Mon. Not. Roy. Astron. Soc.*, 328:1039, 2001.

22. R. R. Caldwell. A Phantom menace? *Phys. Lett.*, B545:23–29, 2002.

23. Sean M. Carroll. Quintessence and the rest of the world. *Phys. Rev. Lett.*, 81:3067–3070, 1998.

24. Douglas Clowe, Marusa Bradac, Anthony H. Gonzalez, Maxim Markevitch, Scott W. Randall, Christine Jones, and Dennis Zaritsky. A direct empirical proof of the existence of dark matter. *Astrophys. J.*, 648:L109–L113, 2006.

25. R. R. Caldwell, Rahul Dave, and Paul J. Steinhardt. Cosmological imprint of an energy component with general equation of state. *Phys. Rev. Lett.*, 80:1582–1585, 1998.

26. Robert R. Caldwell, Marc Kamionkowski, and Nevin N. Weinberg. Phantom energy and cosmic doomsday. *Phys. Rev. Lett.*, 91:071301, 2003.

27. Edmund J. Copeland, Andrew R Liddle, and David Wands. Exponential potentials and cosmological scaling solutions. *Phys. Rev.*, D57:4686–4690, 1998.

28. Takeshi Chiba, Takahiro Okabe, and Masahide Yamaguchi. Kinetically driven quintessence. *Phys. Rev.*, D62:023511, 2000.

29. Michel Chevallier and David Polarski. Accelerating universes with scaling dark matter. *Int. J. Mod. Phys.*, D10:213–224, 2001.

30. Neil J. Cornish, David N. Spergel, and Glenn D. Starkman. Does chaotic mixing facilitate Omega < 1 inflation? *Phys. Rev. Lett.*, 77:215–218, 1996.

31. Edmund J. Copeland, M. Sami, and Shinji Tsujikawa. Dynamics of dark energy. *Int. J. Mod. Phys.*, D15:1753–1936, 2006.

32. Konstantinos Dimopoulos and Leonora Donaldson-Wood. Warm quintessential inflation. *Phys. Lett.*, B796:26–31, 2019.

33. Gia Dvali, Andrei Gruzinov, and Matias Zaldarriaga. A new mechanism for generating density perturbations from inflation. *Phys. Rev.*, D69:023505, 2004.

34. Thibault Damour and Viatcheslav F. Mukhanov. Inflation without slow roll. *Phys. Rev. Lett.*, 80:3440–3443, 1998.

35. Konstantinos Dimopoulos and Charlotte Owen. Quintessential Inflation with α-attractors. *JCAP*, 1706(06):027, 2017.

36. Konstantinos Dimopoulos, Charlotte Owen, and Antonio Racioppi. Loop inflection-point inflation. *Astropart. Phys.*, 103:16–20, 2018.

37. Michael Dine, Lisa Randall, and Scott D. Thomas. Supersymmetry breaking in the early universe. *Phys. Rev. Lett.*, 75:398–401, 1995.

38. Michael Dine, Lisa Randall, and Scott D. Thomas. Baryogenesis from flat directions of the supersymmetric standard model. *Nucl. Phys.*, B458:291–326, 1996.

39. G. F. R. Ellis. Topology and cosmology. *Gen. Rel. Grav.*, 2:7–21, 1971.

40. Katherine Freese, Joshua A. Frieman, and Angela V. Olinto. Natural inflation with pseudo–Nambu-Goldstone bosons. *Phys. Rev. Lett.*, 65:3233–3236, 1990.

41. Joshua A. Frieman, Christopher T. Hill, Albert Stebbins, and Ioav Waga. Cosmology with ultralight pseudo Nambu-Goldstone bosons. *Phys. Rev. Lett.*, 75:2077–2080, 1995.

42. Gary N. Felder, Lev Kofman, and Andrei D. Linde. Instant preheating. *Phys. Rev.*, D59:123523, 1999.

43. Bo Feng and Ming-zhe Li. Curvaton reheating in nonoscillatory inflationary models. *Phys. Lett.*, B564:169–174, 2003.

44. L. H. Ford. Cosmological constant damping by unstable scalar fields. *Phys. Rev.*, D35:2339, 1987.

45. L. H. Ford. Gravitational particle creation and inflation. *Phys. Rev.*, D35:2955, 1987.

46. Yasunori Fujii. Origin of the gravitational constant and particle masses in scale invariant scalar-tensor theory. *Phys. Rev.*, D26:2580, 1982.

47. Bo Feng, Xiu-Lian Wang, and Xin-Min Zhang. Dark energy constraints from the cosmic age and supernova. *Phys. Lett.*, B607:35–41, 2005.

48. Alan H. Guth, David I. Kaiser, and Yasunori Nomura. Inflationary paradigm after Planck 2013. *Phys. Lett.*, B733:112–119, 2014.

49. Dalia S. Goldwirth and Tsvi Piran. Initial conditions for inflation. *Phys. Rept.*, 214:223–291, 1992.

50. Alan H. Guth. The inflationary Universe: A possible solution to the horizon and flatness problems. *Phys. Rev.*, D23:347–356, 1981. [Adv. Ser. Astrophys. Cosmol.3,139(1987)].

51. S. W. Hawking. Black hole explosions. *Nature*, 248:30–31, 1974.

52. S. W. Hawking and I. G. Moss. Fluctuations in the inflationary Universe. *Nucl. Phys.*, B224:180, 1983.

53. Dragan Huterer and Michael S. Turner. Prospects for probing the dark energy via supernova distance measurements. *Phys. Rev.*, D60:081301, 1999.

54. Michael Joyce and Tomislav Prokopec. Turning around the sphaleron bound: Electroweak baryogenesis in an alternative postinflationary cosmology. *Phys. Rev.*, D57:6022–6049, 1998.

55. D. Kazanas. Dynamics of the Universe and spontaneous symmetry breaking. *Astrophys. J.*, 241:L59–L63, 1980.

56. William H. Kinney. Horizon crossing and inflation with large eta. *Phys. Rev.*, D72:023515, 2005.

57. Renata Kallosh, Andrei Linde, and Diederik Roest. Superconformal inflationary α-attractors. *JHEP*, 11:198, 2013.

58. Lev Kofman, Andrei D. Linde, and Alexei A. Starobinsky. Towards the theory of reheating after inflation. *Phys. Rev.*, D56:3258–3295, 1997.

59. Alexander Yu. Kamenshchik, Ugo Moschella, and Vincent Pasquier. An alternative to quintessence. *Phys. Lett.*, B511:265–268, 2001.

60. Edward W. Kolb and Michael S. Turner. *The Early Universe*, volume 69. 1990.

61. Miao Li. A model of holographic dark energy. *Phys. Lett.*, B603:1, 2004.

62. Andrei D. Linde. A new inflationary Universe scenario: A possible solution of the horizon, flatness, homogeneity, isotropy and primordial monopole problems. *Phys. Lett.*, 108B:389–393, 1982. [Adv. Ser. Astrophys. Cosmol.3,149(1987)].

63. Andrei D. Linde. Scalar field fluctuations in expanding Universe and the new inflationary Universe scenario. *Phys. Lett.*, 116B:335–339, 1982.

64. Andrei D. Linde. Chaotic inflation. *Phys. Lett.*, 129B:177–181, 1983.

65. Andrei D. Linde. Initial conditions for inflation. *Phys. Lett.*, 162B:281–286, 1985.

66. Andrei D. Linde. Eternal chaotic inflation. *Mod. Phys. Lett.*, A1:81, 1986.

67. Andrei D. Linde. Hybrid inflation. *Phys. Rev.*, D49:748–754, 1994.

68. Andrei D. Linde. Fast roll inflation. *JHEP*, 11:052, 2001.

69. Eric V. Linder. Exploring the expansion history of the Universe. *Phys. Rev. Lett.*, 90:091301, 2003.

70. Andrew R. Liddle and D. H. Lyth. *Cosmological inflation and large scale structure.* 2000. Cambridge University Press (2000-04-13) (1656) ASIN: B019NEC0Z6.

71. David H. Lyth and Andrew R. Liddle. *The primordial density perturbation: Cosmology, inflation and the origin of structure.* 2009. Cambridge University Press; Revised edition (June 11, 2009) ASIN: B00AKE1VFO.

72. Andrei D. Linde, Dmitri A. Linde, and Arthur Mezhlumian. From the Big Bang theory to the theory of a stationary universe. *Phys. Rev. D*, 49:1783–1826, 1994.

73. F. Lucchin and S. Matarrese. Power law inflation. *Phys. Rev.*, D32:1316, 1985.

74. Andrew R. Liddle and Anupam Mazumdar. Inflation during oscillations of the inflaton. *Phys. Rev.*, D58:083508, 1998.

75. Andrei Linde and Mahdiyar Noorbala. Measure problem for eternal and non-eternal inflation. *JCAP*, 1009:008, 2010.

76. Andrei D. Linde and Antonio Riotto. Hybrid inflation in supergravity. *Phys. Rev.*, D56:R1841–R1844, 1997.

77. David H. Lyth and Ewan D. Stewart. Thermal inflation and the moduli problem. *Phys. Rev.*, D53:1784–1798, 1996.

78. David H. Lyth and David Wands. Generating the curvature perturbation without an inflaton. *Phys. Lett.*, B524:5–14, 2002.

79. Ming-zhe Li, Xiu-lian Wang, Bo Feng, and Xin-min Zhang. Quintessence and spontaneous leptogenesis. *Phys. Rev.*, D65:103511, 2002.

80. David H. Lyth. What would we learn by detecting a gravitational wave signal in the cosmic microwave background anisotropy? *Phys. Rev. Lett.*, 78:1861–1863, 1997.

81. Jerome Martin and Robert H. Brandenberger. The TransPlanckian problem of inflationary cosmology. *Phys. Rev.*, D63:123501, 2001.

82. R. D. Peccei, J. Sola, and C. Wetterich. Adjusting the cosmological constant dynamically: Cosmons and a new force weaker than gravity. *Phys. Lett.*, B195:183–190, 1987.

83. P. J. E. Peebles and A. Vilenkin. Quintessential inflation. *Phys. Rev.*, D59:063505, 1999.

84. P. J. E. Peebles and J. T. Yu. Primeval adiabatic perturbation in an expanding universe. *Astrophys. J.*, 162:815–836, 1970.

85. Bharat Ratra and P. J. E. Peebles. Cosmological consequences of a rolling homogeneous scalar field. *Phys. Rev.*, D37:3406, 1988.

86. N. Suzuki et al. The Hubble Space telescope cluster supernova survey: V. Improving the dark energy constraints above $z > 1$ and building an early-type-hosted supernova sample. *Astrophys. J.*, 746:85, 2012.

87. K. Sato. First order phase transition of a vacuum and expansion of the Universe. *Mon. Not. Roy. Astron. Soc.*, 195:467–479, 1981.

88. Varun Sahni, M. Sami, and Tarun Souradeep. Relic gravity waves from brane world inflation. *Phys. Rev.*, D65:023518, 2002.

89. Alexei A. Starobinsky. A New Type of Isotropic Cosmological Models Without Singularity. *Phys. Lett.*, 91B:99–102, 1980. [Adv. Ser. Astrophys. Cosmol.3,130(1987); ,771(1980)].

90. Y. Shtanov, Jennie H. Traschen, and Robert H. Brandenberger. Universe reheating after inflation. *Phys. Rev.*, D51:5438–5455, 1995.

91. R. K. Sachs and A. M. Wolfe. Perturbations of a cosmological model and angular variations of the microwave background. *Astrophys. J.*, 147:73–90, 1967. [Gen. Rel. Grav.39,1929(2007)].

92. Eva Silverstein and Alexander Westphal. Monodromy in the CMB: Gravity Waves and String Inflation. *Phys. Rev.*, D78:106003, 2008.

93. Paul J. Steinhardt, Li-Min Wang, and Ivaylo Zlatev. Cosmological tracking solutions. *Phys. Rev.*, D59:123504, 1999.

94. Alexei A. Starobinsky and Junichi Yokoyama. Equilibrium state of a selfinteracting scalar field in the De Sitter background. *Phys. Rev.*, D50:6357–6368, 1994.

95. R. A. Sunyaev and Ya. B. Zeldovich. Small scale fluctuations of relic radiation. *Astrophys. Space Sci.*, 7:3–19, 1970.

96. Michael S. Turner. Coherent Scalar Field Oscillations in an Expanding Universe. *Phys. Rev.*, D28:1243, 1983.

97. W. G. Unruh. Notes on black hole evaporation. *Phys. Rev.*, D14:870, 1976.

98. Alexander Vilenkin. Creation of universes from nothing. *Phys. Lett.*, 117B:25–28, 1982.

99. Alexander Vilenkin. Quantum fluctuations in the new inflationary Universe. *Nucl. Phys.*, B226:527–546, 1983.

100. Steven Weinberg. The cosmological constant problem. *Rev. Mod. Phys.*, 61:1–23, 1989. [,569(1988)].

101. C. Wetterich. Cosmology and the fate of dilatation symmetry. *Nucl. Phys.*, B302:668–696, 1988.

Index

R^2 inflation, 156

ΛCDM, *see* concordance model (ΛCDM)

α-attractors, 158, 161, 240

η-problem, 116, 165, 175

$f(R)$ gravity, 156

A-term inflation, 183

accelerated expansion, 18, 23, 28, 73, 74, 124

acceleration equation, 18, 22, 110

acoustic oscillations, 96, 105

acoustic peaks, 95–97, 105, 207

acoustic scale, 100

action (S), 137, 138, 157
  $f(R)$, 156
  Einstein-Hilbert, 153
  Higgs inflation, 154
  scalar-tensor, 153

action principle, 137, 153, 157

adiabaticity condition, 121, 122, 163

age of the Universe, 29, 36, 64
  age problem, 36

anthropic reasoning, 193, 206

Aristotle, 209

attractor, 211, 218
  exponential, 174, 213, 227, 231
  inflationary, 125, 165, 195, 220
  scaling, 213, 215, 227
  tracking, 215

axion decay constant, 141, 147

axion monodromy inflation, 141

axions, 41, 51, 111, 119, 147

axiverse, 221

barotropic parameter (w), 16, 134
  dark energy, 28
  k-essence, 224
  phantom, 222
  scalar field, 111, 209, 213
  variable, 28, 218

baryogenesis, 54, 59, 64, 73

baryon acoustic oscillations (BAO), 26, 100, 207

baryon asymmetry, 54, 62, 64

baryon-photon plasma, 96, 207

Big Bang, 10, 27, 49, 62, 86

Big Bang nucleosynthesis (BBN), 40, 54, 64, 70, 215, 233, 238

Big Crunch, 21–23, 27, 85, 243–246

Big Freeze, 247

Big Rip, 32, 34, 244–247

bispectrum, 261

black holes, 79, 193
  cosmic, 244
  evaporation, 42, 81, 124, 245
  primordial, 42, 72, 77, 124, 151, 176

blackbody spectrum, 47, 64

Boltzmann brains, 194

Boltzmann constant (k_B), 47, 110

bouncing universe, 201

Brans-Dicke theory, 153

Bunch-Davies vacuum, 263

Casimir experiment, 79

Cepheids, 7

chameleon, 227

Chandrasekhar limit, 7

chaotic inflation, 139, 188, 190, 192, 195, 200, 232

chaotic mixing, 199

Chaplygin gas, 229

closed universe, 19–22, 98, 105, 196

COBE constraint, 53, 92, 104, 128

coincidence problem, 29, 232, 242

cold dark matter (CDM), 41, 64, 105, 106
  halos, 106
  thermal, 50

Coleman-Weinberg potential, 150, 183

comoving coordinates, 11, 33

compact Universe, 198

Compton wavelength, 24, 83, 124, 126, 221

concordance model (ΛCDM), 205, 244, 259

conformal factor, 154

conformal time, 68, 263

conformal transformation, 154

continuity equation, 16, 110, 225

Copernican principle, 9

cosmic distance ladder, 6

cosmic egg fallacy, 15, 73

cosmic inflation, *see* inflation

cosmic microwave background (CMB) radiation, 6, 52, 64, 67
  dipole anisotropy, 52
  multipoles, 92, 95, 96
  peak structure, 96, 207
  polarization, 99
  primordial anisotropy, 53, 65, 78, 92, 230

spectrum, 92
temperature today, 52, 92
cosmic neutrino background, 54, 57
cosmic strings, 60, 96
cosmic variance, 93
cosmic web, 107
cosmological constant (Λ), 22, 205, 243, 259
    fine-tuning, 27, 205, 259
    problem, 27, 29, 205, 206, 259
cosmological horizon, *see* horizon
cosmological principle, 14, 52, 64, 72, 76
coupled dark energy, 226
CPL parametrization, 28
critical density (ρ_c), 20
curvaton field, 133
curvaton reheating, *see* reheating
curvature parameter (k), 15, 18, 96, 198
curvature perturbation (ζ), 100, 101, 131, 133,
        260
    adiabatic & isocurvature, 101
    spectrum ($\mathcal{P}_\zeta$), 100, 101, 104, 128, 164,
        166, 185, 191, 261
cycloid curve, 22

dark energy, 2, 64, 100, 105, 111, 205
    coupled, *see* coupled dark energy
    definition, 27
    phantom, *see* phantom dark energy
dark matter, 39–41, 64
DBI inflation, 162
de Sitter instanton, 196
decoupling, matter-radiation, 51, 68, 95, 99,
        207
density contrast (δ), 101, 105
density parameter (Ω), 20, 71
    baryons (Ω_B), 40, 43, 57
    CDM (Ω_{CDM}), 43
    dark energy (Ω_{DE}), 207
    gravitational waves (Ω_{GW}), 238
    massive neutrinos (Ω_ν), 54, 106
    matter (Ω_m), 24, 57, 105
    radiation (Ω_r), 43
    stars (Ω_{stars}), 39
    today (Ω_0), 24, 70
    vacuum density (Ω_Λ), 24, 43
density perturbations, 101
    adiabatic, 84, 97
    Gaussian, 84, 99
    primordial, 40, 53, 65, 72, 78, 82, 84, 126–
        128
deuterium bottleneck, 56

diffusion zone, 189–193, 195, 196
dissipation coefficient (Υ), 164, 167
Doppler effect, 12, 52, 67

e-folds, 87, 113, 133, 179, 236, 237
effective relativistic degrees of freedom (g_*), 48,
        57, 110, 134
Einstein frame, 154–157, 228
electron-positron pair annihilation, 53
electroweak
    energy scale, 59, 216, 221, 232
    phase transition, 59, 135
energy conditions, 246
enhanced symmetry point (ESP), 122, 145, 179,
        200, 222, 233
ensemble average, 92
equality, matter-radiation, 30, 71, 134, 226
equation of state, 16, 110, 229, 246
equation of state parameter, *see* barotropic
        parameter (w)
equivalence principle
    strong, 83
    weak, 221, 227
ergodic theorem, 93
eternal inflation, 188–193, 195, 196, 200
Euler-Lagrange equation, 138
event horizon, *see* horizon
expansion of the Universe, 8, 12, 64, 94, 104,
        131
    origin, 65, 72, 85

fake Universe paradox, 194
false vacuum inflation, 179
fast-roll inflation, 175, 176, 178
fat WIMPzillas, 124
Fermi paradox, 194
fifth force, 221, 227, 238
filaments, 5, 107
fine-structure constant, 222
flat Universe, 20, 21, 29, 70, 87, 96, 98
flatness problem, 65, 70
    resolved by inflation, 75
flaton field, 177
Floquet index, 120
FLRW metric, *see* metric
fluid equation, *see* continuity equation
Fokker-Planck equation, 186
free-streaming, 106
freeze-out, 50, 63
Friedmann equation, 15, 22, 110, 198
Friedmann universes, 20, 243

galaxy clusters, 5, 207
gamma-ray bursts (GRB), 208
geometric units, 110
ghost condensate, 225
ghost field, 224
God, 62, 193, 206
Gospel of John, 196
graceful exit, 161, 172, 174, 226, 227
grand unified theory (GUT), 60, 71, 72, 151
gravitational constant, Newton's (G), 15, 24, 110, 205, 259
gravitational instability, 78, 94, 104
gravitational lensing, 41, 43, 207
gravitational reheating, *see* reheating
gravitational waves, 99, 102, 103, 129, 208, 238
gravitinos, 41, 72, 77, 119, 176
GUT-scale inflation, 77, 87, 118, 151, 200

Harrison-Zel'dovich spectrum, 102, 104
Hawking radiation, 42, 81, 244
Hawking temperature, 81, 83, 88, 110, 124, 126, 245
heavy fields and inflation, 162
helium abundance, 56
Hesiod, 195
Higgs field, 59, 111, 151, 155, 221
    GUT, 60, 151
Higgs inflation, 155, 157, 228
hilltop inflation, 142, 190, 192, 200
    quadratic, 145, 175, 179
    quartic, 144
holographic dark energy, 231
Holographic Principle, 230
homogeneity, 14, 197, 198
horizon, 10, 33, 69, 95
    event, 34, 79, 124, 195, 200, 231
    exit, 89, 126, 127, 134
    particle, 34, 83
    reentry, 89, 95, 96, 134, 240
horizon problem, 65, 68
    resolved by inflation, 74
Hot Big Bang, 2, 49, 57, 60, 64, 73, 76, 116, 195
Hubble constant (H_0), 8, 14, 97, 135, 207, 211
Hubble diagram, 207, 208
Hubble flow, 9, 12, 34, 106, 259
Hubble friction, 112, 115, 117, 149, 164
Hubble horizon, *see* Hubble radius
Hubble parameter (H), 12, 20, 113
Hubble radius, 36, 231
Hubble time, 29, 36, 126, 185, 191
Hubble-Lemaître law, 8, 15

hybrid inflation, 77, 148, 174, 176, 179

inflation, 2, 64, 65, 74–76, 82, 96, 99, 127, 189
    definition, 73
inflationary attractor, *see* attactor
inflationary paradigm, 112, 116, 126, 129
inflationary plateau, 112, 152, 160, 200, 233
inflection-point inflation, 183, 192, 195
initial conditions of inflation, 86, 195, 201
instant preheating, *see* preheating
integrated Sachs-Wolfe effect (ISW), *see* Sachs-Wolfe effect
isotropy, 14

Jeans length (λ_J), 104, 230
Jordan frame, 154, 228

k-essence, 224
k-inflation, 161, 191
kination, 234
Klein-Gordon equation, 112, 126, 139, 209
    kinetically dominated, 181, 209, 234
    phantom, 222
    warm, 164

Lagrangian (L), 137
Lagrangian density ($\mathcal{L}$), 138, 142, 147, 191, 221
    α-attractors, 158
    Higgs inflation, 157
    k-inflation, 161
    phantom, 222
    scalar field, 138, 265
Langevin equation, 185
large scale structure, 5, 15, 42
last scattering surface, 6, 52, 67, 91–95, 207
lightest supersymmetric particle (LSP), 41
lightspeed (c), 17, 32, 33, 73, 79, 83, 109, 172
line element, 33, 257
Little Rip, 248
loop inflection-point inflation, 183
luminosity distance, 7, 208
Lyman-α forest, 208
Lyth bound, 130

MACHOs, 41
Mathieu equation, 120
Maxwell-Bolzmann distribution, 49
measure problem, 192
metric, 32, 130, 154
    FLRW, 33, 257
Milne universe, 21, 198
modified gravity, 153, 228

modular inflation, 175
moduli fields, 72, 77, 176
monopoles, 72, 77, 176, 178
multiverse, 189, 192–194

natural inflation, 145, 173
natural units, 109, 115
Nature, 1, 57, 112, 131, 196
neutralinos, 41
neutrinos, 42, 53, 228
new inflation, 143, 145, 179
no-hair theorem, 86, 125, 195
noncanonical scalar field, 154, 158, 160, 180,
        240
nonconformally invariant fields, 84, 124
nonminimally-coupled scalar field, 155, 228
nonoscillatory inflation, 122, 233
null energy condition, 32, 87, 246

observable Universe, 6, 10
Occam's razor, 131, 220
occupation number (n_k), 120, 123
old inflation, 171
open Universe, 20, 21, 98, 198
oscillating inflation, 180
oscillons, 119

parametric resonance, 119, 151
particle production, 79–82, 120, 124, 185, 238
peculiar velocity, 12
perfect fluid, 111, 161, 258, 266
phantom dark energy, 30, 34, 114, 222, 231, 246
phase transitions, 57, 64, 149, 151, 172
pivot scale, 101, 102, 127, 128, 134, 135
Planck constant (h), 17, 47, 79
        reduced ($\hbar$), 27, 47, 79, 109
Planck density (ρ_P), 27, 62
Planck distribution, 47
Planck energy scale, 27, 60, 188, 196, 200, 205,
        221
Planck length (ℓ_P), 27, 62, 199
Planck mass (M_P), 24, 110, 154, 205
        reduced (m_P), 110, 142
Planck time (t_P), 62, 197
Plato, 139, 230
pocket universe, 189, 192–194
pole inflation, 160
power-law inflation, 174, 192, 195
preheating, 119
        instant, 122, 233, 235
primordial black holes, *see* black holes

primordial curvature perturbation, *see* curva-
        ture perturbation (ζ)
probability distribution, 186
pseudo Nambu-Goldstone boson (PNGB), 147,
        220
Pseudo Rip, 248

Q-balls, 119
quantum decoherence, 80, 186
quantum diffusion, 187, 188
quantum fluctuations, 82, 84, 89, 125, 185, 189,
        230
quark-gluon plasma, 58
quasi-de Sitter inflation, 86, 112, 161, 176, 191,
        210
quintessence, 209
        coupled, 226
        exponential, 213, 240
        freezing, 211
        inverse-power-law (IPL), 215, 232
        PNGB, 220
        scaling-freezing, 215
        thawing, 211, 215, 218, 220, 232, 242
        tracking-freezing, 218
quintessential inflation, 122, 220, 232, 240
quintessential tail, 233
quintom, 223

radiation density, 48, 53, 110
radius of curvature (comoving), 18, 196
Raychadhuri equation, 259
recombination, 51
redshift, 12
reheating, 73, 116, 172
        curvaton, 233, 235
        gravitational, 124, 233, 236, 240
        modulated, 131, 133
        perturbative, 116, 164
reheating temperature, 118
relic problem, 71
        resolution by inflation, 77
rotation curves, 39
running spectral index (n_s'), 102, 129

Sachs-Wolfe effect, 78, 94, 95, 101
        integrated (ISW), 95, 207
Sachs-Wolfe plateau, 95–97
Sakharov conditions, 64
scalar curvature (R), 153, 157, 231, 248, 258
scalar field, 28, 111
        density and pressure, 111, 161, 224

spectrum of perturbations, 126, 166, 264

scalar-tensor theory, 153–156

scalaron, 157

scale factor (a), 11, 15

    Λ dominated Universe, 23, 86

    closed universe, 22

    flat Universe, 29

    phantom, 32

scaling attractor, *see* attractor

Schwarzschild radius, 22, 24, 124, 230, 231

shift-symmetry, 147, 200

Silk scale, 97

singularity, 62, 196, 246

slow-roll

    approximation, 114, 162

    inflation, 114, 190

    parameters, 116, 129, 165

solitons, 119

sound horizon, 96, 99, 162, 207

sound speed (c_s), 16, 96, 104, 162, 163, 207, 224, 230

spacetime foam, 62, 65, 72, 85, 195–197

spacetime interval, 32

spectator field, 131, 187, 209, 233

spectral index

    scalar (n_s), 97, 101, 103, 129, 162, 166, 185

    tensor (n_t), 102, 130

speed of light, *see* lightspeed (c)

standard candles, 6, 207, 208

Standard Model of Cosmology, 2, 99

standard model of particle physics, 57, 59, 63, 111, 135, 192

Starobinsky-type inflation, 152, 159, 190

steady-state universe, 52

stochastic inflation, 185

stochastic resonance, 121

string landscape, 164, 193

strong CP-problem, 41

structure formation, 40, 78, 82, 99, 104, 107, 208, 230

Sunyaev-Zel'dovich effect, 97, 208

superluminal expansion, 15, 34, 73, 79, 85, 126, 172, 189

supernovae type Ia, 7, 207

tensor perturbation spectrum $(\mathcal{P}_h)$, 102, 130, 162

tensor to scalar ratio (r), 103, 130, 162, 166

theory of everything (ToE), 60, 139, 241

thermal bath, 48, 73, 116, 120, 121, 133, 164, 177

thermal equilibrium, 17, 46, 54, 59, 64, 67, 73

thermal inflation, 142, 151, 177

Thomas Aquinas, 62

Thomson scattering, 51, 95

threshold energy, 49

tracker, 211, 216, 232

tracking condition, 212

tracking parameter (Γ), 211

transplanckian effects, 264

Tully-Fisher relation, 8

turnaround, 106

ultra-slow-roll inflation, 182, 191, 195

uncertainty relation, 79, 83, 126, 197

Universe

    age, *see* age of the Universe

    closed, *see* closed Universe

    definition, 1

    flat, *see* flat Universe

    nontrivial topology, *see* compact Universe

    observable, *see* observable Universe

    open, *see* open Universe

Unruh effect, 82

vacuum density (ρ_Λ), 23, 110, 196, 205, 230, 243

virialization, 106

virtual particles, 79, 221, 263

warm inflation, 164, 234

waterfall field, 149, 179

WIMP miracle, 41, 51

WIMPs, 41

X-ray observations, 40, 42, 207, 208

youngness paradox, 193

zero-point energy, 27, 79, 230, 259

Printed in the United States
By Bookmasters